1983

Pesticide Analytical Methodology

Pesticide Analytical Methodology

John Harvey, Jr., EDITOR
E. I. Du Pont de Nemours and Company

Gunter Zweig, EDITOR
U.S. Environmental Protection Agency

ASSOCIATE EDITORS
Richard Cannizzaro
Henry Dishburger
Joseph Sherma

Based on a symposium
jointly sponsored by the
Divisions of Pesticide Chemistry
and Analytical Chemistry
at the 178th Meeting of the
American Chemical Society
Washington, D.C.,
September 9–14, 1979.

ACS SYMPOSIUM SERIES **136**

AMERICAN CHEMICAL SOCIETY
WASHINGTON, D. C.   1980

Library of Congress CIP Data

Pesticide analytical methodology.
  (ACS symposium series; 136 ISSN 0097-6156)

  Includes bibliographies and index.

  1. Pesticides—Analysis—Congresses. 2. Pesticides—
Environmental aspects—Congresses.
  I. Harvey, John, 1925- . II. Zweig, Gunter. III.
American Chemical Society. Division of Pesticide
Chemistry. IV. American Chemical Society. Division
of Analytical Chemistry. V. Series: American Chemis-
try Society. ACS symposium series; 136.

SB960.P47   632'.95'028   80-19470
ISBN 0-8412-0581-7   ASCMC 8   136 1–406 1980

ACS Symposium Series

M. Joan Comstock, *Series Editor*

FOREWORD

The ACS SYMPOSIUM SERIES was founded in 1974 to provide a medium for publishing symposia quickly in book form. The format of the Series parallels that of the continuing ADVANCES IN CHEMISTRY SERIES except that in order to save time the papers are not typeset but are reproduced as they are submitted by the authors in camera-ready form. Papers are reviewed under the supervision of the Editors with the assistance of the Series Advisory Board and are selected to maintain the integrity of the symposia; however, verbatim reproductions of previously published papers are not accepted. Both reviews and reports of research are acceptable since symposia may embrace both types of presentation.

CONTENTS

PREFACE

Analytical methodology for pesticide determinations has greatly advanced during the past 30 years. The emphasis during the first 20 years was placed on increased sensitivity of detection. That these efforts were successful is attested by achieving sensitivity both quantitatively and qualitatively in the femtogram range (10^{-15} g).

In a recent review of pesticide analytical methodology (Zweig, G. "The Vanishing Zero—Ten Years Later," *JAOAC*, **1978**, *61*(2), 229–248) the author pleaded with analytical chemists to improve the selectivity, resolving power, and identification capability for sub-microgram quantities of pesticide residues. The rationale was that the toxicologists lag far behind the analytical chemists and often are not able to assess the toxicological significance of minute levels of pesticide residues.

We believe that this symposium on Recent Advances in Pesticide Analytical Methodology fulfills this plea. High-performance liquid chromatography (HPLC) has made the greatest advances. Chapters on HPLC cover subjects on metabolism studies; automation of HPLC; evaluation of LC columns; the effect of the mobile phase on reversed-phase chromatography; the electrochemical or amperometric detector; and fluorogenic detection.

Quantitative thin-layer chromatography (TLC) has progressed considerably and has been adapted to forensic chemistry of pesticide poisoning incidents. High-performance TLC (HPTLC) with precoated plates is discussed separately.

Fourier transform infrared spectroscopy; negative ion spectroscopy; and immunochemical technology are the subjects of several chapters. A chapter on cleanup presents a general challenge for all types of analytical techniques. Chemical derivatization of pesticides and metabolites enhances the detectability and ability to analyze small quantities of pesticide residues.

Applications of modern analytical techniques also are discussed relative to the analysis of tetrachlorodibenzo-*p*-dioxin; organotin compounds; airborne pesticides; computerized data processing for metabolism studies; and human exposure studies.

The organizers of this symposium wish to thank the contributors for their excellent talks, written chapters, and cooperation in meeting all deadlines. The Division of Pesticide Chemistry also wishes to acknowledge the joint sponsorship of this symposium with the Division of Analytical Chemistry.

Biochemicals Department   JOHN HARVEY, JR.
Experimental Station
E. I. Du Pont de Nemours and Company
Wilmington, Delaware 19898

Hazard Evaluation Division   GUNTER ZWEIG
Environmental Protection Agency
Washington, D.C. 20460

April 17, 1980

Modern High-Performance Liquid Chromatography in Pesticide Metabolism Studies

JOHN HARVEY, JR.

E. I. Du Pont de Nemours & Company, Inc., Biochemicals Department, Experimental Station, Wilmington, DE 19898

This paper is the only one in the liquid chromatography portion of this symposium which will attempt to deal with chromatography specifically from the viewpoint of the pesticide metabolism chemist. A residue analyst knows what compound he must analyze for, and develops his method with the properties of that substance in mind. On the other hand, the pesticide metabolism chemist has a different problem. At the conclusion of the treatment, exposure, and harvest phases of a radiolabeled metabolism study, he divides his material into appropriate samples, and extracts each sample with selected solvents to obtain the radioactive materials in soluble form. Typically these extracts consist of low levels (ppm) of carbon-14 labeled metabolites in a complicated mixture of normal natural products from the plant, animal, or soil source. The identity of each metabolite is unknown, and each must be isolated from the natural background and from other labeled metabolites in sufficient quantity and in adequate purity for identification studies, usually by mass spectrometry. The situation is rather like looking for the proverbial "needle in the haystack" when one does not know the size, shape, or composition of the needle, or even how many needles there are in the stack. At this point a separation technique must be selected with certain important requirements in mind.

First, the technique must be non-destructive. We must have confidence that the material we isolate and identify is the same as the metabolite the bio-system produced. Second, the technique must be powerful enough to effect the desired purification. Third, we need to achieve the required purity and quantity with an economical expenditure of time and resources. High performance liquid chromatography (hereafter referred to as HPLC) is ideally suited to this challenge.

• The conditions of HPLC are mild - usually ambient temperatures - with ordinary solvents from hexane to water. Only in very rare instances have rearrangements or degradations been reported under LC conditions, and when suspected, alternative routes are usually available to provide the separation without the degradation.

● HPLC is a powerful separation technique. Commercially
  available columns of 4.6mm i.d. x 25cm length regularly
  exhibit 8,000 - 12,000 or more theoretical plates.
● HPLC is fast and economical in terms of labor and
  reagents. The effluent fractions come off the column in
  small volumes of solvents readily available for liquid
  scintillation counting, and/or further purification steps.
● In addition, HPLC is extremely versatile, one chromatograph being capable of separating compounds by any of 4-6
  distinct modes of chromatography by simple interchanges
  of columns. When the capacity for infinite variation of
  the composition of the mobile phase is added to the
  available choices in the stationary phase a nearly
  unlimited versatility is displayed.

It is not my purpose to expound chromatographic theory, or
to discuss the fine points of column preparation, solvent
selection, or new advances in detectors. The papers that follow
deal with recent developments in these areas, and what they
report is as applicable to pesticide metabolism analyses as to
residue analyses. Instead, I shall describe a working radiochromatograph for the pesticide research laboratory and discuss
some of the problems associated with this type of instrument.
I shall close by describing some of the possibilities and
procedures open to the analyst in the study of pesticide
metabolism problems.

The Radiochromatograph

A schematic diagram of a radiochromatograph is shown in
Figure 1. In its simplest form any chromatograph consists of a
solvent reservoir, a constant flow high pressure chromatographic
pump, a suitable chromatographic column, and one or more
detectors. An injection valve is the most convenient way of
introducing the sample onto the column, and a recorder produces
the permanent record of detector responses. For the ultimate in
versatility the addition of a device for generating a solvent
gradient is highly desirable. In our own laboratory we have
several radiochromatographs of varying complexity. The one thing
they all have in common is that they are lab-assembled
instruments, with all of the component parts being commercially
available. This approach allows the ultimate in flexibility,
including the capacity to replace individual components at any
time whether because of breakdown, obsolescence, or change in
requirements. In our radio-LC's the general detector is an
ultraviolet absorbance type. This serves, when analyzing crude
mixtures, to indicate where the non-labeled UV absorbing
components may be found, and also enables us to establish the
retention time of model compounds or standards. This type of
detector is sensitive and convenient for many purposes, but any
of the many available LC detectors may be substituted (index of
refraction, infrared, etc.) depending on the user's requirements.

The unique heart of a radiochromatograph is the radioactivity detector. These devices are essentially liquid scintillation counters, simplified by the absence of the usual mechanism for the storage and transport of counting vials. They do, of course, have to retain the necessary shielding which makes them rather bulky as shown in Figure 2. This unit is part of a Model 3021 Tri-Carb Scintillation Spectrometer (Packard Instrument Company) but equivalent results have been obtained with a Nuclear-Chicago Liquid Radiochromatography System Model 4526 (Searle Analytic, Inc.). In these detectors a flow cell is positioned between the photomultiplier tubes of the scintillation counter and the column effluent flows through this cell in contact with an appropriate scintillator.

At this point either of two approaches may be taken. One is to mix the column effluent continuously with a liquid scintillation counting cocktail and pass the homogeneous stream through the flow cell in the detector (homogeneous counting). The other is to pass the effluent stream over an insoluble scintillator which has been packed into the flow cell (heterogeneous counting). The two approaches have been compared critically by Schutte (1). This author concluded, and we concur, that the heterogeneous system is best for preparative chromatography when activities are high and losses are unacceptable. In homogeneous counting all or a portion of the sample is mixed with the cocktail and becomes, in effect, unrecoverable. For our own work in pesticide metabolism, samples are invariably small, hard-to-come-by, and therefore very precious. Consequently, we prefer a system which will enable us to detect the radiolabeled fractions without loss of any material.

Because the photomultiplier tubes and associated circuitry of the scintillation counters have already been developed to a high state of efficiency (2), we must turn our attention to the flow cell itself and the way we operate our LC columns and the controls of the counter in order to get the most benefit out of the system.

Flow Cells for Radiochromatographs

As shown in Figure 3, our detectors came equipped with a cylindrical plastic cell about the size and shape of a scintillation counting vial into which a U-shaped tube had been drilled. When this cell has been packed with scintillation grade anthracene a flow cell quite satisfactory for aqueous systems is obtained. The top of the cell is, of course, totally opaque, to exclude extraneous light from the photomultiplier tubes, and is sized and shaped to fit the well in the scintillation counter being used.

The situation for organic solvents is less satisfactory, however. In my experience, all commercial cells depend at some point on epoxy cement which simply does not hold up to continuous exposure to many organic solvents. As shown in Figure 4, we have

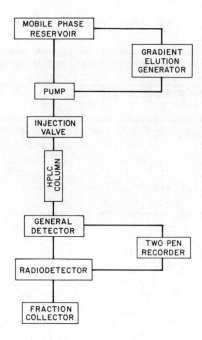

*Figure 1.   Schematic of a radiochromato-
graph*

Figure 2.   Radioactivity detector

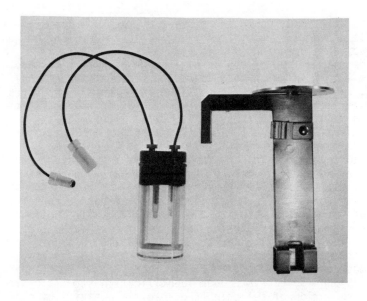

Figure 3.   Scintillation flow cell for aqueous systems

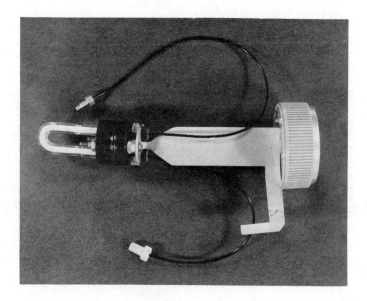

Figure 4.   Scintillation flow cell for organic solvents

found a simple solution to the problem by taking a commercially available 3-inch long glass chromatographic column (Chromatronix, Inc.) and, after bending it into a U-shape, attaching it to hang below the regular cap of the scintillation counter. This solution provides a cell in which only glass and Teflon® contact the solvent, and the cell is able to withstand moderate pressures and any solvent conditions that our chromatographic columns themselves withstand. Packed with a cerium activated lithium glass scintillator (which is available from a number of supply companies) a completely inert and durable cell is readily obtained.

Because photomultiplier tubes can only detect events which occur on or near the surface of the scintillator column exposed to the tubes, the greater the surface to volume ratio, the more efficiently events will be detected. The literature contains a number of suggestions in which the simple U-tube is modified into coils and spirals which present larger surfaces per unit of volume to the photomultiplier tubes (3). While this does increase sensitivity, a practical limit to minimum required sensitivity is set in most metabolism studies by the fact that we must separate enough compound to carry through identification studies. Since we have the option of selecting the specific activity of our original ^{14}C-compound, we have found the U-tubes described entirely adequate.

Operational Problems of the Radiochromatograph

Let us consider some of the special problems encountered in the operation of a radioisotope detector and the compromises that must be considered. Like any chromatographic detector, a carbon-14 detector should have a small volume and a short hold-up time in order to minimize band spreading and loss of resolution. Unfortunately radioisotopes are measured with an inherent time factor - disintegrations per minute. Therefore, the smaller the cell and the shorter the hold-up, the lower will be the sensitivity, a circumstance which is totally at odds with the first requirement. In practice, we have found that a U-tube with a cross-section diameter of 2mm is generally satisfactory. This gives a cell with a void volume of 200-300 µl, which is high compared to the 2-10 µl volumes of many UV flow cells, and may introduce some band spreading when used with the best new HPLC columns.

Regardless of cell size, the faster you push a radioactive peak through the cell, the smaller that peak will appear to the radioactivity detector (Figure 5). Thus low level samples cannot be run at high flow rates. With the usual 4.6mm x 25cm HPLC columns, flow rates of 0.5-1.0ml/min. are used routinely with no difficulty.

The inherent time factor in radioisotope detection can also be partially compensated by the electronic controls. While counting times of 10-20 minutes are commonly used for liquid

scintillation counting, counting intervals on the order of seconds
are all that is available for chromatographic effluents. These
effects are shown in Figure 6. If we select a relatively long
counting interval (e.g., 25 seconds) we smooth out the base line,
but also cause an electronic apparent band spreading of low level
samples. If we select a shorter counting interval the apparent
band spreading is reduced, lower level peaks are detectable, but
base-lines and peaks are definitely rougher.

Solvent quenching effects, which are of importance in liquid
scintillation counting, seem to have little effect with a solid
scintillator. However, the glass scintillators used with organic
solvents are only about 1/10th as effective as anthracene.
Consequently, we detect chromatographic peaks with a minimum of
about 300 disintegrations per minute using anthracene and
approximately 3000 dpm with the glass.

HPLC Versatility

Leaving the radioisotope detector at this point, let us
examine the versatility inherent in the liquid chromatographic
process and develop a strategy for HPLC separations from the
viewpoint of the metabolism chemist. Liquid chromatography is
not an exact science. However, a great deal of practical
knowledge has been accumulated, and a tested theory has evolved
to relate many chromatographic parameters to the very practical
business of separating of chemical compounds.

The versatility of HPLC is so great, that a hit or miss
approach even to the selection of the initial column and conditions is unlikely to give worthwhile results (4). Indeed, a
working knowledge of chromatographic theory is a prerequisite to
appreciate the factors involved in choosing even the initial
conditions for a chromatographic separation. A knowledge of
theory is further necessary in determining from the results of
that initial experiment what modifications are required to
improve the separation. Fortunately both courses and books are
available on HPLC; they deserve considered attention.

Basically, the liquid chromatographer has four modes of
chromatography to work with (5): adsorption, size exclusion, ion
exchange, and partition. Simplistically, that means the ability
to separate compounds by four different chemical mechanisms:
adsorptive power, molecular size, ionic attraction, and partition
between two liquid phases. Partition chromatography itself has
undergone extensive change and no longer refers to an immobile
phase with a physically absorbed liquid layer, but instead
utilizes a solid immobile phase, usually a silica, to which a
layer of long chain (C_8-C_{18}) organic molecules has been chemically
bonded. This bonded surface layer has many properties like a
liquid and may itself contain functional groups (e.g., -CN) which
profoundly alter its properties (e.g., polarity and selectivity).
However, this layer possesses the distinct virtue of being

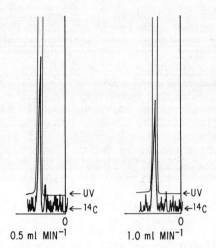

*Figure 5.   Effect of flow rate on ¹⁴C peak
(Porasil A, 2.8 mm × 1000 mm/CHCl₃)*

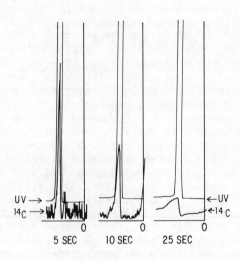

*Figure 6.   Effect of counting interval on ¹⁴C peak (Porasil A, 2.8 mm × 1000 mm
CHCl₃, 0.50 mL min⁻¹)*

chemically bonded to the support, and hence cannot be washed off. This type of chromatography can be divided into reversed phase and normal phase depending on whether the mobile phase is more polar (reversed) or less polar (normal) than the stationery phase.

A decade ago packings for LC columns (Figure 7) were more or less naturally occurring materials of irregular shape and considerable range in size (Figure 7C). Five years ago, the best packings were synthetically produced, spherical in shape, moderately sized (35-75μ), and sometimes covered with a chemically bonded layer of organic material (Figure 7A). Today, advances in technology have nearly caught up with theory, and the most efficient particles are totally porous micro-particles (Figure 7B) of 10 microns or less in size (6). The first two types were usually sold in bulk, and each user packed his own columns. The micro-particulates require special packing techniques, and are usually available already packed in standard-sized, stainless steel columns. These micro-particulate columns exert considerable resistance to flow, necessitating the use of pressures of 1,000-5,000 psi to obtain a satisfactory flow rate. Thus, HPLC sometimes is considered to mean High Pressure Liquid Chromatography. These micro-particulate columns also have high capacities because of their totally porous nature, and all in all present a very attractive combination of properties for analysis of pesticide metabolite extracts.

Separation Strategy

Separation of extremely complex mixtures of radiolabeled metabolites in natural products generally requires the use of more than one chromatographic step. The sequence in which the steps are carried out is furthermore crucial to a satisfactory result, and is itself dependent upon the initial prechromatographic separations (7). We might, for example, expect that our initial biological system had been extracted to give three fractions, any or all of which might contain radioactivity: non-polar (hexane-soluble), moderately polar (ethyl acetate-soluble), and very polar (alcohol/water soluble).

Non-polar and moderately polar solutes generally present the least difficulty, and these extracts are amenable to reversed phase partition chromatography followed by final purification of separated fractions by adsorption chromatography. Reversed phase chromatography is a practical first step because it is effective for a very wide range of compounds, and secondly because it has less tendency to be "fouled" by irreversible absorption of highly polar contaminants.

The polar extracts are more difficult because they contain conjugates of pesticides together with normal natural products, like sugars, which are closely related to the conjugates. Furthermore, these solutions are likely to contain natural products which have become radiolabeled as a result of reincorporation of radiolabeled carbon dioxide resulting from complete breakdown of the radiolabeled pesticide. For these polar extracts, size exclusion chromatography on a soft gel like BioGel P-2 (BioRad Laboratories) is an excellent first step. Bio-macromolecules come off the column immediately, and small inorganic molecules are retained. Little fractionation of the middle sized molecules occurs, but the clean-up has been considerable. Following this, ion exchange chromatography will identify the ionic components, while the non-ionic species may be isolated by reversed phase and adsorption chromatography.

Applications of HPLC to Pesticide Metabolism Problems

In the time that is left, let us look at several examples of the use of a ^{14}C-detector in practical situations. Figure 8 shows the resolution of two ^{14}C-metabolites from each other in a mixture of naturally-occurring compounds in river water. This is an early stage purification step on BioGel P-2, a gel filtration medium. Each of these peaks is now ready to be purified further by chromatography in some other mode.

Figure 9 shows the purification of another ^{14}C-compound in a partially purified extract of goat milk. At this stage at least seven other anions are visible to the ultraviolet detector as the components are separated on Permaphase AAX anion exchanger. Even now the strong UV band almost directly above the radioactive peak appears to belong to a contaminant rather than the ^{14}C-compound itself, indicating that further work will be required.

Figure 10 shows the happy result which occurs when a ^{14}C-unknown has been isolated, purified, and identified. Following synthesis of the proposed structure, the synthetic model and the purified ^{14}C-compound are co-chromatographed. We can adjust the sensitivity of the UV detector so that UV from the ^{14}C-compound is more or less invisible, and the UV seen is due only to the non-radioactive model. When the two coincide as neatly as shown here we have further evidence for the validity of our identification.

Finally, we have encountered situations where carbon-14 from a labeled pesticide has been incorporated into normal natural products including amino acids. We were interested in seeing if we could separate the normal amino acids from a protein hydrolysate on a single column with sufficient resolution to identify separate amino acids. Figure 11 shows a radiochromatogram of the separation of sixteen standard ^{14}C-amino acids. Were this a product of a metabolism study, we could then isolate any of these fractions, convert the residual amino acid to the N-trifluoroacetyl O-butyl derivative for gas chromatography and further confirmation of structure.

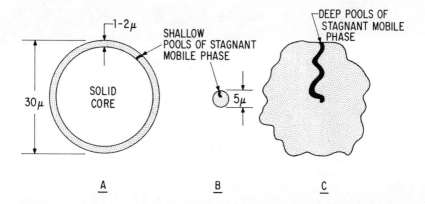

John Wiley & Sons, Inc.

Figure 7.   Types of particles for modern liquid chromatography (6): (A) super-ficially porous particle; (B) very small totally porous particle; (C) totally porous particle

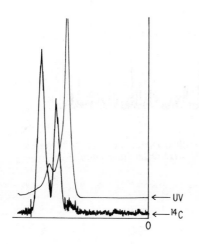

Figure 8.   Separation of ^{14}C metabolites from naturally occurring compounds in river water (Bio-Gel P-2, 6 mm $\times$ 1000 mm/H_2O, 0.8 mL min^{-1})

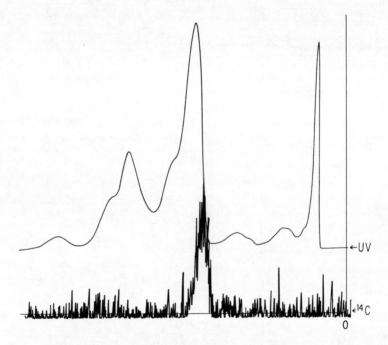

Figure 9. Separation of a ^{14}C anion from naturally occurring acids in milk (Perma-phase AAX, 2.8 mm $\times$ 1000 mm/.01N HNO$_3$, 0.50 mL min^{-1})

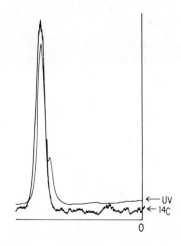

Figure 10.   Comparison of a ^{14}C metabolite with a model compound (Porasil A, 2.8 mm $\times$ 1000 mm/THF, 0.4 mL min^{-1})

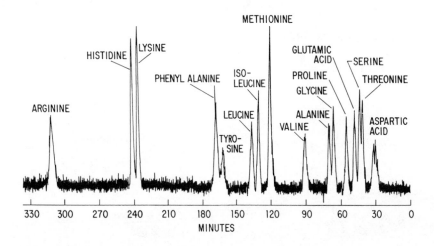

Figure 11.   Resolution of a synthetic mixture of ^{14}C amino acids

Literature Cited

1. Schutte, L., J. Chromatogr., 1972, 72, 303.

2. Rapkin, E., "Liquid Scintillation Counting", Vol. 2,
    Heydan, London, 1972.

3. Bransome, E. D., Jr., Ed., "The Current Status of Liquid
    Scintillation Counting", Grune and Stratton, New York,
    1980, pp 100-102.

4. Synder, L. R. and Kirkland, J. J., "Introduction to Modern
    Liquid Chromatography, 2nd Ed., John Wiley & Sons, Inc.,
    New York, 1979, pp 16-21.

5. Roberts, T. R., "Radiochromatography", Elsevier Scientific
    Publishing Co., New York, 1978, p 128.

6. Synder, L. R. and Kirkland, J. J., "Introduction to Modern
    Liquid Chromatography, 2nd Ed., John Wiley & Sons, Inc.,
    New York, pp 169-183.

7. Synder, L. R. and Kirkland, J. J., "Introduction to Modern
    Liquid Chromatography", 2nd Ed., John Wiley & Sons, Inc.,
    New York, 1979, pp 753-762.

RECEIVED December 28, 1979.

FAST–LC Concepts for Automated Pesticide Analysis

DONALD A. BURNS

Technicon Industrial Systems, 511 Benedict Ave., Tarrytown, NY 10591

The planet earth, fortunately for us, is blessed with conditions for "life as we know it" in the form of two kingdoms: animals and plants. One of these animals, man, doesn't always approve of everything else that grows, so he takes positive action to limit the growth of some organisms by creating large machines which spew forth mists of chemicals called pesticides. Since these toxic substances may not disappear from the environment as rapidly as they were dispensed, man is relegated to the laboratory to monitor the residues left in or on the food he plans to eat. Various regulating agencies require ever-increasing numbers of tests at ever-rising costs, so it behooves those who must cope with this situation to study carefully the alternatives. The laboratory director may hire more analytical chemists, or provide more automated instruments for his existing staff. The automation approach nearly always results in higher precision and lower costs per test.

One analytical technique especially suited to pesticide analysis is high performance liquid chromatography (HPLC), and one particular version of it is called FAST-LC (an acronym for Fully Automated Sample Treatment for Liquid Chromatography). It is the various aspects of this automated sample treatment which will be considered in this report.

Unit Operations

Just as a voyage from one point to another consists of many individual steps, so also does an analytical procedure involve many individual operations. These are often referred to as "unit operations", and they may be classified in several ways (eg. obtaining a sample, getting it ready for analysis, the analysis itself, or the detection step for quantitation). The sample may be solid, liquid, or gas, and its pre-treatment could include such operations as extraction, filtration, dialysis, distillation, concentration, or even chromatography.

The analysis step could be quite direct (as colorimetry often is) or a chromatographic separation may be required prior to detection. If chromatography is employed, it may be necessary to derivatize the sample, and this can be done either before or after the separation on the column. And in addition to colorimetry, other means of detection are available to the analyst: ultraviolet (UV) spectrophotometry, fluorescence, electro-chemical, etc.

In this report it will be shown how many of these unit operations can be combined into partially- or fully-automated systems. Although several types will be described, the emphasis will be on HPLC-based systems. And that is probably as it should be, because HPLC is still the most popular analytical technique, according to Thomas & Mosbacher (1).

The Automation of HPLC

Handling Liquids. Let's take a look at how HPLC came to be fully automated. One begins with a column and some sort of detector and readout for quantitation. Next we add a valve with a sample loop and a relatively high pressure pump to move the mobile phase. By adding a sampler and a second pump to aspirate the samples sequentially into the loop of the valve, we've gone a long way toward automation. We can complete this simplest of systems by adding an actuator for the valve and some type of programmer to maintain synchronization between sampler and valve (Figure 1).

Although these modules provide a basic automated system, some serious limitations prevail. Samples must be aqueous or high boiling organic solutions, since volatile samples would be partially or wholly lost before analysis. Also, the samples must be liquids, and we know that many will be solids. Finally, the samples must be ready to run, requiring no pre-treatment of any kind, and we know that this is generally not the case. How can these limitations be overcome?

Some samples lend themselves to small column cleanup such as that provided by Waters' Sep-Pak (a miniature chromatographic column). And several suppliers offer processing turntables which will perform various cleanup operations which are frequently required prior to chromatography. And when samples have been cleaned up (whether manually or by one of these automated devices), they can then be transferred to an HPLC unit by any one of several available auto-injectors.

Many of these manual operations can be incorporated into a properly-designed automated system by adding a few more components to Figure 1. By going from a single-channel to a 3-channel proportioning pump, one can use the top two channels to bring together an aqueous sample and the organic solvent into which he wishes to extract the analyte. Extractions can be accomplished continuously in coils, and the two phases separated in appropriately-shaped fittings, using the third channel of the pump to pull the organic phase through the loop of the sample valve.

But in the real world, one doesn't ordinarily perform the extraction immediately. By going to a 6-channel pump (Figure 2), one can expand the usefulness of the cleanup system by adding more unit operations. One can now start with a diluent (to get the sample into the correct concentration range), add the sample to this air-segmented stream and mix it in coil "M", then add the organic solvent and perform the extraction in coil "E".

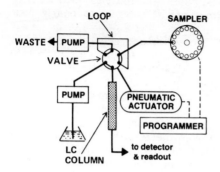

Figure 1. Basic automated HPLC system (18)

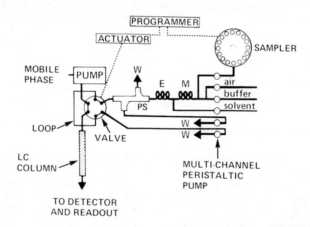

Figure 2. Automated HPLC system with on-line extraction (18)

In place of the diluent (or in addition to it, using more channels) one can add a buffer to obtain the optimum pH for the most efficient extraction. Or, one can add a reagent (or two, or more) and perform a pre-column derivatization, if that is required to improve separation or sensitivity. Multi-channel proportioning pumps are available with up to 28 channels, with pump tubes which can handle acids, bases, and most solvents at flow rates ranging from about 50 microliters per minute up to nearly four milliliters per minute.

Handling Solids. When samples are solid, one generally performs some sort of disintegration or dissolution operation to get the sample into a liquid for greater ease in handling. A SOLIDprep sampler is available which can accommodate up to 20 samples (up to about ten grams each) in glass or plastic containers, dumping them individually into a homogenizer for rapid grinding and dispersion in a liquid phase. Any number of washes can be programmed to eliminate cross-contamination.

Filtration. When samples emerge from the SOLIDprep sampler, they usually require filtration. A Continuous Filter module is available which employs a long roll of one-inch-wide filter paper which is kept in continuous motion so that a new surface is presented for each sample, again eliminating cross-contamination.

Evaporation. Another unit operation is evaporation to dryness and re-dissolution in another solvent. Figure 3 is a diagram of an Evaporation to Dryness Module (EDM). An inert matrix (in the form of a teflon wire of circular cross section) is threaded through a glass evaporator tube and wound around two pulleys. The evaporator tube has a side-arm going to a source of vacuum, so there is a continuous stream of air being drawn into the distal end of the tube. The sample (in solvent "A") is deposited on the moving wire at a point where the in-rushing air pushes the solution along the wire. The sample is not carried into the evaporator tube by the movement of the wire, but is pushed in, flowing as a sheath stream over the surface of the inert matrix. The solvent evaporates and is carried away by the moving air, leaving the solute as a residue on the surface of the matrix.

When the residue emerges at the left end of the evaporator tube, it is re-dissolved in solvent "B" inside the take-off fitting near the pulley. The solution is removed faster than solvent "B" is supplied; thus, the emerging stream is air-segmented so that it can be pumped a considerable distance without loss of sample integrity. When a keeper is used on the matrix, even some volatile pesticides can be transferred from one solvent to another. Also, by stopping the matrix for (say) five minutes while continuing to load the right-hand section of it, then re-starting the matrix and allowing it to pass through the take-off fitting in (say) one half minute, the analyst can obtain a concentration improvement of an order of magnitude. The EDM is destined to open up new automation procedures for the analytical chemist, as it performs both solvent exchanges (which are so often necessary for HPLC) and sample concentraton (which is frequently required in residue analysis).

When an on-line heating step is called for, reaction coils can be mounted in temperature-controlled modules which can be mounted inside so-called analytical cartridges. The top surfaces of these cartridges also serve as mountings for mixers, extractors, phase separators, dialyzers, and a host of other devices for various unit operations.

Combining Unit Operations

Let us now see how some of these unit operations can be combined into automated systems. The first combination doesn't employ chromatography, but it is shown (Figure 4) because it is probably the first totally automated pesticide analyzer ever to operate, having been published by Winter (2) in 1959. It determined carbamates by their inhibition of the enzyme cholinesterase. When no pesticide was present, the enzyme decomposed its substrate (acetyl choline) into a product which reacted with an indicator to give a detectable color change. The presence of the pesticide, however, inhibited the enzymatic reaction and recorder tracings were lower. Employing both continuous dialysis and heated incubation coils, this indirect method was capable of determining carbamates in the range 0.! to 1.0 PPM at the rate of 20 samples per hour.

The rest of the examples in this report involve liquid chromatography and either pre- or post-column sample treatment. Table I will give the reader some idea of the analyses which can be done and what kinds of treatment and detectors are in use. Pre-column treatment, for example, may include cleanup by dialysis, by chromatography, by extraction, by distillation, or by precipitation of protein. Pre-column concentration may be done by chromatographic column or by evaporation. And pre-column derivatization may involve digestion or simpler reactions. At the other end of the table, post-column derivatives have been made for use with both fluorescent and colorimetric detectors.

Pre-Column Reactions. Some pre-column cleanups are as simple as the dilution shown in the analysis of a fermentation broth (Figure 5). The dialyzer is used here not in its usual way, but rather as a filter, thus insuring a particle-free stream for the loop of the injection valve.

The pre-column derivatization shown in Figure 6 involves ion-pairing reactions carried out on some alkaloids. Gfeller, et al (3) formed the ion pairs with picric acid in the mixing coils, then extracted them into chloroform. Detection was at 330 nm.

Post Column Derivatization. The carbamate analyzer described by Moye (4) has been modified to run on the FAST-LC system diagrammed in Figure 7. This arrangement permits displaying both the UV trace and the more sensitive fluorometric trace after a fluorescing label has been introduced by a post-column reaction. The two tracings shown in Figure 8 were obtained on a 2-pen recorder following an injection of 15 ng of lannate. The band broadening of 25% is quite acceptable in view of the increased sensitivity. When the sample level was decreased to 0.4 ng, the UV tracing disappeared into the background noise (Figure 9), while the tracing of the fluorescent derivative showed that still lower levels could be quantitated.

Although there is some evidence to support operation of continuous flow systems without air segmentation, all but the very shortest hydraulic paths show improved performance when bubbles are introduced at regular intervals. The series of tracings in Figure 10 clearly demonstrates the

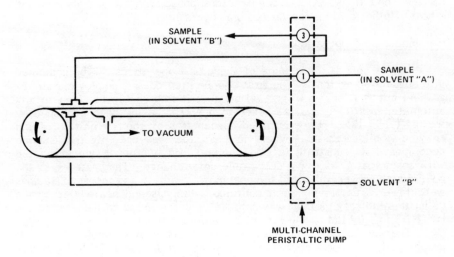

Figure 3.   Evaporation to dryness module (EDM)

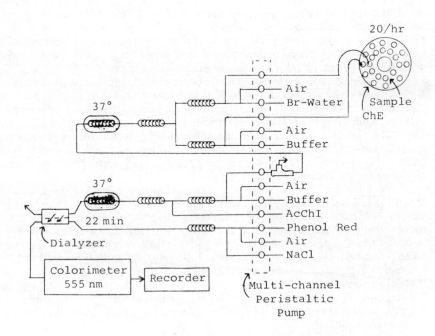

Figure 4.   First fully automated pesticide analyzer

Table I.   Examples of various pre- & post-column
           treatments

Pesticides	Pre-Column	Post-Column	Detector	Sample	Ref
Org-Cl cpds	CU-Chro		EC	Liq	10
Herbicide	CU-Chro	Der	Fl	Sol	7
* Triazine	{ CU-Extr { Der-Dig				12
General	Con-Chro			Liq	11
Carbamates		Der	Fl	Liq	4
Nitrosamides		Der	Vis	Liq	6
Parathion		Der	Vis		13
Org-P cpds		Der	Vis		14
2,4-D			UV	Liq	8
* 2,4-D	CU-Dist		Vis	Liq	15
* Org-P cpds	CU-Dial		Vis		2
Aldrin			e	Liq	16

Non-Pesticides

Vitamins	CU-Extr		UV	Sol	17
Drugs	{ CU-Ppt { CU-Extr { Con-Evap		UV	Liq	9
Alkaloids	{ CU-Extr { Der	Der			3

KEY:

Chro	Chromatography		EC	Electro Chemical
Con	Concentration		Extr	Extraction
CU	Cleanup		Evap	Evaporation
Der	Derivatization		Fl	Fluorescence
Dial	Dialysis		Liq	Liquid
Dig	Digestion		Ppt	Precipitation
Dist	Distillation		Sol	Solid
e	electron capture		UV	Ultraviolet
			Vis	Visible

* Non-LC method

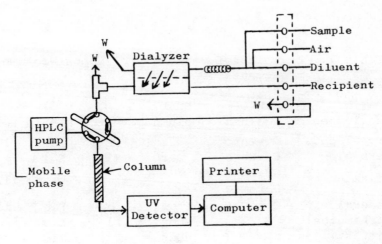

American Laboratory

Figure 5.   Pre-column sample treatment for fermentation broth (19)

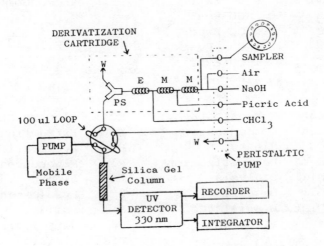

American Laboratory

Figure 6.   Pre-column derivatization for alkaloids (19)

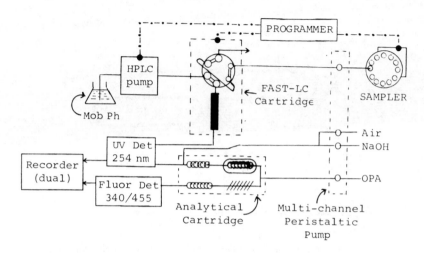

Figure 7.   Post-column derivatization for carbamates

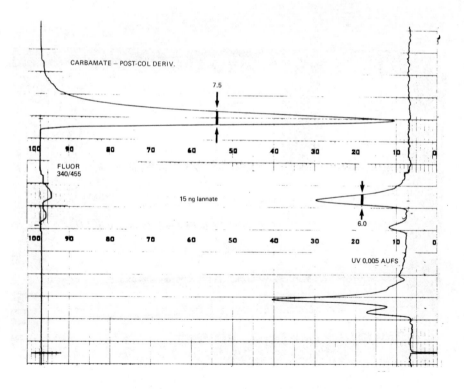

Figure 8.   Comparison of UV and fluorescent tracings from carbamate chromatogram (15 ng)

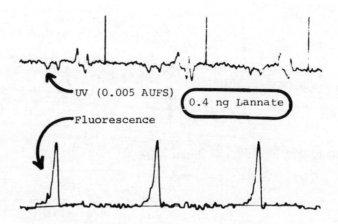

*Figure 9.   Comparison of UV and fluorescent tracings from carbamate chromato-
gram (0.4 ng)*

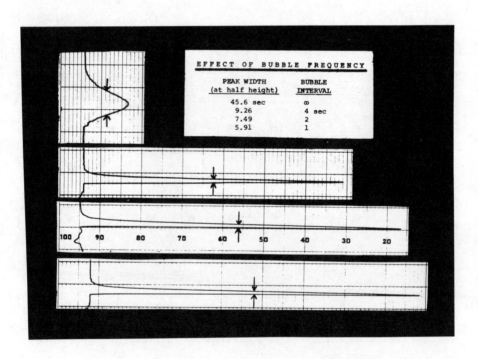

Figure 10.   Effect of air segregation on band broadening

band broadening which can be expected in a typical 5-10 minute post-column reaction when bubbles are omitted; it also shows the improvement possible with increasing bubble frequency.

Gfeller, et al (5) have similarly demonstrated how sample integrity can be retained with air-segmentation. In his analyzer for digitalis glycosides via post-column derivatization, he employed a ten minute reactor at elevated temperatures to make a fluorescent product, improving sensitivity 100-fold while limiting band broadening to only 15% and maintaining CV's of about 1%.

The next pesticide analyzer is described by Singer, et al (6) and used post-column derivatization for a nitrosamide-specific reaction. Following chromatographic separation and UV detection, the stream was air-segmented and treated with the Griess reagent (Figure 11). The nitrosamide was cleaved in dilute acid at 90° C, and the released nitrous acid diazotized the sulfanilic acid which then coupled with the amine to form a highly colored dye. The total time in the derivatization cartridge was about three minutes, giving sensitivities of the order of one nanomole.

Total Automation. In a joint effort between Technicon and Monsanto, Cowell (7) has fully automated the analysis of a water-soluble herbicide. Five-gram samples of animal feeds were ground up with the SOLIDprep sampler and the suspension was filtered with the Continuous Filter module. The solution was then passed through an ion exchange column, and an aliquot drawn into a sample loop. After switching the valve, the mobile phase pushed the sample through a pre-column and into an analytical column for separation. Since there was a sensitivity problem, the column effluent was air-segmented so that a fluorogenic label could be added. The post-column reaction with o-phthalaldehyde took place in a reaction cartridge, and the resulting stream was directed to a fluorometer for detection and recording. The complete flow diagram is shown in Figure 12. At this writing, proprietary considerations preclude giving additional details.

Stevens (8) has described a fully-automated analysis for the pesticide 2,4-D, albeit starting with liquid samples and an auto-injector. His instrumentation (shown in Figure 13) is especially interesting since it conserves the mobile phase by recycling it. The column effluent, after leaving the UV detector, is directed back to the mobile phase reservoir where it is stirred magnetically and used again. The very slight rise in the baseline throughout the day is a small price to pay for this saving of often expensive solvents.

A final example of totally automated HPLC (although it isn't for pesticides) will demonstrate how many different unit operations can be done in a single system to take the tedium out of repetitive analyses. The drug analyzer depicted in Figure 14 was designed to determine therapeutic levels of theophylline in human serum (8). The sampler (in the center) aspirates 50 uL of serum into the analytical cartridge, then to the EDM, and finally to the LC module. The following series of operations takes place at the rate of 20 samples per hour without operator intervention: unmeasured, untreated sample is aspirated, diluted with buffer, and mixed with an internal standard; the system then precipitates the protein, removes the particulates, extracts the analyte (and internal standard) into

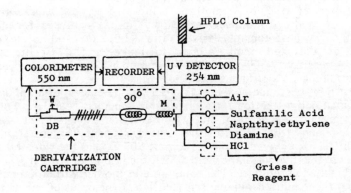

American Laboratory

Figure 11. Post-column derivatization for nitrosamides using a colorimetric detector (19)

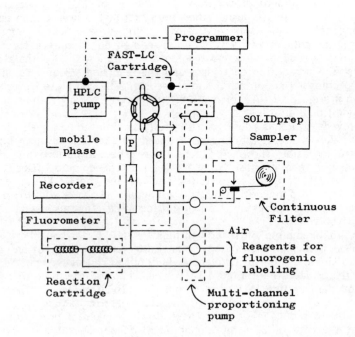

Figure 12. Totally automated herbicide analyzer

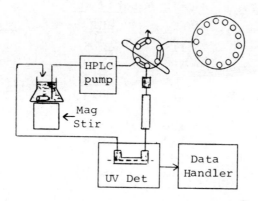

Figure 13. 2,4-D analyzer with mobile phase recycling

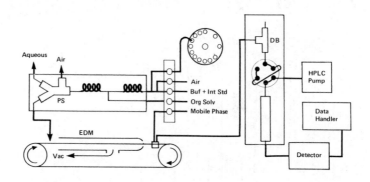

American Laboratory

Figure 14. FAST–LC analyzer for drugs in biological fluids (19)

Table II.   Precision study on drug analyzer.

Sample No.	Peak Heights (mm)		Ratios
	T	BHT	T/BHT
1	117	62.5	1.87
2	106	57.0	1.86
3	118	63.5	1.86
4	128	68.5	1.87
5	133	72.0	1.85
6	120	65.0	1.85
7	113	60.0	1.88
8	107	56.5	1.89
9	115	61.0	1.89
10	116	63.0	1.84
11	108	59.0	1.84
12	120	64.0	1.88

	Average	1.87
	SD	0.018
	CV	0.95%

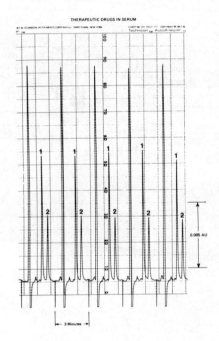

Figure 15.   Repetitive chromatograms from drug analyzer—therapeutic drugs in serum:(1) theophylline, 10µg/mL; (2) internal standard (β-hydroxytheophylline), 30 µg/mL; column: FAST · LC-8, 4.6 × 150 mm (5µ), 3.5 mL/min, 2000 psi (19)

an organic solvent, separates the aqueous and organic phases, and evaporates the organic solution to dryness; finally, it takes up the residue in a second solvent (the mobile phase), makes an automatic injection into the LC column, detects the separated peaks, and produces a finished chromatogram (Figure 15). As shown in Table II, CV's are limited to an average of under 1% despite the small sample size and relatively high analytical rate.

The Future

This brief summary of the state of current instrumentation for the total automation of pesticide analyses will hopefully entice the reader to apply these principles of mechanized unit operations to his own analytical problems. There are enough creative scientists in this field, each trying to lighten his workload, that it seems safe to make this prediction: by 1984 we'll have a pesticide analyzer which will take our unmeasured, untreated sample into one end of the instrument and give us final answers, printed in correct concentration units, twice as fast as today, with one tenth the sample size, and with half the CV's. If an analyst makes up his mind to do so, he can likely automate anything.

Acknowlegement

Thanks are due Mr. Jose Fernandez, Mr. Anthony Pietrantonio, and Dr. J. Russel Gant, without whose dedicated efforts and considerable talents the FAST-LC system described might still be on the drawing board.

Literature Cited

1. Thomas, E J; Mosbacher, C J; Ind Res/Dev, 1979, (2), 97.
2. Winter, G D, Ann NY Acad Sci, 1960, 87, 629.
3. Gfeller, J C; Huen, J M; Thevenin, J P, J Chromat, 1978, 166, 133.
4. Moye, H A; Scherer, S J; St John, P A, Anal Letters, 1977, 10 (13), 1049.
5. Gfeller, J C; Frey, G; Frei, R W, J Chromat, 1977, 142, 271.
6. Singer, G M; Singer S S; Schmidt, D G, J Chromat, 1977, 133, 59.
7. Cowell, J, personal communication, 1979.
8. Stevens, T S, personal communication, 1978.
9. Industrial Method #MA-1001, Technicon Industrial Systems, 511 Benedict Ave, Tarrytown NY 10591
10. Dolphin, R J; Willmott, F W; Mills, A D; Hoogeveen, L P J, J Chromat, 1976, 122, 259.
11. Euston, C B; Baker, D R, Amer Lab, 1979, (3), 91.
12. Hormann, W D; Formica, G; Ramsteiner, K; Eberle, D O, JAOAC, 1972, 55, 1031.
13. Ott, D E, Bull Environ Contam & Tox, 1977, 17, 261.
14. Ramsteiner, K A; Hormann, W D, J Chromat, 1975, 104, 438.
15. Ott, D E; Freistad, H O, JAOAC, 1977, 60 (1), 218.
16. Willmott, F W; Dolphin, R J, J Chromat Sci, 1974, 12, 695.

17.   Industrial Method #MA-1002, Technicon Industrial Systems,
      511 Benedict Ave, Tarrytown NY 10591.
18.   Burns, D A; Res/Dev, 1977, (Apr), 22.
19.   Burns, D A; Fernandez, J I; Pietrantonio, A L; Gant, J R;
      Amer Lab, 1979, (Oct), 80.

RECEIVED December 28, 1979.

Performance Evaluation of Liquid Chromatographic Columns

JOSEPH J. DESTEFANO

E. I. Du Pont de Nemours & Company, Inc., Instrument Products Division, Wilmington, DE 19898

The measurement of column performance criteria to compare columns, column packings, and column loading procedures has long suffered from the non-standardization of test procedures and disagreement as to which of the many performance criteria are most significant. It is the purpose of this paper to point out several practical aspects which must be considered when designing experiments to test column performance.

When one attempts to evaluate LC columns, the first question that must be answered is, "What is good?". In other words, which of the many measurable performance parameters are most significant in terms of the ultimate use of the column--chromatographic analysis? Most experts in HPLC recognize the value of measuring the reduced column performance parameters, such as reduced plate height (h) and column resistance parameter (ϕ), because these dimensionless parameters make comparisons of columns much easier(1). Therefore, experts in chromatography might define a good column as one which had a reduced plate height of between 2 and 5 at the minimum of a reduced plate height versus reduced linear velocity plot. This plot would rise very gently from the minimum as velocity increased. In addition, specific permeability or column resistance parameter values would be consistent with theoretical values for the particle size of the packing and mobile phase viscosity. On the other hand, most users of HPLC columns might define a good column as one which separates their particular sample mixture in relatively short analysis times, one that remains stable for a significantly long period of time, and one which, at the end of its lifetime, can be readily replaced with a column of similar performance.

0-8412-0581-7/80/47-136-031$05.00/0

Both of these definitions of column "goodness" have their
strengths. The experts' definition has the important advantage
of being able to compare results obtained from columns containing
different packings, packed in different ways, using different test
systems, with the testing being performed in various laboratories.
But the users' definition is readily visualized in practical
terms which relate directly to the use of the column in the
individual laboratory. The manufacturers of HPLC columns must of
necessity, be sensitive to the needs of the chromatographers, but
that sensitivity should not mean that quantitative column
evaluation techniques can be ignored. Therefore, the two
definitions of column "goodness" can be combined and simplified
to describe the following performance criteria:

- Column efficiency
- Peak symmetry
- Column permeability
- Column stability
- Column reproducibility

It must then be decided which of the various measures of these
criteria, shown in Table I, are the most generally useful.

COLUMN EFFICIENCY

The most commonly used criterion for judging column
performance is efficiency as measured by the number of theoretical
plates or column plate count (N) exhibited by the column during
the separation of a test mixture. The larger the number of
theoretical plates, the more likely it is that the column will
produce the desired separations. However, while popular, N is not
a complete performance parameter for making comparisons. For
example, N does not take into account particle size as does the
reduced plate height, h. Another measurement, h_{min}, accounts for
all of these factors as well as the mobile phase linear velocity
and sample diffusion. However, N is the term most commonly
recognized as being related to resolution (2), as shown in
Equation 1:

$$\text{Resolution} = \frac{1}{4} \frac{(\alpha - 1)}{(\alpha)} \frac{(k')}{(1 + k')} (\sqrt{N}) \quad \text{(Eq. 1)}$$

Hence, resolution, a key criterion for HPLC column users, is
maximized for a given mobile phase/stationary phase system by
maximizing N. Therefore, even if researchers are inclined to
measure h_{min} as their column efficiency parameter, manufacturers
and users of LC columns can be forgiven for continuing to use N
for internal laboratory purposes.

TABLE I

PERFORMANCE CRITERIA

PERFORMANCE PARAMETER	MEASURED QUANTITY
Column efficiency	• Theoretical plates (N) • Height equivalent to a theoretical plate (HETP) • Reduced plate height (h) • Minimum reduced plate height (h_{min})
Peak Symmetry	• Skew • Tau/Sigma ratio • Asymmetry factor
Column Permeability	• Specific permeability • Column resistance parameter • Pressure required for given flow and mobile phase • Flowrate achieved with given pressure and mobile phase
Column Stability	• Long-term testing • Accelerated testing
Column Reproducibility	• Solute retention (k', capacity factor) • Selectivity (alpha) • Packing characteristics (ie., surface area of adsorbents, surface coverage of bonded phases, etc.)

Plate number by itself is not a sufficiently reliable specification. The plate numbers exhibited by a column can be varied by adjusting the test conditions which are usually designed to maximize those values. For example, very low viscosity mobile phases such as pentane are often used in spite of their practical difficulties such as high volatility to take advantage of the superior mass transfer properties of such solvents. Also, test solutes often are chosen for their low polarity to minimize band tailing, and for their low molecular weight to maximize mass transfer effects rather than for their relationship to real-life samples. In addition, the test is often conducted at unusually slow flowrates to take advantage of the fact that most column packings are more efficient under these conditions.

It should be stressed that there is nothing wrong with these practices. In fact, column performance is best compared under ideal thermodynamic conditions (1). Hence, test systems should be chosen to produce the best column performance, since most workers like to see how well a column really can perform. However, it should be recognized that when considering the number of plates specified for a column, it is necessary to examine the test conditions used to generate that number.

PEAK SYMMETRY

Another problem associated with using the number of theoretical plates (N) as a performance criterion is that there are several equations which can be used to make this calculation, as shown in Figure 1. All of these equations are equivalent for gaussian peaks. However, for tailing peaks, all of the equations are subject to error, some more than others. Therefore, a measure of peak symmetry is required to be able to determine the validity of plate height measurements.

Several methods can be used to measure the degree of tailing exhibited by a peak. One method is a computer-calculated value for the mathematical peak skew based on a novel peak analysis method (3). Briefly summarized, Figure 2 shows that the contour of a peak can be described as having two components: a Gaussian component having the standard deviation, sigma, and an exponential modifier having the time constant, tau. Peak tailing increases with the tau/sigma ratio as shown in Figure 3. This computer-measured ratio can be used to calculate peak skew (4) using Equation 2:

$$\text{skew} = \frac{2(\text{tau/sigma})^3}{[1 + (\text{tau/sigma})^2]^{3/2}} \qquad (\text{Eq. 2})$$

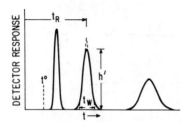

TANGENT METHOD $N = 16 \left(t_R / t_W \right)^2$

AREA METHOD $N = 2\pi \left(h' t_R / A \right)^2$

HALF-WIDTH METHOD $N = 5.54 \left(t_R / W_{1/2} \right)^2$

Figure 1. Equation for calculating the number of theoretical plates (number may vary with: mobile-phase viscosity–column temperature; test solute type and molecular weight; mobile-phase velocity)

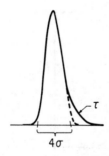

Figure 2. Graphical depiction of τ (exponential tailing component) and σ (gaussian component) features of an asymmetrical peak (τ/σ ratio related to peak skew)

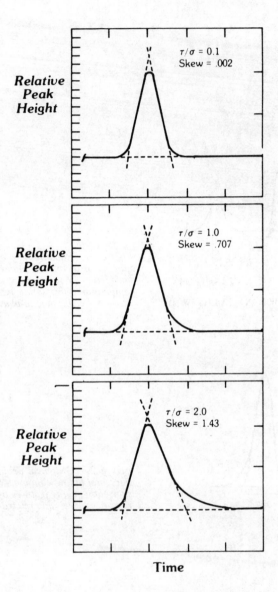

Figure 3.   Computer-generated peaks having different skew (τ/σ) values (constant σ²; variable τ²)

With previous methods, peak skew could not be determined with
sufficient precision to be meaningful (3). However, with the
tau/sigma method, peak skew is easily calculated to give precise,
useful, quantitative information on peak shape. The tau/sigma
method for calculating peak skew is preferred over other methods
since this approach has a sound theoretical basis and values can
be measured with greater precision and accuracy.

However, using peak skew as a performance criterion does
suffer from one drawback: it is necessary to have a sophisticated
computer to make the calculation. Therefore, most users of HPLC
columns will have to use hand-calculated values for peak asymmetry
measurement (Figure 4). However, these hand-calculated values can
be correlated to a first approximation to the computer-generated
values of peak skew using the graph shown in Figure 5.

COLUMN PERMEABILITY

Measurement of column permeability is not clear-cut. The
permeability of a column can be determined in several different
ways. For example, it can be measured as the specific permeability,
K^o (EQ. 3), which corrects for mobile phase viscosity and column
length (5):

$$K^o = \frac{v\eta L}{\Delta P} \qquad (EQ. \ 3)$$

where: v = mobile phase linear velocity (cm/sec)
         η = mobile phase viscosity (poise)
         ΔP = column pressure (atm x 10^6)
         L = column length (cm)

Because of these corrections, specific permeability (K^o) is a
measurement for making comparisons among columns of various types.
Unfortunately, it is difficult to convey what this measurement
means when one is attempting to use a column for a given application. A more easily visualized and more practical way to
describe column permeability is to specify a flowrate measured
for a given column with a given mobile phase at a given pressure
and temperature. Alternatively, the pressure required to produce
a given flowrate with a given mobile phase and temperature can be
specified. These latter two specifications are not absolute
measurements as is the specific permeability, but they do suggest
to the column user the pressure required to carry out a particular
separation.

COLUMN STABILITY

The stability of the column packed bed and the packing bonded

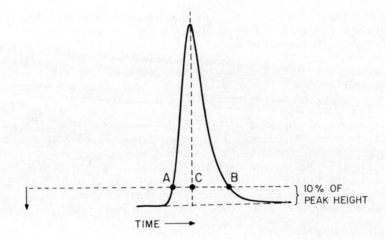

Figure 4.   Method for hand-calculating peak asymmetry (peak asymmetry factor
CB/AC)

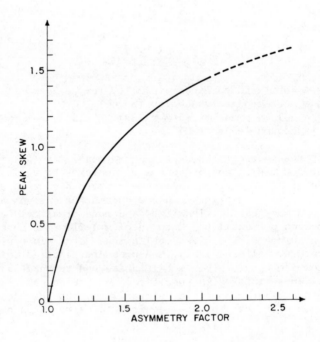

Figure 5.   Graph of first-approximation relationship between peak skew and the
peak asymmetry factor

phase are difficult to measure. Usually, these parameters are measured through accelerated column operation tests. These tests normally utilize auto-samplers, computer data handling systems, high operating pressures and mobile phase flowrates, and elevated temperatures. In addition, difficult mobile phases such as concentrated buffers controlled at very high or very low pH values are often used to examine column stability.

The results of such tests are most useful for determining differences in stability among columns from various manufacturers (ie., they are good <u>relative</u> tests). However, these tests should not be considered absolute measures of column stability since their relationship to real-life use conditions is not known.

COLUMN REPRODUCIBILITIES

Lastly, column-to-column reproducibility experiments must be designed to be sensitive to variations in particle surface area and pore structure, differences in the amount of coverage of bonded phases, and variations in the packed bed which may result during column loading. The retentivity (capacity factor, k') of a test solute under standard conditions is one indicator of column to-column reproducibility. This is especially true for bonded-phase packings since the amount and chemical characteristics of the bonded-phase have large effects on the retention of a solute. This fact is best illustrated by the data presented in Table II. Three different batches of an experimental bonded-phase packing were prepared using three different synthetic approaches. The resulting three packings contained about the same amount of organic material, but the retention of benzyl alcohol on each packing was greatly different under identical chromatographic conditions. The point is clear: simply knowing that a packing is "an octadecyl bonded phase with 15% organic" is not sufficient to determine whether column-to-column variations will degrade a difficult separation.

While retentivity specifications are useful, an even better and more sensitive test for column-to-column reproducibility is the selectivity factor, alpha. This factor measures the ratio of the retentivity or capacity factors (k') of two solutes and determines if there are any differences between columns which would result in the two peaks moving closer together or further apart.

Unlike the retentivity measure, k', relatively minor changes in alpha can have very major effects on resolution as shown in Table III. The variances in alpha that can be tolerated from column-to-column to maintain a given separation ($R_s > 1.25$) is dependent on the alpha value involved. Table IV shows these allowable variances for several values of alpha. This table

TABLE II

EFFECT OF BONDED PHASE ON RETENTION

Packing Lot	% Organic	k' of Benzyl Alcohol*
#1	13.7	4.7
#2	12.6	7.7
#3	13.5	11.2

*Conditions: Bonded phase - 3 experimental batches of propyl nitrile, mobile phase - 2/98 isopropanol/hexane.

TABLE III

EFFECT OF k' AND α ON Rs

FOR: • Constant   N = 4900
       • Constant α = 1.24
       • Varying   k' $\pm$ 10% from 2.0

$$R_s \text{ for k' of } 1.8 = 2.18$$
$$R_s \text{ for k' of } 2.2 = 2.33 \quad \text{or} \quad 2.25 \pm 3.5\%$$

FOR: • Constant   N = 4900
       • Constant   k' = 2.0
       • Varying α $\pm$ 10% from 1.24

$$R_s \text{ for } \alpha \text{ of } 1.116 = 1.21$$
$$R_s \text{ for } \alpha \text{ of } 1.364 = 3.11 \quad \text{or} \quad 2.16 \pm 44\%$$

TABLE IV

ALLOWABLE VARIANCE IN ALPHA

ALPHA	ALLOWABLE VARIANCE*	ALLOWABLE VARIANCE**
1.2	6.7%	10.0%
1.3	13.8%	16.9%
1.4	20.0%	22.9%

*To maintain R_s >1.25 assuming N = 4900 and k' = 2.0
**To maintain R_s >1.25 assuming N = 10,000 and k' = 2.0

demonstrates that the specification of a column selectivity factor
within narrow limits can indicate if the difficult separation
achieved on one column will be accomplished on another similar
column, assuming constant N.

There is a practical difficulty, however, in designing
selectivity tests which will provide this kind of information. It
is difficult to find two solutes which are sufficiently different
so that their retentivities will react differently to changes in
column conditions but which will still elute fairly close to one
another in a single test run. Generally, it is best to use test
compounds with differing functional groups to maximize the chances
of the test system being truly diagnostic.

A final important reproducibility specification should be
considered which applies specifically to bonded-phase packings.
First, the bonded-phase should be specified as being polymeric or
monomeric. If polymeric, information on the % organic or % carbon
for the packing and the chemical structure of the bonded phase
should be provided. However, as shown before, this information
is often not sufficient to determine lot-to-lot chromatographic
reproducibility. If the bonded phase is monomeric, data on the
% organic or % carbon and chemical structure are also useful, but
in addition, the surface coverage calculated from these values (6)
should also be provided (EQ. 4).

$$\text{Coverage} \atop (\mu\text{moles/m}^2) \quad = \quad \frac{W}{M \times S} \quad \text{x}10^6 \quad (\text{EQ. 4})$$

where: W = weight of organic layer (g/g packing)

        M = molecular weight of bonded group (g/Mole)

        S = the specific surface area of the adsorbent,
             corrected for the weight of bonded phase
             (m^2/g packing)

Coverage values should be maintained constant from packing lot to
packing lot to standardize the number of unreacted acidic silanol
groups still present on the silica surface. These active, acidic
groups may interact with solutes to cause a mixed retention
mechanism. This effect will contribute significantly to peak
tailing, especially for basic solutes. Because of steric
considerations, it is not possible to react all of the silanol
groups with the silane reagent. However, by maximizing the number
of groups reacted, the remaining silanols are often made inaccessible for interaction with most solutes because of steric
shielding. Therefore monomeric bonded phases, with maximum,
reproducible surface coverage are preferred to assure reproducible
columns.

TESTING PROCEDURES

Once a decision is made regarding the performance criteria to be measured, it then becomes necessary to design proper tests. The testing of column performance, whether in the laboratory or by column manufacturers, should be carried out with the following points in mind:

- Since it is not possible to use test solutes which would be of interest to all users of HPLC columns, test mixtures should be composed of simple, stable, pure, and readily available organic compounds.

- Column performance is best compared under ideal kinetic conditions. Hence, test systems should be chosen which will produce the best column performance (e.g., low viscosity mobile phases, low molecular weight solutes, etc.).

- Testing apparatus should be designed to minimize band spreading external to the column (e.g., short, narrow connecting tubing between the column and injector and detector, low dead-volume detector flow cell, etc.).

- Test procedures should be designed and described so that they are readily repeatable in different laboratories.

SUMMARY

Using a standard test system, a minimum of the following specifications should be determined by the user when a column is obtained:

- Theoretical Plates
- Peak Symmetry
- Permeability
- Selectivity (or Retentivity)

These column performance criteria should be specified by the manufacturer and users should measure them whenever a new column is received. These column specification measurements should be kept for future reference and re-checked whenever it is suspected that the column has deteriorated.

ABSTRACT

Traditionally, HPLC column performance has been based on theoretical plate measurements. These measurements, while useful, do not provide sufficient data for complete column evaluation. In addition, when calculating theoretical plates, errors caused by peak tailing can lead to an overestimation of the resolving power of a chromatographic column. To more fully evaluate a modern, high-performance liquid chromatographic column, several performance specifications should be monitored in addition to theoretical plate measurements. These specifications include: column permeability, peak symmetry, and sample retention characteristics. All performance parameters should be measured under a standard set of well-documented chromatographic conditions. The factors affecting the choice of performance measures and test conditions are given.

LITERATURE CITED

1. Knox, J.; Bristow, P.; Chromatographia, 1977, 10, 279.

2. Snyder, L.R.; Kirkland, J.J.; "Introduction to Modern Liquid Chromatography"; John Wiley and Sons, New York, 1974; pp. 35, 36, 37, 38.

3. Kirkland, J.J.; Yau, W.W.; Stoklosa, H.J., Dilks, Jr., C.H.; J. Chromatogr. Sci., 1977, 15, 303.

4. Grushka, E.; Anal. Chem., 1972, 44, 1733.

5. Anonymous, Du Pont Instruments, L. C. Column Report - Column Performance Criteria, No. E18181.

6. Unger, K.K.; Becker, N.; Roumeliotis, P.; J. Chromatogr. Sci., 1976, 125, 115.

RECEIVED December 28, 1979.

New Techniques for Improving Mobile-Phase Selectivity in Reversed-Phase Chromatography

STEPHEN R. BAKALYAR

Rheodyne, Inc., 2809 Tenth St., Berkeley, CA 94710

This paper reviews recently developed techniques for improving the selectivity of reversed phase high performance liquid chromatography (RPLC) by manipulating mobile phase chemistry. The discussion is limited to packings with bonded hydrocarbon coatings. That is, it does not deal with "partition" chromatography, where there is a deliberate attempt to coat the packing with a bulk organic liquid phase. Under some conditions the packing may be coated with nonpolar mobile phase modifier species. But we will consider the stationary phase to be a nonpolar, hydrocarbonaceous surface.

The chemical interactions that impart a large range of selectivities to RPLC can be divided into two types. The first type are primary equilibria between solutes and the two phases. The second type are secondary equilibria between solutes and complexing agents in the mobile phase. These two types can in turn be subdivided into various specific intermolecular interactions. Figure 1 summarizes these. The following discussion of RPLC selectivity improvement techniques will use figure 1 as an outline, discussing each mechanism in turn.

Primary Equilibria

Hydrophobic Effect. The primary retention force in normal phase adsorption is the attraction of solute polar moieties to the polar stationary phase. In contrast, the retention force in RPLC is repulsion from the mobile phase. The stationary phase is a relatively passive surface, the solute attraction for the hydrocarbon stationary phase being weak and non-selective. How does this come about, and what are the resultant selectivity characteristics?

0-8412-0581-7/80/47-136-045$05.00/0

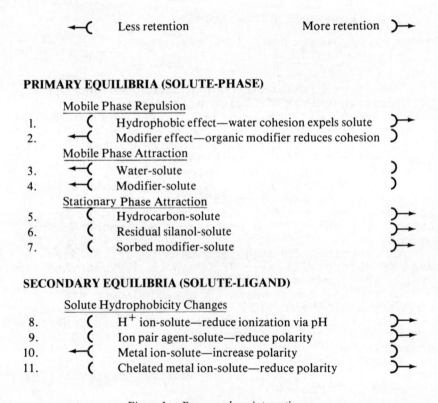

Figure 1.   Reverse-phase interactions

   Water has a very high cohesive energy density,
primarily caused by intermolecular hydrogen bonds.
This cohesiveness produces a high surface tension, and
all but the most polar or ionic solutes experience a
hydrophobic force that drives them onto the stationary
phase and causes retention. Most chemists are not as
familiar with hydrophobic effects as with polar interactions, since the former have only recently been discussed in the chromatographic literature (1-6). The
hydrophobic effect results from the strong attractive
forces between water molecules. The "structure" of
the water must be distorted or disrupted when a solute
is dissolved. We can think of the solute as forming a
cavity in the water. Highly polar or ionic solutes can
themselves interact strongly with water, compensating
for this distortion, and are thus easily solvated.
But nonpolar solutes are relatively insoluble and are
squeezed out of the water and onto the apolar stationary phase surface. In this association the total number of water molecules in contact with the solute is
reduced, and the water cavity area decreases, an ener-
getically favorable event.
   Although the detailed retention mechanisms are as
yet unclear (see for example 7,8), there is a building
consensus that reversed phase chromatography is dominated by the hydrophobic effect. Retention is therefore
primarily a function of solution phenomena in the
mobile phase, and it is not surprising that RPLC has
many ways to modify selectivity by manipulating the
chemical nature of the mobile phase.
   The hydrophobic effect imparts to RPLC an inherently high selectivity for differences in the hydrocarbon backbone of solutes. The addition of a methylene group or other bulky, relatively nonpolar moiety
such as a chloro group causes significantly increased
retention.

   Modifier Effect. The hydrophobic force can be
reduced by decreasing the mobile phase cohesiveness.
Since water is the most cohesive common liquid, adding
any miscible organic solvent will do. The effect will
be larger with less polar solvents and greater concentrations. Adjusting organic modifier ("B" solvent)
concentration is the primary means of changing retention.

   Water-Solute. Polar moieties of the solute cause
reduced retention because they can hydrogen bond and
dipole interact with the water, partially mitigating
the hydrophobic repulsive force. In more familiar

terms, we simply say that greater solute polarity imparts greater water solubility, and this favors an
equilibrium shift to the mobile phase.

Modifier-Solute. Polar groups of the solute can
interact with polar groups of solvent modifiers. A
particular solvent modifier will exhibit specific
solute interactions which another solvent of approximately similar strength or polarity does not have.
Although we speak of a solvent's "polarity" as if it
were a fixed inherent property, solvent polarity does
not accurately reflect the eluting strength for all
solutes. This is because strength is the sum total of
three types of molecular interactions acting concur-
rently, e.g., dispersion, orientation, and hydrogen
bonding. Each solvent has these interactive components
in a unique ratio. Thus two solvents of approximately
equal polarity can have different interactive profiles.
Figure 2 shows the profiles for three common solvents,
methanol, acetonitrile, and tetrahydrofuran. Depending
on how one defines the various interactive strengths
the numerical values differ. But this illustration
will serve our purpose. Solutes also have a profile,
and when there is a good interactive match between
solvent and solute the solvent strength is particularly
high, and the retention decreases.
     A simple example of modifier-solute selectivity
effects is shown in figure 3. The retention of five
different functional groups is compared for three different binary solvent mixtures with water. The acetonitrile and tetrahydrofuran concentrations have been
choosen so as to provide the same retention for benzene, a k' of 4.7. Different solvent modifiers are
strongest for different solutes. The shifting of peaks
relative to each other is quite pronounced among the
different solvents.
     Polar group selectivity also occurs in ternary
solvent systems (5,10). For example, the addition of
5% to 25% of a third solvent to a water-acetonitrile
mixture can alter the relative retention of peaks, and
often resolve overlapping peaks. Dolan et al (11) have
employed ternary mobile phases of water, methanol and
tetrahydrofuran to analyze vitamin tablets where inter-
ferring peaks could not be resolved with binary mixtures. See Figure 4.

Hydrocarbon-Solute. As mentioned previously, the
attraction of the bonded hydrocarbon stationary phase
for the solute is weak and unselective, and probably is
not a significant direct contributor to selectivity.

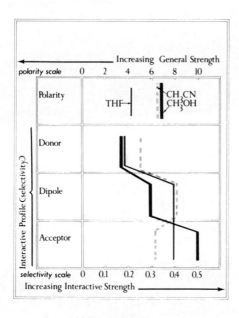

Figure 2.   Solvent interactive profiles (the three bottom scales show the interactive strengths of the solvents: hydrogen donor, hydrogen acceptor, and dipole)

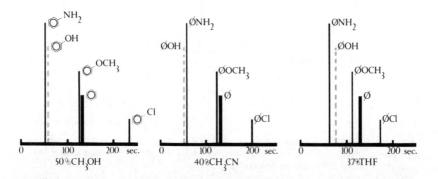

Figure 3.   Selectivity of modifier solvent (column: 250 × 4.6 mm I.D. 10-μm Lichrosorb RP-8; other columns provide differing degrees of selectivity effects with these solvents)

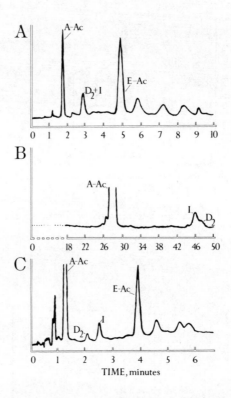

Journal of Chromatographic Science

Figure 4.   Separation using ternary mobile phase (Conditions: (A) 95% methanol in water at 4 mL/min on 30-cm column; (B) 95% methanol in water at 0.5 mL/ min on four 30-cm columns (increased number of plates); (C) 12% water/25% THF/63% methanol at 6 mL/min on 30-cm column (increased selectivity); column was Microbondapak C_{18}, 10 μm) (11)

Residual Silanol-Solute. Even after bonding
hydrocarbon chains to the silica, some residual unreacted silanol groups remain. These polar groups can
interact with polar functional groups of solutes, i.e.,
normal phase attraction. The precise role they play in
retention and selectivity is not well understood, but
the different selectivities exhibited by various packings probably result at least in part from their different amounts of unreacted silanol groups.

Sorbed Modifier-Solute. Organic modifier added to
the aqueous mobile phase can perhaps sorb onto the
stationary phase. This could have various effects,
including shielding the bonded hydrocarbon layer and/or
the residual silanol groups. It is possible that the
selectivity effects mentioned in the modifier effect
section are actually the result of stationary phase
interactions. Whatever the precise mechanism, the fact
remains that addition of different organic modifiers
provides a powerful selectivity tool.

Secondary Equilibria

Secondary selectivity interactions will now be
discussed. The idea of secondary interactions is to
change the effective polarity of the solute, hopefully
to different degrees for different solutes so as to
improve the selectivity. Making the solute more hydrophobic increases its retention; making it less hydrophobic decreases its retention.

Hydrogen Ion-Solute. These interactions are
simply the well known effects of pH on acidic, basic or
zwitterionic solutes. The reactions below show the
mechanisms with a weak base and a weak acid.

Hydrophilic solute		Hydrophobic solute
RNH_3^+	$\rightleftharpoons$	$RNH_2 + H^+$
$H^+ + RCO_2^-$	$\rightleftharpoons$	RCO_2H

The work of Twitchett and Moffat (12) provides a good
example. They investigated 30 compounds selected as
representative of a wide variety of drug substances and
claimed that drugs of any lipid solubility, molecular
weight, chemical structure and acidity/basicity can be

chromatographed by RPLC if an appropriate eluent (solvent composition and pH) is chosen.

   Ion Pair Agent-Solute. The polarity of solutes which have ionic character can also be adjusted by complexing the solute with an oppositely charged ion. Termed ion pair chromatography, it is particularly useful for increasing the retention of fully charged species, such as quaternary amines and sulfonic acids, where pH control has little effect. This technique has received much attention in the recent literature (13-16). Gloor and Johnson have written a useful practical guide (17).
   When an ion pair agent is added to the mobile phase, the agent conjugates with the oppositely charged solute, and the hydrophobic end of the agent contributes to the shift toward hydrophobicity. The following reactions show a quaternary amine pairing with a weak acid, and an alkyl sulfate pairing with a weak base; both are under pH conditions where the solute is charged.

Hydrophilic solute	Ion pair agent		Hydrophobic paired solute

$$RCO_2^- \quad + \quad {}^+N(C_4H_9)_4 \quad \rightleftharpoons RCO_2N(C_4H_9)_4$$

$$RNH_3^+ \quad + \quad {}^-O_3SO(CH_2)_{11}CH_3 \rightleftharpoons RNH_3O_3SO(CH_2)_{11}CH_3$$

   There is debate over whether the ion pair forms in the mobile phase to form a new species, which then sorbs to the stationary phase, or whether the agent sorbs to the stationary phase by itself, forming in effect an ion exchange surface (18-20). In any case, the effect is roughly the same - an increase in retention of solute species that are charged oppositely to the ion-pairing agent.

   Metal Ion-Solute. Transition metals have been loaded onto classical LC columns and used for selective separations (21). More recently successful attempts have been made to control selectivity by metal additions directly to the mobile phase (22). This is a form of ion pairing, but the pairing agent is polar instead of hydrophobic. Silver ion has been used for the separation of olefins (23). M. deRuyter and A. deLeenheer have employed argentation chromatography to resolve difficult mixtures of Retinyl esters (24).
   In these cases where the metal is added to the

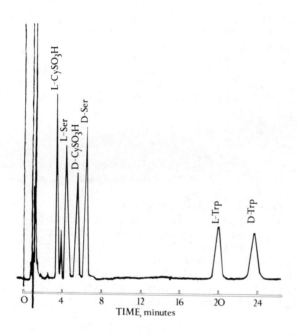

Analytical Chemistry

Figure 5.   Optical isomer separation using optically active metal chelate mobile phase (26)

mobile phase, the effect is to increase solubility, and thereby reduce retention. In the next technique the opposite effect is achieved.

Chelated Metal Ion-Solute. If the metal is first chelated with a relatively hydrophobic chelating agent, solute interactions will increase retention. Cooke et al (25) have developed such a technique using 4-dode-cyldiethylenetriamine and Zn(II). Not only does this chelated metal greatly increase retention for certain anionic solutes, presumably by an ion pairing inter-action, but the relatively rigid conformation of the metal chelate imparts marked selectivities.

An impressive application of the chelated metal technique was recently reported by LePage et al (26). Optical isomers of dansyl amino acids were resolved using an optically active metal chelate. High alpha values and high efficiency conditions were achieved, as shown in figure 5.

Summary

Selectivity in reversed-phase liquid chromato-graphy separations can be controlled by a multitude of techniques. Reversed-phase separations are inherently selective for the hydrocarbon structure of solutes, owing to the nature of hydrophobic effects. Selectiv-ity for the polar structure can be controlled by choice of organic solvent modifiers, and/or choice of complex-ing agents. The complexing agents take advantage of the wealth of classical chemistry available, such as charge control via pH adjustment, as well as sophisti-cated metal chelating techniques using optical isomers.

"Literature Cited"

1. Horvath, E.; Melander, W.; and Molnar, I. J. Chromatogr., 1976, 125, 129.
2. Karger, B.L.; Gant, J.R.; Hartkopf, A.; and Weiner, P.H., J. Chromatogr., 1976, 128, 65.
3. Horvath, C.; Melander, N.; Molnar, I., Anal. Chem., 1977, 49, 142.
4. Molnar, I.; Horvath, C., J. Chromatogr., 1977, 142, 623.
5. Bakalyar, S.R.; McIlwrick, R.; Roggendorf, E., J. Chromatogr., 1977, 142, 353.
6. Bakalyar, S.R., American Laboratory, 1978, 10, 43.
7. Scott, R.P.W.; Kucera, P., J. Chromatogr., 1977, 142, 213.

8. Colin, H.; Guiochon, G., J. Chromatogr., 1977, 141, 289.
9. Bakalyar, S.R., Spectra-Physics Chromatography Review, 1977, 3, (2),6.
10. Tanaka, N.; Goodell, H.; Karger, B.L., J. Chromatogr., 1978, 158, 233.
11. Dolan, J.W.; Gant, J.R.; Tanaka, N.; Giese, R.W.; Karger, B.L., J. Chromatogr. Sci., 1978,16, 616.
12. Twitchett, P.J.; Moffat, A.C., J. Chromatogr., 1975 111, 149.
13. Wahlund, K.G., J. Chromatogr., 1975, 115, 411.
14. Knox, J.H.; Jurand, J., J. Chromatogr., 1978, 149, 297.
15. Johansson, I.M.; Wahlund, K.G.; Schill, G., J. Chromatogr., 1978, 149, 281.
16. Tilly-Melin, A.; Askemark, Y.; Wahlund, G.; Schill, G., Anal. Chem., 1979, 51, 976.
17. Gloor, R.; Johnson, E.L., J. Chromatogr. Sci., 1977 15, 413.
18. Kissinger, P.T., Anal. Chem., 1977, 49, 883.
19. Tomlinson, E.; Riley, C.M.; Jefferies, T.M., J. Chromatogr., 1979, 173, 89.
20. Scott, R.P.W.; Kucera, P., J. Chromatogr., 1979, 175, 51.
21. Walton, H.F., Separ. Purif. Methods, 1975, 4, 189.
22. Sternson, L.A.; DeWitte, W.J., J. Chromatogr., 1977, 137, 305.
23. Vonach, B.; Schomburg, G., J. Chromatogr., 1978, 149, 417.
24. de Ruyter, M.G.M.; deLeenheer, A.P., Anal. Chem., 1979, 51, 43.
25. Cooke, N.H.C.; Viavattene, R.L.; Eksteen, R.; Wong, W.S.; Davies, G.; Karger, B.L., J. Chromatogr., 1978, 149, 391.
26. LePage, J.N.; Linder, W.; Davies, G.; Sietz, D.E.; Karger, B.L., Anal. Chem., 1979, 51, 433.

RECEIVED December 28, 1979.

Electrochemical Detection of Picomole Amounts of Oxidizable and Reducible Residues Separated by Liquid Chromatography

PETER T. KISSINGER, KARL BRATIN, WILLIAM P. KING, and JOHN R. RICE

Department of Chemistry, Purdue University, West Lafayette, IN 47907

Liquid chromatography with electrochemical detection (LCEC) is coming into widespread use for the trace determination of easily oxidizable and reducible organic compounds. Detection limits at the 0.1 picomole level have been achieved for a number of oxidizable compounds. Due to problems with dissolved oxygen and electrode stability, the limit of detection for easily reducible substances is currently about tenfold less favorable. The modern interest in electrochemical detectors for liquid chromatography was stimulated by the recognition that this technique was ideal for the study of aromatic metabolism in the mammalian central nervous system. Most of the papers published during the past eight years have focused on the applications of the LCEC technique to neurochemical problems. Since the first commercial detectors became available in 1974, a number of other areas of application have been explored. A running bibliography of LCEC applications is frequently updated and provides a useful overview of current applications (1). The basic concepts of LCEC have been recently reviewed in several places (2-5). Thus far there have been few applications to pesticide or herbicide residues. The purpose of this chapter is to briefly introduce the technique and explore its potential utility for the determination of some additives and residues of agricultural interest. It is assumed that the reader is informed about liquid chromatography but has little knowledge of organic electrochemistry.

Basic Concepts

Liquid chromatography (LC) and hydrodynamic electrochemistry are, for the most part, very compatible technologies which in combination yield important advantages for a number of trace determinations. In order of decreasing importance, the three major advantages are selectivity, sensitivity, and low cost. The use of modern LC for residue determinations requires a selective detector with a rapid response time, wide dynamic range, and low

active dead volume (<20 µL). Because electrochemistry is a surface technique, small volume transducers (<1 µL) can easily be
constructed. Since there is no need for a light beam to indirectly convert the chemical information into an electric current,
electrochemical detectors can be simpler and considerably less
expensive than popular ultraviolet absorption and fluorescence
detectors.

A commercial thin-layer amperometric detector is illustrated
in Figure 1. The cell consists of two blocks of inert fluorocarbon separated by a thin fluorocarbon gasket (typ. 50 µm thick)
which is slotted to form a rectangular channel. A working
electrode is positioned in the lower block (typ. 3 mm diameter)
and an auxiliary electrode of the same size is located across
the channel in the upper block. A reference electrode probe is
positioned downstream opposite the exit port. The potential
difference between the working electrode material and the bulk of
the solution film is controlled relative to the reference electrode. This applied potential is selected so that the analytes
of interest will rapidly react at the working electrode. All
charge passing through the cell (due to the electron transfer at
the working electrode) flows between the working and auxiliary
electrodes. Essentially no charge passes along the thin-layer
of solution to the reference electrode. This geometry is ideal
from an electrochemical point of view since the solution resistance is very small even at low ionic strengths (due to the
nearness of A and W) resulting in very little difficulty with
the iR drop problem common to many thin-layer electrochemical
experiments. A more detailed discussion of the critical features
of cell geometry is presented elsewhere (6). For those not
familiar with three-electrode electrochemical instrumentation,
a chapter on this subject is available (7). The cell depicted
in Figure 1 will typically exhibit a linear dynamic range of six
orders of magnitude.

Three working electrode materials are commonly used in thin-
layer amperometric detectors: glassy (or "vitreous") carbon,
carbon paste, and gold amalgamated with a thin film of mercury.
The selection of the proper working electrode for a given application has been described (8). In general, carbon electrodes
are used for oxidation and mercury electrodes for reductions.
While carbon paste electrodes are often advantageous for use
with aqueous mobile phases, they are eroded by non-aqueous solvents and therefore glassy carbon is preferred in many applications.

The compatibility of electrochemical detection with the
various modes of liquid chromatography is limited. For all
practical purposes, electrochemical detection is not suitable for
use with normal phase adsorption or partition chromatography due
to the solvents of low dielectric constant used as the mobile
phase. On the other hand, reverse-phase adsorption and partition
(including ion-exchange or ion-pairing systems) are highly com-

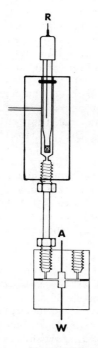

R

A

W

Bioanalytical Systems

Figure 1. Thin-layer electrochemical detector ((A) auxiliary electrode; (R) reference electrode; (W) working electrode)

patible with the LCEC technique since the mobile phase is commonly
an aqueous buffer with or without one or more polar organic
modifiers (typically methanol or acetonitrile). Amperometric
detection requires the presence of ions in the mobile phase.
While ionic strengths as low as 1 mM have been used with the cell
illustrated in Figure 1, most LCEC experiments are carried out
with a minimum of 0.05 M buffer salts in the mobile phase. Post-
column mobile phase changes (pH, ionic strength, solvent content)
and post-column reactions (redox cross reactions, derivatizations, enzyme catalyzed reactions) can expand the utility of
electrochemical as well as other detectors. These subjects have
recently been treated in some detail (9). Suffice it to say that
direct detection, without post-column chemistry, is always
preferable (more reliable, more sensitive, less expensive).

What type of compounds are the most suitable for an LCEC
method? Easily oxidized compounds (phenols, aromatic amines,
mercaptans, etc.) are almost always good candidates. Easily
reducible compounds (nitro compounds, quinones, etc.) are also
a good possibility. It is important to recognize that the
analytes must be oxidized or reduced at an electrode. Many
compounds can undergo rapid chemical redox reactions in homogeneous solution, but behave very poorly at electrode surfaces.
There is no substitute to obtaining a voltammogram (current-
voltage characteristic curve) when evaluating the possibility of
solving a problem with LCEC.

Cyclic voltammetry (CV) permits the very rapid evaluation of
a compound's electrochemical reactivity. This technique is
performed at a stationary electrode placed in a stationary solution. A typical response is illustrated in Figure 2 for the
oxidation of hydroquinone at a carbon paste electrode in a medium
suitable for reverse phase chromatography. Although the $2e^-$
oxidation to benzoquinone is chemically reversible (both the
oxidized and reduced form of the couple are stable during the
time for this experiment to be completed), the electron transfer
kinetics are slow, as indicated by the shape of the voltammogram.
To select a potential suitable for LCEC, a value 50 mV beyond the
oxidation peak potential is a good first choice. The hydroquinone nucleus may be thought of as an electroactive functional
group or an "electrophore" and substituted hydroquinones will
behave similarly. Thus, it is not necessary to obtain CV information for all compounds prior to making a judgment about the
applicability of LCEC.

Figure 3 illustrates several reverse-phase chromatograms for
a synthetic mixture of hydroquinone (peak 1), phenylalanine
(peak 2), 2-methylhydroquinone (peak 3), 2,5-dimethylhydroquinone
(peak 4), and phenol (peak 5). In Figure 3A an electrochemical
detector was used at a potential of +1.07 volts vs. Ag/AgCl and
the chromatogram is plotted as a "hydrodynamic chronoamperogram"
using the jargon of electrochemists. Of the material injected
(1,4 ng; 2,10 ng; 3,8 ng; 4,20 ng; 5,4 ng) all were detected with

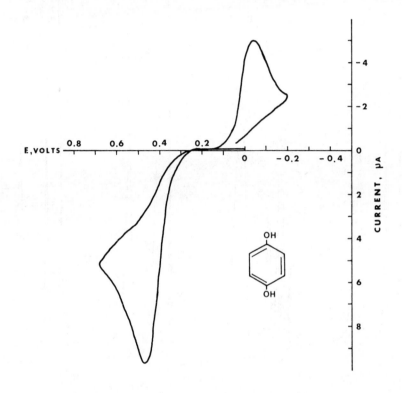

Figure 2. Cyclic voltammogram of hydroquinone (1 mM) at a carbon paste electrode (pH 5.0 McIlvane buffer)

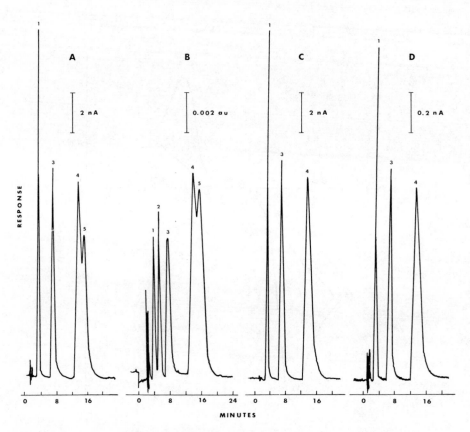

Bioanalytical Systems

*Figure 3.   Reverse-phase (C_{18}) chromatograms of hydroquinone (cmpd 1), phenyl-
alanine (cmpd 2), 2-methylhydroquinone (cmpd 3), 2,5-dimethylhydroquinone
(cmpd 4), and phenol (cmpd 5).*

*Mobile phase pH 5.0 McIlvane buffer, 1 mL/min: (A) Electrochemical detector, carbon
paste electrode, +1.07 volts; (1) 4 ng; (2) 10 ng (not detected); (3) 8 ng; (4) 20 ng; (5)
4 ng; (B) UV detector, 254 nm; (1) 2 μg; (2) 5 μg; (3) 4 μg; (4) 10 μg; (5) 2 μg;
(C) electrochemical detector, +0.500 volts (same amounts injected as for A); (D) electrochemical detector, +0.500 volts; (1) 0.40 ng; (2) 1.0 ng; (3) 0.80 ng; (4) 2.0 ng; (5)
0.40 ng.*

good signal-to-noise except phenylalanine, which is not oxidizable at the electrode potential used. Figure 3B illustrates the corresponding chromatogram obtained with a conventional 254 nm UV absorption detector (500-fold more material was injected in this case compared with Figure 3A!). Note that phenylalanine is now easily detected. This points to one of the most important advantages of LCEC for biomedical and environmental samples: it is almost always highly selective compared to UV absorption at 254 nm. This feature is of obvious importance to residue work, where a very large number of UV absorbing natural products can become probable interferences.

The selectivity of electrochemical detection is further illustrated by Figure 3C (same amounts injected as per Figure 3A) which was obtained at a significantly lower potential (+ 0.500 volts), where only the three hydroquinones can be detected. When the amount of material injected was reduced by a factor of ten, the chromatogram in Figure 3D was obtained. While only 0.4 ng of hydroquinone is detected, the signal-to-noise ratio remains excellent and the amount injected could be further reduced by a factor of ten. In some ideal cases, minimum detectable quantities (S/N = 2) of 1 pg (typically 5 femtomoles) have been achieved with standard solutions.

Voltammetric information can be obtained directly from the electrochemical detector as well as from cyclic voltammetry experiments. The recommended experiment is "normalized hydrodynamic voltammetry" in which a sample is repeatedly injected onto the LC column with different potentials applied to the amperometric detector. The peak heights (oxidation or reduction current) for one or more components are tabulated as a function of potential and divided by the peak height for each compound found at the highest potential utilized. At sufficiently large potentials, the electrochemistry of most substances becomes mass transport limited and the peak height will no longer depend on potential. When the normalized peak height values ($\emptyset$) for each component are plotted as a function of the applied potential, a conventional S-shaped voltammogram will result, as illustrated in Figure 4 for the hydroquinone and phenol peaks in Figure 3. It is important to recognize that it is not necessary to operate an electrochemical detector in the "plateau" region of the voltammogram. In fact, some substances do not reach a plateau prior to the background limit imposed by the mobile phase.

It should now be clear why phenol was "tuned out" in Figures 3C and 3D. Such curves can be very useful to insure the identity of a given component by comparison with the voltammogram obtained for repeated injection of an appropriate standard. Note that the relative oxidation currents are plotted "down" in Figure 4, in keeping with the convention of American electroanalytical chemists. The process of obtaining a normalized hydrodynamic voltammogram is similar to recording a UV spectrum by repeated injection of a sample at different wavelengths. In both UV

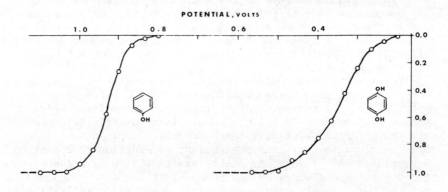

Bioanalytical Systems

Figure 4.   Normalized hydrodynamic voltammograms obtained from liquid chromatograms (conditions as per Figure 3A)

spectrophotometry and electrochemistry it is often sufficient for qualitative purposes to measure relative peak heights at only 2 or 3 energies (i.e., a complete spectrum or voltammogram is not always necessary).

Phenolic Residues

An obvious application of LCEC in residue analysis is to pesticides capable of yielding a phenolic group. As described above, phenols are well suited to electrochemical detection since they are readily oxidized at a carbon electrode. About 25% of the pesticides in use today possess a substituted phenol moiety which may be cleaved from the molecule. Among such compounds are a few herbicides, fungicides and several organo-phosphorous insecticides. Roughly two-thirds of all carbamates are based on phenols since their anticholinesterase activity is increased by a substituted phenyl ring (10). Many phenols are also used directly as pesticides. For example, chlorophenols have many applications as insecticides, fungicides, antiseptics and disinfectants. Pentachlorophenol alone is the second most widely used biocide in North America with 80 million pounds produced in 1977. Phenols and polyphenols also result from the metabolism of many pesticides containing an aromatic nucleus. This connection between pesticides and phenols has often been utilized in residue analytical methodology owing to the large body of analytical chemistry available for phenols.

The usable anodic range of various carbon electrodes extends to about +1.2 to +1.4 volts. The limiting factor is background current due to the oxidation of mobile phase components (e.g. water, methanol, impurities). Carbon paste based on mineral oil (CPO) was used as the electrode for the voltammetric data reported here. Glassy carbon, due to its structural stability, is also used in LCEC where separations require a high proportion of organic modifier. Glassy carbon works well as an electrode although it exhibits higher noise and background currents than CPO. For most compounds voltammograms on glassy carbon are comparable to those on CPO, however, the waves of certain compounds are broadened and shifted to higher potentials relative to CPO. Most phenol oxidation potentials lie within the accessible range. Phenol itself oxidizes at roughly intermediate potentials. The oxidation potential of a substituted phenol is strongly influenced by inductive, resonance, and steric effects of the substituents. Groups capable of donating electron density to the aromatic ring decrease the oxidation potentials (shifting the bands in a negative direction) while electron withdrawal has the opposite effect. An earlier study (11) demonstrated that half-wave potentials ($E_{1/2}$) of monosubstituted phenols and anilines are linearly related to Hammett's substituent constants. Reasonable success was achieved in predicting the half-wave potentials for disubstituted compounds.

For those unfamiliar with the "Kissinger Charts" (Figure 5) for tabulating electrochemical information, a detailed explanation is available on request to the authors. In brief, the box for each compound is plotted between the cyclic voltammetric peak potential (E_p) and the potential at half the peak current ($E_{p/2}$). The width of the box is indicative of the electron transfer rate constant. An empty box indicates that the electrochemical process results in a product which cannot be reduced on the reverse CV trace. A cross-hatched box indicates that a reducible product is produced but that it is not the fundamental oxidation product. A filled box means that the electron transfer reaction is chemically reversible (e.g. a hydroquinone would be typical of such a compound).

The role oxidation potential plays in selectivity is demonstrated in the determination of pentachlorophenol (PCP) and one of its metabolites tetrachlorohydroquinone (TCHQ) (12). When a simple ether extraction was used for isolation of these phenols from acidified urine, the extracts also contain considerable acid and neutral dietary components. Figure 6A illustrates a chromatogram obtained at sufficient detector potential such that the response from electroactive coextractants obscures the presence of low levels of TCHQ. TCHQ can be oxidized at quite low potentials (+0.2 V) where the response from interfering substances is considerably reduced, Figure 6B. PCP, on the other hand, is only detected at higher potentials (>+0.85V). A higher k' helps separate PCP from interferences but the baseline and lower response makes the lower limit of detection for PCP about an order of magnitude above that for TCHQ. This example also points up the increased selectivity towards phenolic metabolites. Oxidation potentials are significantly reduced by the commonly encountered hydroxylation process. Furthermore, low operating potentials give low background and noise, allowing the high sensitivity to be easily realized for secondary metabolites.

Carbamate esters are in widespread use as agricultural pesticides. Several phenolic parents of carbamates are listed in Figure 5. Of the carbamates tested by cyclic voltammetry none were oxidized at useful potentials. If the pH is increased to 11 or above, oxidation of the phenol formed by hydrolysis is detected. Derivatization of the liberated phenol is often used in gas chromatographic procedures to circumvent problems with chromatography of the labile carbamate (13). LCEC can be adapted to many existing procedures for carbamates by detection of the phenol. Liquid chromatography avoids the need for derivatizing many compounds and offers the advantages of automated pre-column trace enrichment (14). Such an approach permits a very rapid method for determination of phenolic carbamates in relatively simple matrices such as groundwaters by direct sample injection after treating the sample with base (Figure 7). Since the free phenol is the primary degradation product it must be measured beforehand. Post-column reaction schemes for carbamates

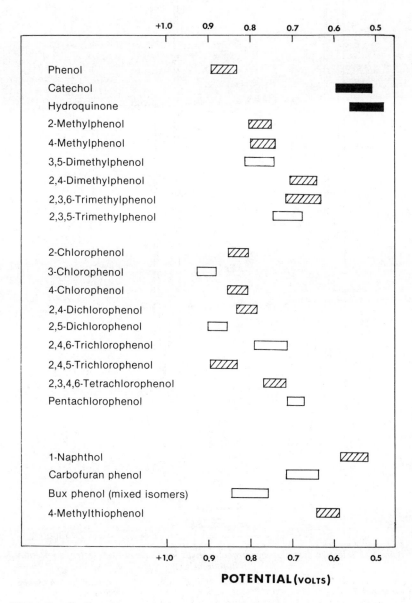

Figure 5. Cyclic voltammetric data for selected phenols of environmental interest (carbon paste electrode, 200 mv/sec, pH 4.0 citrate buffer, 10% ethanol (v/v))

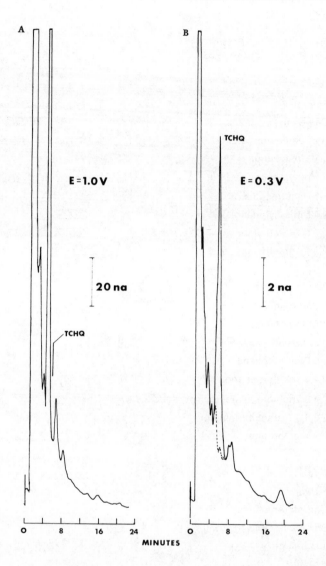

Figure 6.   Selective detection of tetrachlorohydroquinone at low electrode potential in human urine.

Sample preparation: pooled urine (pH 2) spiked to 50 ppb tetrachlorohydroquinone, hydrolyzed 30 min at 90°–100°C; 5-mL volume extracted three times with ether, combined extracts evaporated and redissolved in 0.5 mL 40% methanol, 0.1 mL injected. Column: 15 cm × 4.6 mm Waters μ Bondapak C_{18}; mobile phase: 0.1M ammonium acetate pH 6, 38% methanol, 1.0 mL/min; detector electrode: glassy carbon, TL-5; 2 mil gasket; potentials vs. Ag/AgCl/3M NaCl reference electrode.

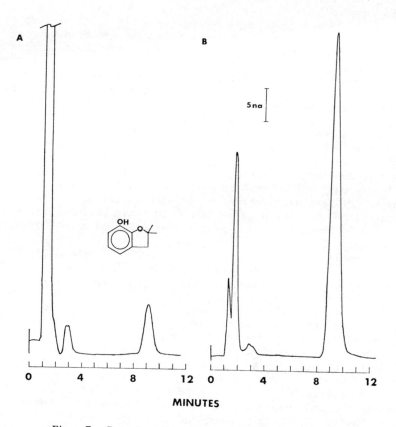

Figure 7.   Detection of a carbamate by hydrolysis to phenol.

*Chromatographic conditions are the same as in Figure 6; working electrode potential
+0.9 V—(A) 0.1 mL of acidified (pH 2) river water spiked with 1.0 ppm carbofuran and
0.1 carbofuran phenol; (B) same sample after adjustment to pH 12.*

utilizing fluorescence detection have been described by several
groups (15,16). Detection involves hydrolysis of the eluted
carbamate by addition of base at elevated temperature followed by
the addition of o-pthalaldehyde which in the presence of mercap-
toethanol reacts with methylamine from the carbamate to give a
fluorescent product. Electrochemical detection used in a similar
fashion has the advantage in that post-column addition of reagents
is not needed.

A few reports have appeared on the application of reverse
phase LCEC to trace phenols. Ott (17) used a dual detector
approach (UV and EC) to evaluate the latter for the determination of 2-phenyl-phenol in orange rind. EC was found to be more
selective as none of several UV interferences were oxidized. The
40% ethanol in the mobile phase was incompatible with the carbon
paste working electrode. The stability was improved at the
expense of sensitivity by post-column dilution with aqueous
buffer. The use of glassy carbon to circumvent this problem in
the determination of phenolic preservatives has been described
(18). Armentrout et al. (19) describe the determination of trace
phenolic compounds in wastewater. A polymeric cation-exchange
resin was used for separation with acidic acetonitrile-water
eluent. Of several electrode materials studied 50% carbon black/
polyethylene gave the highest signal-to-noise ratios. Detection
limits were below 1 ppb. The detector employed a tubular working electrode.

A complete, general mechanism has not been elucidated for
the electrochemical oxidation of phenols. This is because the
process consists of several distinct yet overlapping events
which in turn have resulted in fragmentary and sometimes con-
flicting results. There is strong evidence for at least two
primary products which can be formed: a phenoxy radical and
phenoxy cation. Both are highly reactive species which undergo
reactions to form multiple products. Some representative products are included in the scheme in Figure 8. Many of these
secondary products also oxidize at the electrode creating more
reactive species which in turn result in larger, more diverse
products. A common product is an insoluble polymeric material.
The mechanism is strongly dependent on several factors, including phenol structure and concentration, solvent composition and
electrode potential.

In electrochemical detection the main consideration is
the initial electron transfer process which produces the measured
signal. As shown above, substitution influences the electrode
potential required. A second controlling factor is the pH of the
medium. Phenolic oxidation potentials generally shift to more
negative values with increasing pH, and become pH independent
where the phenol exists wholly as the phenolate ion. Thus, it
is important to obtain voltammograms under conditions close to
the chromatographic system used. Moreover, it is ideal to
ascertain the oxidation potential versus pH relationship within
the useful chromatographic range.

Secondary reactions are also a consideration in certain instances. As often occurs in conventional electrochemical experiments with phenols, insoluble polymeric products are formed which adhere to the electrode surface, diminishing the rate of the electron transfer step. The result is a decrease in efficiency of the electrode in oxidizing the phenol. This is observed as a decrease in current response. Since the rates of the coupling reactions are second order or greater with respect to the phenol, filming problems are minimized in thin-layer hydrodynamic amperometry due to the very low concentrations ($<10^{-6}$M) of phenol normally involved. Eventually, as greater amounts are injected, the deactivation will appear and become progressively worse. This filming is also dependent on the polarity of the medium. Mobile phases with high organic modifier content lessen the tendency of products to interfere because they are more soluble. It is important to remember that those phenols considered here under normal LCEC conditions show only a slight tendency to deactivate electrodes. Poorly behaving examples are the esters of p-hydroxybenzoic acid (18) and hydroxyindoles (14).

Secondary products capable of further oxidation can also increase the unit response for certain compounds. The principle was recently demonstrated in the detection of catecholamines where a post-column addition was used to adjust effluent pH increasing the rate of the chemical step leading to electroactive secondary products. Oxidation of secondary products of phenolic oxidations probably occurs to a degree under all conditions.

Aromatic Amine Residues

Aromatic amines, molecules containing one or more aromatic rings bearing a primary or substituted amine function, constitute a second important class of electrochemically oxidizable compounds. Several specific reports dealing with the determination of aromatic amines by LCEC have recently appeared.

Loss of one electron to yield a cation radical is the predominant oxidation process for aniline, the simplest primary aromatic amine, and for the majority of its ring-substituted analogs (20). The specific potential at which this occurs is influenced by the electron donating or withdrawing effects of the ring substituent(s). Figure 9 illustrates the variation in oxidation potential with structure for several representative compounds. Extensive delocalization of the unpaired electron occurs throughout the ring system, and both head-to-tail and tail-to-tail coupling reactions rapidly consume the electrogenerated radicals. For example, both 4-aminodiphenylamine and benzidine (4, 4′ - diaminobiphenyl) are recovered after oxidation of aniline in aqueous solution, as shown at the top of the next page.

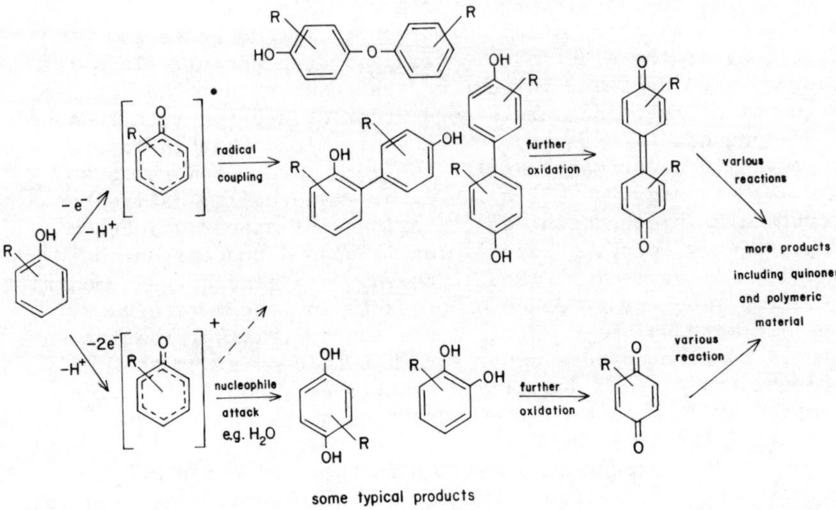

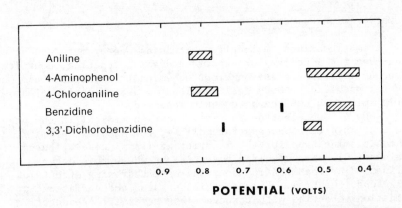

some typical products

Bioanalytical Systems

Figure 8.   A brief outline of the electrochemical oxidation of phenols showing some typical reaction products

Aniline

4-Aminophenol

4-Chloroaniline

Benzidine

3,3'-Dichlorobenzidine

0.9   0.8   0.7   0.6   0.5   0.4

POTENTIAL (VOLTS)

Figure 9.   Cyclic voltammetric data for selected amines of environmental interest (carbon paste electrode, 200 mV/sec, pH 4.0 citrate buffer, 10% ethanol (v/v))

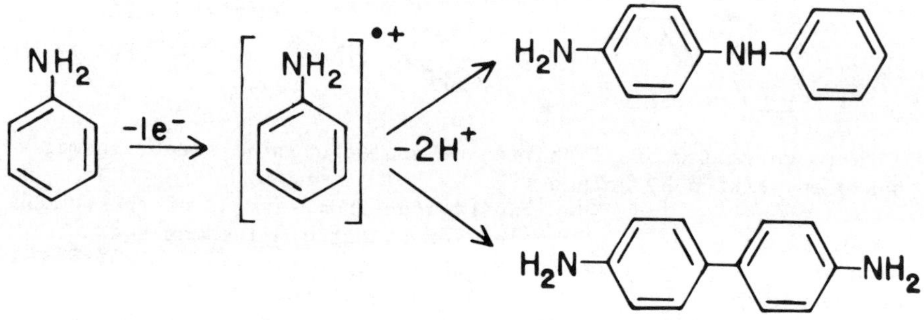

Depending on the identity and position of substituent groups, some of these coupling products may be subject to a subsequent 2 electron oxidation, as described below. Consequently, most anilines may be expected to yield either one or two electrons per molecule oxidized. If further subjected to exhaustive electrolysis, particularly at extremes of pH, the ultimate product is a dark, intractable polymeric solid.

Secondary or tertiary aromatic amines, those being further substituted at the nitrogen, behave in qualitatively similar ways, although their oxidation potentials are usually shifted more positive. Bulky substituent groups may also inhibit the extent of subsequent coupling reactions.

A special class of substituted anilines are those having a hydroxyl group or a second amino groups positioned ortho or para to the amine function. These molecules are easily oxidized in a two electron process to the corresponding quinoneimine or quinonediimine. The ease of oxidation reflects the high stability of the extended conjugation of double bonds in the oxidized form. Here, also, other ring substituents may slightly alter the value of the oxidation potential. In addition, the imine groups of such species are susceptible to hydrolysis, such that the ultimate product of the oxidation of a p-phenylenediamine would be the corresponding p-benzoquinone, as shown below.

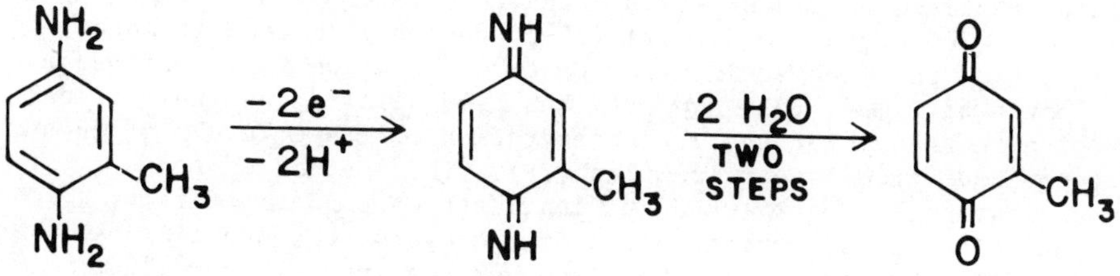

This characteristic ease of oxidation is also exhibited by benzidine and similar multi-ring molecules capable of forming an oxidation product with extended conjugation:

$$\text{H}_2\text{N}\!-\!\!\langle\bigcirc\rangle\!-\!\langle\bigcirc\rangle\!-\!\text{NH}_2 \xrightarrow[-2\text{H}^+]{-2e^-} \text{HN}\!=\!\langle\bigcirc\rangle\!=\!\langle\bigcirc\rangle\!=\!\text{NH}$$

A thorough review of aliphatic and aromatic amine electrochemistry has been written by Nelson (21).

A number of individual pesticides from several of the common classifications produce single-ring aromatic amines as degradation products. For example, Propham and carbetamide, members of the carbamate class, yield aniline as a residue while Monuron and Dimilin, classified as ureas, degrade to p-chloroaniline. Lores, Bristol, and Moseman of the EPA developed an LCEC assay for pesticide-derived aniline and a number of its halogenated derivitives (22). Their scheme employed two sets of isocratic operating conditions in order to effectively quantify all of the compounds examined. Reported detection limits for the seven amines determined were 200-400 pg.

Amine-substituted biphenyl and related compounds, including benzidine and its congeners, are also of considerable analytical interest. Although not generally considered to be pesticide residues per se, their determination in environmental samples further illustrates the applicability of LCEC. Many of these are highly toxic and are the subject of stringent governmental regulation (23). Chromatographic determinations have recently appeared for several compounds in this class, such as MOCA (4,4´-methylenebis-(2-chloroaniline)) (24) which behaves electrochemically as a substituted aniline. Analyses utilizing LCEC have been developed for benzidine in urine (25), for benzidine and 3,3´-dichlorobenzidine in wastewater (26) and for benzidine and the toxic polyaromatic 2-naphthylamine, among others, in waste water (27). Detection limits for actual samples with these methods are approximately 100 pg.

The second aromatic ring contained in these molecules significantly increases their hydrophobic character, resulting in larger capacity factors on reverse-phase chromatographic material. In practice, this has necessitated the use of substantial amounts of organic modifiers (30-50 percent) in the chromatographic mobile phase in order to effect elution in a reasonable length of time. This property has also allowed the exploration of a technique for the preconcentration of trace levels of these compounds from aqueous samples (28). This method of trace enrichment, recently used by Ogan, Katz, and Slavin for the determination of several polyaromatic hydrocarbons utilizing fluorescence detection (29), involves passing a large volume (2-50 mL) of the aqueous sample through a short (3 cm) column packed with reverse-phase material which is contained in the sample loop of the injection valve of a conventional LCEC instrument. Since the aqueous sample solution contains no organic solvent, the capacity factors of the amines of interest

are very large. Consequently, these molecules are very strongly
retained as the sample passes through and out of the sampling
precolumn. Simultaneously, a second pump passes mobile phase
containing the proper amount of organic solvent through the
analytical column. A schematic diagram of the complete system
is shown in Figure 10. After a known amount of sample has been
pumped through the sample enrichment column, the valve position
is switched and the retained compounds flushed off in a narrow
band by the mobile phase onto the analytical column. A second
virtue of this configuration is that the hydrophilic compounds
present in the sample are not concentrated, and hence contribute
little to the chromatogram. This approach is illustrated in
Figure 11, which shows a chromatogram of a standard solution
containing benzidine at 50 parts per trillion, and that of a
ground water sample also containing approximately 50 parts per
trillion benzidine which was collected some distance from an
industrial chemical dumping area where benzidine was known to have
been released.

Electroreducible Residues

Many electrochemical techniques have been applied to envir-
onmentally important reducible molecules. DC polarography has
been applied to the determination of nitroaniline herbicides
(30), bipyridylium herbicides (31), parathion and other nitrophenols (32,33), and dithiocarbamate fungicides (34-36). AC
polarography has been used for the determination of explosives
(37), dithiocarbamate fungicides (34,35,38), and nitrophenols
(39). Differential pulse polarographic (DPP) methods have found
widespread application to the determination of electroactive
agrochemicals, where sample interferences are minimal. DPP
methods have been used for heterocyclic pesticides (atrazine,
ametryne, and terbutryne) in natural waters (40), Vacor (rodenti-
cide) residues (41), nitrophenyl pesticides (42), and azomethine
insecticides in grain formulations (43). These methods compare
favorably with the usual gas chromatographic methods; however,
they lack selectivity for molecules with similar electrochemical
behavior. For example, the insecticides, cytrolane and cyolane,
could not be measured with DPP when both were present in a sample
(43). Reductive LCEC, which combines a chromatographic separation with electrochemical detection, easily permits simultaneous
quantitation of cyolane and cytrolane, as well as many others.
The development of the reductive mode LCEC technique has
been slow because of difficulties in preparing convenient and
reliable working electrodes for use with a high efficiency
chromatographic separation. In addition, problems are encountered
with dissolved oxygen and heavy metals. Solid electrodes have
been used with limited success for reductive LCEC. Mercury
pool electrodes (44-47), the DME (48-53), and platinum wire
electrodes coated with mercury (49) are generally not satis-

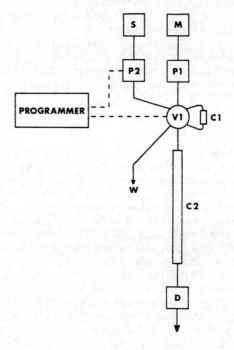

Figure 10.   Components of a liquid chromatograph used for the trace enrichment
of aromatic amines by column preconcentration

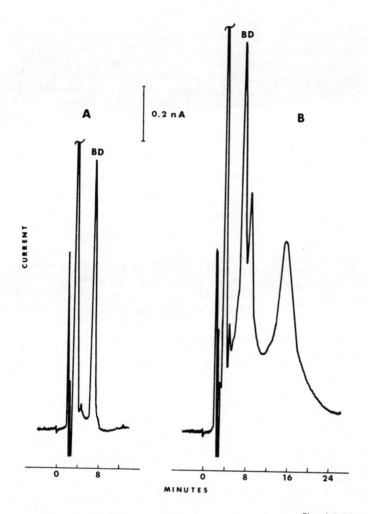

Figure 11. Determination of 50 parts per trillion benzidine (BD) using the instrumental configuration shown in Figure 10.

A 20-mL sample was preconcentrated on a 3-cm column containing C_2 reverse-phase material and eluted and chromatographed with a mobile phase of 30:70 methanol:0.1M ammonium acetate, pH 6.2 on a 15 cm × 4.6 mm analytical column packed with C_2 material; electrochemical detector operated at +500 mV vs. Ag/AgCl—(A) standard aqueous solution; (B) ground water sample collected near an industrial chemical dumping area.

factory. Mercury pool and DME electrodes suffer from vibrational
effects, edge effects due to creeping of solution between the
mercury and its container, and the time dependent surface area.
Platinum electrodes coated with mercury have a small negative
range. Our laboratory (6,54,55) and MacCrehan et.al. (56-58)
have used mercury amalgamated gold electrodes for detecting
reducible compounds which were previously inaccessible with
carbon electrodes. These electrodes have the advantages of being
inexpensive, easily prepared, and mechanically rigid.

The reductive detector cell resembles the oxidative cell
illustrated in Figure 1 except that gold is used for the working
electrode. The mercury film surface is prepared by first
polishing the gold to a mirror-like finish. Triple-distilled
mercury is placed on the gold disk and after ca. 5 minutes the
excess is removed. The lifetime of the electrode surface is
dependent on a number of variables including the composition of
the mobile phase, the concentration of heavy metals, the applied
potential, and the nature of the sample. Unlike carbon electrodes,
the mercury film electrode is more susceptible to fouling as the
result of irreversible adsorption of contaminants such as traces
of metals and sulfur compounds present in the mobile phase. Under
typical conditions, the electrode surface is stable for 4-5 days.
When working at the extremes of the useful negative range (-1.25
volts vs. SCE), the electrode surface may need reconditioning on
a daily basis. Detailed descriptions of the mercury-gold cell
and its care are presented elsewhere (54,55).

Dissolved oxygen seriously limits the range of the reductive
mode detector if not removed from the mobile phase and sample.
Even at relatively low negative potentials, very large residual
currents are produced by the reduction of oxygen.

Purging of oxygen from the mobile phase using high purity
nitrogen, helium, or argon gas is preferred by most electrochemists. Michael and Zatka (53) removed dissolved oxygen from
an LCEC system by continuously refluxing the mobile phase. Only
stainless steel tubing is used in the chromatograph because teflon
is permeable to oxygen.

Heavy metals also interfere with the reductive mode detector. The presence of dissolved heavy metals such as Pb^{+2} has a
two-fold effect on the detector residual current. First, the background current increases as the result of the reduction of
dissolved metal ions. Second, the reduced metal changes the
composition of the surface of the working electrode. During the
operation of the detector, the surface of the working electrode
is no longer pure gold-mercury amalgam, but also contains other
metal-mercury amalgams having lower hydrogen overpotentials than
gold-mercury alone. By lowering the hydrogen overpotential of
the working electrode, metal amalgams significantly increase the
background current. The problem can be diminished by using ultra-
pure salts when preparing the mobile phase and keeping the ionic
strength low. Heavy metals can also be removed from mobile

phases by electrolysis, if ultra trace determinations are to be attempted.

Many pesticides are suitable for reductive LCEC. Examples are listed in Table 1. The reduction of aromatic nitro compounds has been widely studied (59). Mononitroaromatic compounds are reduced to hydroxylamines in a four-electron process (see below). At low pH, hydroxylamines can be further reduced to amines at more negative potentials than the reduction of the nitro group.

$$ArNO_2 + 4e^- + 4H^+ \longrightarrow ArNHOH + H_2O$$

$$ArNHOH + H^+ \longrightarrow ArNH_2OH^+$$

$$ArNH_2OH^+ + 2e^- + 2H^+ \longrightarrow ArNH_3^+ + H_2O$$

Polynitroaromatic compounds are reduced more easily than mononitro compounds. Additional nitro groups facilitate the reduction; however, the reduced functional groups, hydroxylamine and amine, hinder the reduction of other nitro groups. At pH < 3 the reduction of dinitroaromatic compounds follows two four-electron steps:

$$NO_2-Ar-NO_2 + 4e^- + 4H^+ \longrightarrow NO_2-Ar-NHOH + H_2O$$

$$NO_2-Ar-NHOH + 4e^- + 4H^+ \longrightarrow NHOH-Ar-NHOH + H_2O$$

The hydrodynamic voltammograms of picric acid and p-nitrophenol depicted in Figure 12 illustrate the influence of mechanism on the detection process. If the detector potential is set at the potential on the plateau of the voltammogram of p-nitrophenol, both compounds are detected. If the potential is set at the plateau of the first wave of picric acid, it alone is detected. At more negative potentials the unit response for the polynitroaromatic compounds will be greater than the response for mononitroaromatic compounds since more electrons are involved in the electrochemical reduction of polynitroaromatics.

Molecules which are not electrochemically active can be derivatized using an appropriate "electrophore", making the derivative eligible for electrochemical detection (9). It is fortunate that the nitrophenyl group, commonly used for UV and GC derivatization, is easily reducible. Methods based on the determination of intact carbamate residues by GC have been very disappointing as the result of thermal instability of a large majority of carbamate compounds. Therefore, a number of derivatization methods have been proposed. Carbamates, which produce an aromatic amine or phenol when hydrolyzed, were derivatized using 2,4-Dinitrofluorobenzene (DNFB) to form 2,4-

Table I

Easily Reducible Compounds of Environmental Interest

Dinitroaniline Herbicides

Benefin
Chlornidine
2-(1-methylbutyl)-4,6-dinitrophenol
2-sec-Butyl-4,6-dinitrophenol
4,6-Dinitrobutyl-o-cyclohexylphenol
Nitroaniline
Oryzalin
Trifluralin
4,6-Dinitro-o-cresol

Nitrophenyl Insecticides

Parathion
Methylparathion
Nitrofen

Bipyridylium Herbicides

Paraquat
Diquat

Azomethine bond - insecticides

Cytrolane
Cyolane

Azomethine bond - fungicide

Drazoxolon

Dinitrophenol Pesticides

Dinocap
Medinoterb
Dinobuton

Organometallic Compounds

Methyl mercury
Phenyl mercury
Methyl antimony
Phenyl lead
Methyl lead

Nitrosamines

N-nitrosodiethanolamine
Dimethylnitrosamine
Diethylnitrosamine
Dipropylnitrosamine
N-nitrosoatrazine
N-nitrosomorpholine
N-nitrosopiperidine
N-nitrosopyrrolidine

Explosives

Picric acid
Trinitrotoluene (TNT)
Nitroglycerin

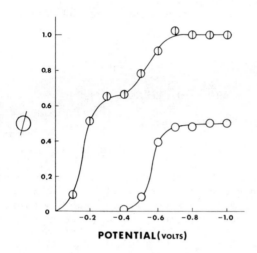

Figure 12. Hydrodynamic voltammograms of picric acid (⊕) and p-*nitrophenol (○) (∅ is the normalized peak current response to repeated injection of 20 ng of each compound; mobile phase: pH 3.7 citrate/phosphate at 0.5 mL/min)*

dinitrophenyl (DNP) derivatives (60-63). Several investigators
have used derivatization with DNFB to confirm the identity of
triazine, urea, and carbamate residues (64-66).

In the past seven years N-nitrosamines have attracted wide
attention due to their carcinogenic properties. In 1977, Ross
et al. (67) reported the presence of N-nitrosodipropylamine in
the formulation of dinitroaniline herbicide, trifluralin, at the
154 ppm level. Subsequent studies by other laboratories showed
that nitrosamine impurities are common to all dinitroaniline-
based herbicides. Nitrosamines contamination resulted from
the nitration of dipropylamine by the excess nitrosating agent.

The electrochemistry of N-nitrosamines has been studied in
detail by numerous investigators (68-72). At low pH, N-nitrosamines are reduced to hydrazines via a four-electron process.
At neutral or alkaline pH, the N-nitroso bond is cleaved in a
two-electron process to produce an amine and dinitrogen oxide.

Low pH:

$$RR'N-NO + H^+ \longrightarrow RR'N-\overset{+}{N}OH$$

$$RR'N-\overset{+}{N}OH + 4e^- + 4H^+ \longrightarrow RR'N\overset{+}{N}H_3 + H_2O$$

Neutral and alkaline pH:

$$RR'N-NO + 2e^- + 2H^+ \longrightarrow RR'NH + 1/2N_2O + 1/2H_2O$$

Bipyridylium herbicides and insecticides containing activated azomethine group are also suitable for LCEC methods of
analysis. The bipyridylium herbicide, methyl viologen (MV,
paraquat), undergoes two, one-electron processes.

$$MV^{+2} + e^- \rightleftarrows MV^{+1} + e^- \longrightarrow MV$$

Insecticides containing an activated azomethine group such
as cytrolane and cyolane are reduced in a single four-electron
process at pH < 8.

$$RR'C=N-\overset{\overset{O}{\|}}{P}(OR)_2 + H^+ \longrightarrow RR'C=\overset{+}{N}H-\overset{\overset{O}{\|}}{P}(OR)_2$$

$$RR'C=NH-\overset{\overset{O}{\|}}{P}(OR)_2 + 4e^- + 3H^+ + H_2O \longrightarrow RR'CH-NH_2 + HO-\overset{\overset{O}{\|}}{P}(OR)_2 + H_2$$

Cytrolane and cyolane were determined in spiked run-off
water to demonstrate the applicability of the thin-layer mercury
amalgamated gold detector for the determination of insecticide
residue. A chromatogram of a sample spiked with 0.94 μg/ml and

0.84 µg/mL of cytrolane and cyolane, respectively, is illustrated
in Figure 13.   The pesticides were isolated according to Paschal
et al. (73) with minor modifications.

Dithiocarbamates and thioureas are included in this section
because of their useful electrochemical behavior at mercury and
mercury amalgam electrodes.   The formation of mercury complexes
results in an easy oxidation at the mercury electrode.   On
the other hand, carbon electrodes are not well suited for the
detection of these compounds because the oxidation occurs beyond
the usual scope of carbon detector cells.

The electrochemical behavior of thiocarbamates has been
studied by several investigators (35,36,38,74-78).   At mercury
electrodes, thiocarbamates are oxidized in a one-electron process
to form insoluble mercury (II) salts.   Thioureas undergo a
similar process.

$$R_2NC\overset{\text{S}}{||}SH \longrightarrow R_2NC\overset{\text{S}}{||}S^- + H^+$$

$$R_2NC\overset{\text{S}}{||}S^- + Hg^o \longrightarrow R_2NC\overset{\text{S}}{||}SHg + e^-$$

$$2\ R_2NC\overset{\text{S}}{||}SHg \longrightarrow (R_2NC\overset{\text{S}}{||}S)_2Hg + Hg^o$$

Cyclic voltammetry is extremely helpful in determining
the potential of the working electrode necessary to detect the
analyte of interest.   Figure 14 illustrates CV data for selected
bipyridylium herbicides, insecticides containing an activated
azomethine group, nitroaromatic pesticides, and herbicides and
their metabolites.   Figure 14 also points out the ease of
oxidation of the dithiocarbamate functional group.   The oxidation
occurs at low negative potentials, whereas typical oxidations
take place at positive potentials as shown in Figure 4.

In conclusion, we have tried to present the principles of
LCEC, describe the present applications that have been made, and
survey the areas of potential utility by reviewing pertinent
chemistry and related methods.   It is our opinion that the success
of LCEC in neurochemistry can carry over to pesticides owing not
only to the advantages of electrochemical detection but also to
the tremendous potential of its adjunct, liquid chromatography.

Acknowledgement

This work was supported by grants from the National Institute
for General Medical Science and the National Science Foundation.

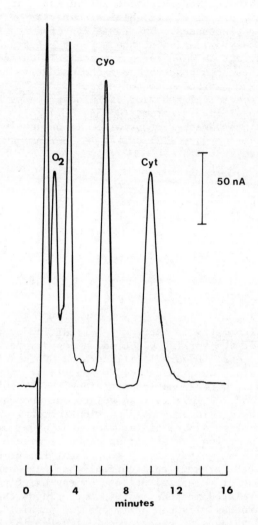

Figure 13. Chromatogram of run-off water spiked with 0.94 µg/mL of cytrolane (Cyt) and 0.84 µg of cyolane (Cyo) (mobile phase: 0.02M acetate buffer, pH 4.15 containing 30% methanol (v/v) at a flow rate of 0.5 mL/min; TL-9A electrochemical transducer was set at −1.25 V vs. Ag/AgCl reference electrode)

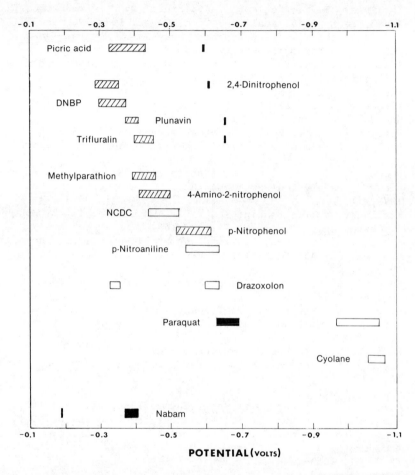

Figure 14. Reductive cyclic voltammetry data for selected compounds of environmental interest obtained using a mercury film electrode (on gold) at scan rates of 200 mV/sec in 0.1M citrate buffer, pH 4.0 and 10% ethanol (v/v) (concentrations of compounds were in the range of 1.8 to 2.3 mg/20 mL)

Literature Cited

1. Bibliography of Recent Reports on Electrochemical Detection";
   Bioanalytical Systems, Inc., W. Lafayette, Indiana.
2. Kissinger, P. T. Anal. Chem., 1977, 49, 447A-456A.
3. Heineman, W. R.; Kissinger, P. T. Anal. Chem., 1978, 50,
   166R-175R.
4. Kissinger, P. T.; Felice, L. J.; Miner, D. J.; Preddy, C. R.;
   Shoup, R. E. Adv. Anal. and Clin. Chem., 1978, 2, 55-175.
5. Kissinger, P. T. "Trace-Organic Analysis by Reverse-Phase
   LC with Amperometric Detection" in Methodological Surveys
   in Biochemistry, Vol. 7, E. Reid, Ed., Ellis Horwood, Ltd.,
   Chichester, 1978, pp. 213-226.
6. Kissinger, P. T.; Bruntlett, C. S.; Bratin, K.; Rice, J. R.
   "Trace Organic Analysis"; National Bureau of Standards
   Special Publication 519; 1979, 705-711.
7. Kissinger, P. T., Ed. "Laboratory Techniques in Electroanalytical Chemistry"; Marcel Dekker, New York, in preparation.
8. Shoup, R. E.; Davis, G. C.; Bruntlett, C. S. Current Separations, Bioanalytical Systems, Inc., W. Lafayette, Indiana,
   1979, 1(1), 4.
9. Kissinger, P. T.; Bratin, K.; Davis, G. C.; Pachla, L. A.
   J. Chromatog. Sci., 1979, 17, 137-146.
10. Kuhr, R. J.; Dorough, H. W. "Carbamate Insecticides: Chemistry,
    Biochemistry and Toxicology"; CRC Press: Cleveland, OH, 1976;
    Chap. 4.
11. Suatoni, J. C.; Snyder, R. E.; Clark, R. O. Anal. Chem., 1961,
    33, 1894-1897.
12. King, W. P.; Kissinger, P. T., manuscript in preparation.
13. Dorough, H. W.; Thorstenson, J. H. J. Chrom. Sci., 1975,
    13, 212-224.
14. Koch, D. D.; PhD Thesis, Purdue University, 1979.
15. Moye, H. A.; Scherer, S. J.; St. John, D. A. Anal. Lett., 1977,
    10, 1049-1073.
16. Krause, R. T. J. Chrom. Sci., 1978, 16, 281-284.
17. Ott, D. E. J. Assoc. Off. Anal. Chem., 1978, 61, 1465-1468.
18. King, W. P.; Joseph, T. K.; Kissinger, P. T. J. Assoc. Off.
    Anal. Chem., accepted for publication.
19. Armentrout, D. A.; McClean, J. D.; Long, M. W. Anal. Chem.,
    1979, 51, 1039-1045.
20. Bacon, J., and Adams, R. N. J. Amer. Chem. Soc., 1-68, 90,
    6596-6599.
21. R. F. Nelson, "Anodic Oxidation Pathways of Aliphatic and
    Aromatic Nitrogen Functions," in Technique of Electroorganic
    Synthesis, N. Weinburg, ed., Wiley, New York, 1974, pp. 535-
    792.
22. Lores, E. M.; Bristol, D. M.; Moseman, R. F. J. Chrom. Sci.,
    1978, 16, 358-362.

23. Fed. Regist., 1974, 39, 23551.
24. Rappaport, S. M. and Morales, R. Anal. Chem., 1979, 51, 19-23.
25. Rice, J. R. and Kissinger, P. T. J. Anal. Toxicol., 1979, 3, 64-66.
26. Riggin, R. M. and Howard, C. C. Anal. Chem., 1979, 51, 211-214.
27. Mefford, I.; Keller, R. W.; Adams, R. N.; Sternson, L. A.; Yllo, M. S. Anal. Chem., 1977, 49, 683.
28. Rice, J. R. and Kissinger, P. T. Environ. Sci. Technol., in preparation.
29. Ogan, K.; Katz, E.; Slavin, W. J. Chrom. Sci., 1978, 16, 518-522.
30. Southwick, L. M.; Willis, G. H.; Dasgupta, P. K.; Keszthelyi, C. P. Anal. Chim. Acta, 1976, 82, 29-35.
31. Engelhardt, J; McKinley, W. P. J. Agric. Food Chem., 1966, 14(4), 377-380.
32. Zietek, M. Mikrochim. Acta, 1976, 463-470.
33. Zietek, M. Mikrochim. Acta, 1976, 549-557
34. Budnikov, G. K.; Toropova, V. F.; Ulakhavich, N. A.; Viter, I. P. Z. Anal. Khimii, 1974, 29(6), 1204-1209.
35. Halls, D. J.; Townshend, A.; Zuman, P. Anal. Chim. Acta, 1968, 41, 51-62.
36. Brand, J. D.; Fleet, B. Analyst, 1970, 95(1136), 905-909.
37. Hetman, J. S. Z. Anal. Chem., 1973, 264, 159-164.
38. Budnikov, G. K.; Supin, G. S.; Ulakhavich, N. A.; Shakurova, N. K. Z. Anal. Khimii, 1975, 30(11), 2275-2277.
39. Burgschat, H.; Netter, K. J. J. Pharm. Sci., 1977, 66(1), 60-63.
40. McKone, C. E.; Byast, T. H.; Hance, R. J. Analyst (London), 1972, 97, 653-656.
41. Whittaker, J. W.; Osteryoung, J. G. Anal. Chem., 1976, 48(9), 1418-1420.
42. Smyth, M. R.; Osteryoung, J. G. Anal. Chim. Acta, 1978, 96, 335-344.
43. Smyth, M. R.; Osteryoung, J. G. Anal. Chem., 1978, 50(12), 1632-1637.
44. Rabenstein, D. L.; Saetre, R. Anal. Chem., 1977, 49, 1036-1039.
45. Saetre, R.; Rabenstein, D. L. J. Agric. Food Chem., 1978, 26, 982-983.
46. Rabenstein, D. L.; Saetre, R. Clin. Chem., 1978, 24, 1140-1143.
47. Saetre, R.; Rabenstein, D. L. Anal. Chem., 1978, 50, 276-280.
48. Buchta, R. C.; Popa, L. J. J. Chromatog. Sci., 1978, 14, 213-219.

49. Wasa, T.; Musha, A., Bull. Chem. Soc. Japan, 1975, 48, 2176-2181.
50. Stillman, R.; Ma, T. S. Mikrochim. Acta, 1973, 491-506.
51. Stillman, R.; Ma, T. S. Mikrochim. Acta, 1974, 641-648.
52. The Model 310 polarographic LC detector, Princeton Applied Research Corp.
53. Michael, L.; Zatka, A. Anal. Chim. Acta, 1979, 105, 109-117.
54. Shoup, R. E.; Bruntlett, C. S.; Bratin, K.; Kissinger, P. T.; "Principles and Applications of Liquid Chromatography with Electrochemical Detection"; Bioanalytical Systems, Inc. W. Lafayette, Indiana, 1979.
55. Bratin, K.; Bruntlett, C. S.; Kissinger, P. T., submitted to J. Liquid Chromatogr.
56. MacCrehan, W. A.; Durst, R. A.; Bellama, J. M. Anal. Lett., 1977, 10, 1175-1188.
57. MacCrehan, W. A.; Durst, R. A. Anal. Chem., 1978, 50, 2108-2112.
58. MacCrehan, W. A.; Durst, R. A.; Bellama, J. M. "Trace Organic Analysis", National Bureau of Standards Special Publication 519; 1979, 57-63.
59. Lund, H. Cathodic Reduction of Nitro Compounds, in M. M. Baizer, "Organic Electrochemistry"; Marcel Dekker: New York, 1973; p. 315.
60. Cohen, I. C.; Wheals, B. B. J. Chromatog., 1969, 43, 233-240.
61. Cohen, I. C.; Norcup, J.; Ruzicka, J. H. A.; Wheals, B. B. J. Chromatog., 1970, 49., 215-221.
62. Holden, E. R. J. Assoc. Off. Anal. Chem., 1973, 56, 713-717.
63. Caro, J. H.; Freeman, H. P.; Turner, B. C. J. Agric. Food Chem., 1974, 22, 860-863.
64. Lawrence, J. F. J. Agric. Food Chem., 1974, 22, 936-938.
65. Holden, E. R-; Jones, W. N.; Beroza, M. J. Agric. Food Chem., 1969, 17, 56-59.
66. Seiber, J. N.; Crosby, D. G.; Fouda, H.; Soderquist, C. J. J. Chromatog., 1972, 73, 89-97.
67. Ross, R. D.; Morrison, J.; Rounbehler, D. P.; Fan, S.; Fine, D. H. J. Agric. Food Chem., 1977, 25, 1416-1418.
68. Lund, H. Acta Chem. Scand., 1957, 11, 990-996.
69. Zahradnik, R.; Svatek, E., Chvapil, M. Chem. Listy, 1957, 51, 2232-2242.
70. Pulidori, F.; Borghesani, G.; Gibhi, C.; Pedriali, R. J. Electroanal. Chem., 1970, 27, 385-396.
71. Borghesani, G.; Pulidori, F.; Pedriali, R.; Bighi, C. J. Electroanal. Chem., 1971, 32, 303-308.
72. Iversen, P. E. Acta Chem. Scand., 1971, 25, 2337-2340.
73. Paschal, D. C.; Bicknell, R.; Dresbach, D. Anal. Chem., 1977, 49, 1551-1554.
74. Brand, J. D.; Fleet, B. Analyst, 1968, 93, 498-506.
75. Halls, D. J.; Townshend, A.; Zuman, P. Anal. Chim. Acta, 1968, 41, 63-74.
76. Smyth, M. R.; Smyth, W. F. Analyst, 1978, 103, 529-567.
77. Jensovsky, L. Chem. Listy, 1955, 49, 1267-73.
78. Fedoronko, M.; Manousek, O.; Zuman, P. Chem. Listy, 1955, 49, 1494-1498.

RECEIVED December 28, 1979.

A Critical Comparison of Pre-Column and Post-Column Fluorogenic Labeling for the HPLC Analysis of Pesticide Residues

H. A. MOYE

Pesticide Research Laboratory, Food Science and Human Nutrition Department, University of Florida, Gainesville, FL 32611

P. A. ST. JOHN

American Instrument Company, Silver Spring, MD 20910

The concept of "fluorogenic labeling" had its beginnings in the early 1950's when dyes containing a naphthalene nucleus were adsorbed to proteins through weak molecular attractions (1). In an effort to better understand the effects of proteins on naphthalene dye fluorescence Hartly and Massey (2) in 1956 reacted 1-dimethylaminonaphthalene-5-sulfonyl chloride (dansyl chloride) with various amino acids. Other authors have detected dansylated amino acids by thin-layer chromatography (3-8). Dansyl chloride reacts with phenols as well and was used by Frei, Lawrence, Hope and Cassidy to derivatize hydrolyzed N-methyl carbamate pesticides prior to HPLC analysis on silica columns with fluorometric detection (9). Early work by Chen demonstrated that dansyl derivatives of amino acids suffered from severe quenching in protic solvents (10). More recently Froehlich and Murphy (11) have shown two to three fold enchancement in fluorescent intensity of dansylated amino acids over pure water by employing 30% dimethyl sulfoxide as solvent.

Primary and secondary amines have been quantitated by thin layer chromatography after derivatization with 7-chloro-4-nitro-benzo-2-oxa-1,3-diazole (NBDCl, 12). The reagent has also been used for analysis of reduced nitrosamines employing adsorptive mode HPLC (13).

We have reported on the use of 9-fluorenylmethyl chloroformate (FMOCCl) as a pre-column fluorogenic labeling reagent for primary and secondary amines, and found it to be particularly suitable for aqueous based HPLC systems such as ion exchange where both dansyl chloride and NBDCl derivatives extensively lose quantum efficiencies with concurrent losses in limits of detection (14). As well, we have also reported on a post-column fluorogenic labeling HPLC arrangement for the analysis of N-methylcarbamate pesticides employing the primary amine specific o-phthalicdi-carboxaldehyde-mercaptoethanol (OPA-MERC) reagent (15). This system was extensively studied by Krause (16), and found to be sensitive, selective, reproducible and stable. By substituting an oxidative calcium hypochlorite reagent for the hydrolytic sodium

0-8412-0581-7/80/47-136-089$05.00/0

hydroxide reagent in this system we have discovered that glyphosate herbicide (n-phosphonomethylglycine) can be readily cleaved
to produce a primary amine which rapidly reacts with the OPA-MERC
reagent to produce a fluorophore (15).

We chose the FMOCCl reagent and the OPA-MERC system to study
the relative merits of pre-column and post-column fluorogenic
labeling when utilized in the development of HPLC based residue
procedures for glyphosate herbicide (GLYPH) and its major metabolite, aminomethylphosphonic acid (AMPA). Some of the conclusions
presented here are based upon experimental data while others arise
from conceptual examination of the two approaches. Additionally,
some of the conclusions recorded here are not definitive but result from a limited number of observations and are included in
order to assist the reader in understanding why certain choices
were made during the development of the procedures.

Parallel, and infrequently duplicative, efforts of the type
described here were judged by the authors to be justifiable in
light of the recent worldwide interest in the herbicide glyphosate
in addition to the fact that the only previously published residue
procedure for the herbicide (17) has generally become to be re-
garded as lengthy, cumbersome and subject to low recoveries.

Apparatus and Reagents

Pre-column Labeling (FMOCCl). A fluorometric HPLC was constructed from two Waters Associates model 6000A pumps, a Waters
Associates model 660 solvent programmer, a Rheodyne model 7010
sample injection valve equipped with a 20 μl sample loop and an
American Instrument Co. Aminco-Bowman spectrophotofluorometer
model 4-8202, equipped with a model B16-63019 flow through cell.
Chromatograms were recorded on a Sargent model MR strip chart
recorder. A Corning CS #0-54 cutoff filter (290 nm) was placed
before the photomultiplier tube (IP28) to reduce the scattered
light from the excitation monochromator. Excitation was at 270
nm and emission at 315 nm; a 150 W. xenon arc lamp was used as a
source. Slit program was set at 3,3,3,3,3,5. Separations were
achieved on Waters Associates μ Carbohydrate (4 mm x 30 cm) or
μ NH_2 (4 mm x 30 cm) columns operated in the anion exchange mode.
Isocratic operation, unless otherwise specified, was conducted at
1.0 ml/min with pH 4 phosphate buffer (0.1 M) containing 25% acetonitrile by volume. Various solvent programs were also attempted.

Acetone was pesticide grade; all other reagents were reagent
grade.

Cation exchange sample cleanup was performed on a 2.4 cm x
50 cm glass column having an integral 500 ml reservoir and a re-
movable stopcock. Exactly 190 g of hydrogen form Dowex 50W -X8,
100-200 mesh, equilibrated with 1 liter of 0.1N HCl, was used to
pack each column and discarded after use.

Water extracts of crops were concentrated on a Büchi Roto-
vapor model R under minimum pressue from a water aspirator;

condensor coolant was kept at 5°C by a circulating refrigerated water bath.

Post-Column Labeling (OPA-MERC). An arrangement (Fig. 1) similar to that previously reported for the determination of N-methylcarbamate pesticides (15) was constructed from a Waters Associates model 6000 pump, a Rheodyne 7010 injector with a 20 µl loop, a 4 mm x 25 cm stainless steel column packed with 13.5 µ Aminex A-27 (Biorad Laboratories), two Milton Roy Model 196-0066-001 reagent pumps, an American Instrument Co. Fluoromonitor equipped with OPA filters (360 and 455 nm) or a Gilson Spectra/glo fluorometer similarly equipped. Chromatograms were recorded on a Varian model A20 strip chart recorder (100 mv). Oxidant (calcium hypochlorite) was pumped and mixed with the HPLC column eluent via model CJ3031 Kel-F "T"s (Laboratory Data Control) and 1.6 mm O.D. x 0.5 mm I.D. Teflon tubing. A 10.6 m delay coil of similar tubing (0.29 ml volume) was used to provide a delay time of about 1 min. 40 sec. before entering another Kel-F "T" into which the OPA-MERC reagent was pumped; a 0.6 m length of tubing carried the mixture to the fluorometer. The OPA reagent was made up as previously described (15). Exactly 1 g was dissolved in 10 ml of dioxane and diluted to 1 liter with pH 10, 0.125 M borate buffer to which 1 ml of MERC was added.

Thermostatting of the columns was accomplised at 62°C with a small oil filled bath (model PY1, Bench Scale Equipment Co.).

Mobile phase for the analytical HPLC ion exchange column was 0.1 M H_3PO_4 at 1.0 ml/min. Calcium hypochlorite was prepared by dissolving 12 mg of HTH (Olin Co.) and 11.6 g NaCl in 1 liter of 0.1 M KH_2PO_4 and adjusting the pH to 9.0 with 10 M KOH.

Experimental

Pre-column Labeling (FMOCC1). FMOCC1 reacts via an Sn2 mechanism with the amino nitrogen of both primary and secondary amines producing a carbamate having a fluorenyl group as the fluorophore (14). Kinetics of the reaction with the primary amine AMPA were impossible to measure by HPLC, however, that of GLYPH showed a rapid reaction (Fig. 2). The micro-scale derivatizations were performed by placing 0.1 ml of 10^{-7}-10^{-3} M GLYPH or AMPA along with 0.9 ml of 0.025 M pH 9 sodium borate, 0.9 ml of acetone and 0.1 ml of 10^{-2} M solution of FMOCC1 in acetone, into a Teflon capped 16 mm x 125 mm culture tube. The solutions were incubated at 23°C for 20 min. without shaking or stirring after which 3 ml portions of ethyl ether were used to wash away excess reagent. Appropriate dilutions were made with water before injection into the liquid chromatograph. A typical chromatogram using the µ Carbohydrate column is shown in Fig. 3. When 10^{-5} M GLYPH was derivatized and compared to an authentic standard, as previously described (14), 109% conversion was calculated, which could be explained by a somewhat impure authentic standard.

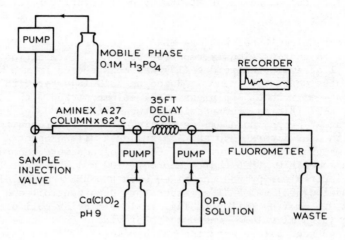

Figure 1.   Schematic of post-column fluorogenic OPA–MERC HPLC arrangement

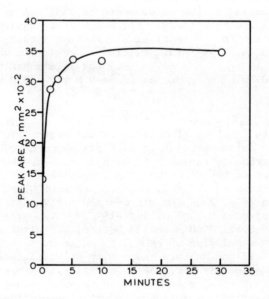

Figure 2.   Kinetics curve of GLYPH–FMOCCl reaction as monitored by HPLC
(5×10^{-5}M glyphosate, 5×10^{-4}M FMOCCl in 1:1 acetone:0.025M sodium
borate, pH 9.0, 23°C)

Crop Extraction and Cleanup (FMOCCl). The extraction procedure found in the Pesticide Analytical Manual (17) was utilized throughout. In this procedure 100 g of chopped crop, 100 ml of chloroform and 200 ml of water were added to a 1 quart Mason jar and blended at medium speed for 15 min. The jar was rinsed with 2 x 20 ml of H_2O and the combined contents and rinses distributed equally between three 60 mm x 120 mm polypropylene centrifuge bottles. After centrifugation at 10,000 rpm for 20 min. the aqueous layers were combined and rotary evaporated to 50 ml at which time the pH was adjusted to 1.0 with concentrated HCl.

All of the sample (50 ml + 10 ml wash) was placed at the top of the 50W - X8 column and eluted at 3.5 ml/min with 0.1 N HCl. GLYPH appeared in the 280 to 400 ml fraction and AMPA appeared in the 580 to 800 ml fraction. These fractions were rotary evaporated separately to approximately 4 ml which were then transferred to a 5 ml volumetric and made to volume with a 1 ml rinse of the flask.

Crop Derivatization (FMOCCl). Derivatization of crop fractions collected from the 50W - X8 column was performed with a proportionately larger amount of FMOCCl to accommodate reagent scavenging by co-extractives that were not fully isolated by the column. In addition, K_2CO_3 (150 mg) had to be added to the extremely acidic (pH 0) concentrate so that the borate buffer was not consumed. Exactly 1 ml of either the GLYPH or AMPA fraction was placed in a Teflon capped culture tube and approximately 150 mg of K_2CO_3 was added with shaking to bring the pH to 11. Along with 4 ml of H_2O 5 ml of 0.1 M FMOCCl in acetone was added to the tube which was capped and reacted at 23°C for 20 min. The reaction mix was washed 3 times with 5 ml of ethyl ether, diluted to 10 ml with H_2O and injected onto the HPLC. Comparisons were made to standards in 0.1 M HCl which were similarly derivatized.

Crop Extraction and Cleanup (OPA-MERC). GLYPH and AMPA were extracted and cleaned up prior to post-column fluorogenic labeling HPLC determination in exactly the same manner as for the FMOCCl procedure with the exception that the concentrated highly acidic fractions from the 50W - X8 columns were adjusted with 10 M KOH to pH 3-8. This was necessary in order to prevent adverse shifting of the HPLC mobile phase pH and subsequent shifts in retention and deterioration of peak shape.

Results and Discussion

HPLC Separations. The columns chosen for the FMOCCl and OPA-MERC experiments were a result of several considerations and observations. Although all four columns used operated in the anion exchange mode only the silica particle columns (μ Carbohydrate and μNH_2) were successful in chromatographing GLYPH and AMPA, ostensibly due to the interaction of the fluorenyl moiety of the

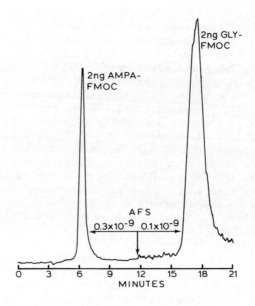

Figure 3.   Typical chromatogram of AMPA and GLYPH derivatives (μ carbo-hydrate column, 25% acetonitrile:0.1M KH₂PO₄, pH 4; 1 mL/min flow, 20 μL injection (14)

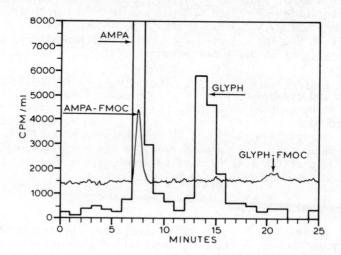

Figure 4.   Chromatogram and elution pattern, as determined by collecting 1-mL fractions following injection of radiolabeled AMPA and GLYPH, showing enhanced retention of GLYPH–FMOC (μ carbohydrate column, 25% acetonitrile/0.025M KH₂PO₄, pH 4; 1 mL/min flow)

derivative with the polystyrene–divinylbenzene polymer forming the plastic bead. Even 0.1 M H_3PO_4 was unsuccessful in eluting either derivative. These columns, however, in contrast to the plastic bead type (HA-X10 and Aminex A-27) could be readily programmed by both pH and ionic strength. The quartenary ammonium plastic bead type column was necessary in the OPA-MERC post-column fluorogenic labeling arrangement, however, due to continued column bleed from the silica particle columns which caused extremely high background fluorescence. In sharp contrast to the easily programmable silica particle columns the plastic bead column could not be programmed at all and took several hours to equilibrate when even small changes were made in ionic strength or pH.

Efficiencies for the two types of columns were both somewhat low at about 1500 theoretical plates for the plastic bead column and only about 1000 for the silica particle columns; the latter were more prone to deterioration from crop co-extractives, dropping to only about 600 after months of use, while the plastic bead column was not measurably affected.

While it is generally recognized that pre-column derivatization frequently leads to improved resolution in a multicomponent mixture our observations with the FMOCCl derivatives of GLYPH and AMPA indicate that it can be sometimes detrimental, as seen in Fig. 4 where ^{14}C labeled GLYPH and AMPA were chromatographed underivatized; 1 ml fractions were collected and counted in a scintillation spectrometer. When GLYPH and AMPA were derivatized and chromatographed a shift in GLYPH retention is observed, from 14 min. to 20.5 min.; the AMPA retention remained unchanged at 7.5 min. Since, of six amino acids, all eluted before AMPA and presented no interferences the increased retention of GLYPH appears to only increase analysis time, a definite detriment.

Post Column Derivatization (OPA-MERC) Optimization. GLYPH response was measured as a function of the OPA-MERC reagent flow and $Ca(ClO)_2$ reagent flow. Flows were varied from 0.2 to 0.6 ml/min; while each was being varied the other was held at 0.3 ml/min, which was observed to be the optimum flow for both of them when column mobile phase (0.1 M H_3PO_4) was held at 1.0 ml/min.

The $Ca(ClO)_2$ cleavage reaction of GLYPH was not studied extensively. By exposing GLYPH to $Ca(ClO)_2$ at pH 2 and then pre-column derivatizing with FMOCCl followed by chromatography on the μNH_2 column it was apparent that glycine was one reaction product that was produced which could then react with OPA-MERC. By lengthening the delay coil from 5 ft to 35 ft, giving an increase in reaction time from 14 sec. to 1 min. 38 sec., an increase in GLYPH peak area by a factor of 2 was realized. Even though it was apparent that GLYPH was not being completely converted to a primary amine the sensitivity was adequate for residue studies and reproducibility was good (see Crop Recoveries section).

Column Cleanup Studies. One of the primary goals of this study was to achieve a significant reduction in time required for sample extraction and cleanup, as well as to eliminate one, or preferably both, of the derivatization steps used in the PAM method. Many and varied attempts were made at single column cleanup of fruits and vegetables, including cation exchange, anion exchange, adsorption, reverse phase and molecular size. Of these, cation exchange on a 50 x 2.4 cm column with 190 g of 100-200 mesh 50W - X8 appeared to be the most satisfactory. Retention of glyphosate was strongly dependent upon eluent pH; maximum retention was observed at 0.05 - 0.1 M HCl and was the key to the single column cleanup since it allowed for maximum separation of GLYPH from sugars, etc., which were not retained by the column. With 0.1 M HCl as eluent AMPA began coming off the column at 580 ml, well separated from crop interferences. GLYPH began eluting at 280 ml; its fraction contained all of the significant crop peaks which were apparent on the HPLC chromatogram but which did not interfere with the GLYPH peak since they were early eluters. The elution patterns are illustrated in Fig. 5.

Crop Recoveries, Pre-Column Derivatization (FMOCCl). As illustrated in Fig. 6 it was possible to recover GLYPH at 0.1 ppm essentially quantitatively from cantaloupe using the FMOCCl precolumn derivatization procedure. However, upon changing HPLC chromatographic conditions for AMPA by lowering the mobile phase ionic strength there appeared to be several interferences which were unresolvable from AMPA-FMOC. Since such interferences seemed insurmountable and it was necessary to quantitate AMPA, no recoveries were attempted for GLYPH alone.

Crop Recoveries, Post-Column Derivatization (OPA-MERC). The good to excellent recoveries realized for all crops studied resulted from the efficiencies of both the cation exchange cleanup and anion exchange analytical columns, the care taken in the rotary evaporation of the samples (5°C condenser temp., maximum aspiration and 40°C water bath), and the selectivity of the OPA-MERC reagent. As seen in Table I, recoveries at 0.1 ppm for AMPA ranged from 61 to 82% and for GLYPH from 70 to 96%.

Typical chromatograms are illustrated in Figs. 7 and 8.

The overall reproducibility of the post-column fluorogenic labeling HPLC as well as the proposed procedure for AMPA and GLYPH is illustrated by the replicate recoveries from cucumber which were measured over a two day period (Table II):

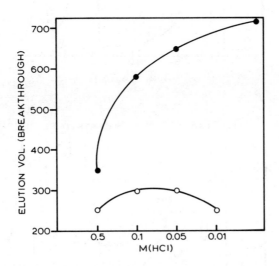

*Figure 5.   Elution pattern of GLYPH (–○–) and AMPA (–●–) from 50 cm ×
2.4 cm 50W − X8 column as a function of eluent pH*

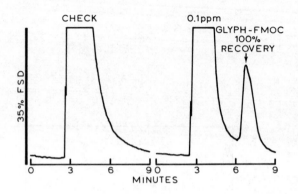

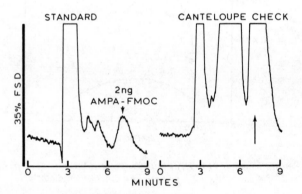

Figure 6. *Chromatogram showing (top) recovery of GLYPH from cantaloupe (μNH₂ column, 1 mL/min flow, 20 μL injection, 25% acetonitrile/0.1M KH₂PO₄, pH 4) and (bottom) cantaloupe coextractives interfering with AMPA derivative (μNH₂ column, 1 mL/min flow, 20 μL injection, 25% acetonitrile/0.025M KH₂PO₄, pH 4)*

Table I. Crop Recoveries at 0.1 ppm[a]

Crop	AMPA[b]	GLYPH[c]
Cantaloupe	68%	92%
Cranberries	61	76
Jalapeño peppers	65	70
Pumpkin	68	90
Cucumber[d]	82	96

a. Single recovery sample; average of duplicate injections.
b. Calculated from peak heights.
c. Calculated from peak areas.
d. Average of five replicate samples.

Table II. Reproducibility of Recoveries at
at 0.1 ppm from Cucumber[a]

Sample No.	AMPA[b]	GLYPH[c]
1	81%	108%
2	83	89
3	81	96
4	83	97
5	84	92
	$\bar{x} = 82.4\%$	$\bar{x} = 96.4\%$
	$s = 1.3$	$s = 7.2$

a. Average of duplicate injections.
b. Calculated from peak heights.
c. Calculated from peak areas.

Standard deviations for AMPA recoveries were less than 2%, excellent for replicate recoveries while the percentage remained high (82.4%). While GLYPH recoveries have always been higher than AMPA for all crops studied the variability has also been higher ostensibly due to the incompleteness of the $Ca(ClO)_2$ cleavage reaction. Still, a standard deviation of less than 8% appears acceptable.

A time and cost analysis for the determination of two residue

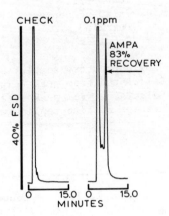

Figure 7. Chromatograms of (left) untreated (check) cucumber and (right) AMPA-fortified cucumber at 0.1 ppm (Aminex A-27 column, 0.1M H_3PO_4, 1 mL/min flow, 20 µL injection, atten 2X)

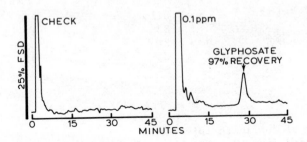

Figure 8. Chromatograms of (left) untreated (check) cucumber and (right) GLYPH-fortified cucumber at 0.1 ppm (Aminex A-27 column, 0.1M H_3PO_4, 1 mL/min flow, 20 µL injection, atten 10X)

samples including a three point analytical curve for AMPA and
GLYPH is shown in Table III:

Table III.   Time and Cost Analysis for AMPA and
GLYPH Residue Determination (OPA-MERC)[a]

Step	Hours	% of 8 hour day	$/sample
1. Extraction	0.5	6	0.22
2. Centrifugation	0.5	6	--
3. Rotary evaporation (50 ml)	1.0	13	--
4. Column prep. (50W - X8)	---[b]	--	12.62
5. Column elution	2.7	34	--
6. Rotary evaporation (5 ml)	1.3	16	--
7. HPLC	2.0[c]	25	0.10
Totals	8.0	100	12.94

a.   Two samples
b.   Columns prepared previous day.
c.   Includes 3 point analytical curve.

Conclusions

For the application described here a post-column fluorogenic
labeling approach was devised which was demonstrated to be sufficiently sensitive (0.1 ppm), reproducible, economical and
relatively rapid compared to the existing PAM procedure. By nor-
malizing on an anion exchange HPLC separation two fluorogenic
reagents were chosen that had demonstrated high quantum yields
and thus were capable of producing high sensitivities for standards. It was necessary to choose a pre-column fluorogenic
labeling reagent (FMOCC1) which derivatized both primary (AMPA)
and secondary amines (GLYPH). This reagent also derivatizes
alcohols under the conditions that were used and consequently
could be expected to be scavanged by crop coextractives while
producing possible interferences; this was indeed observed for
the cantaloupe AMPA fraction. Conversely, since GLYPH is
cleaved by $Ca(ClO)_2$ to produce a primary amine which reacts with
OPA-MERC, a primary amine specific reagent, fewer potential interferences ought to be expected, as was observed. Amino acids,
as well as several naturally occurring phosphonic and sulfonic
acids did not interfere with the post-column fluorogenic labeling determination of the early eluting AMPA. A peak eluting
after AMPA, which did not interfere at the 0.1 ppm level and
occurred in several crops, was not identified.

It was observed that the quartenary ammonium plastic bead
column packings were rugged and less prone to being fouled by
crop coextractives than were the chemically bonded silica particle columns. After over a year of use they retained their efficiency and sensitivity. Their operation with acidic buffers
(pH 1-4) was perfectly compatible with the need to perform the
$Ca(ClO)_2$ cleavage of GLYPH under acidic conditions. Fortunately,
the buffering action of the OPA-MERC reagent was adequate to
shift the pH to 10, an optimum for the ring formation to occur.

The procedure described here takes full advantage of the
water solubility of the two analytes, their anionic and cationic
behavior, the speed of HPLC separations and its suitability for
the analysis of non-volatile compounds.

Literature Cited

1. Weber, G.; Laurance, D. J. R. Biochem. J., 1954, 51, xxxi.
2. Hartley, B. S.; Massey, V. Biochim. Biophys. Acta, 1956,
    21, 58.
3. Gros, C.; Labousse, B. Europ. J. Biochem., 1969, 7, 463.
4. Schmer, G.; Kreil, G. J. Chromatog., 1967, 28, 458.
5. Seiler, N.; Weichmann, J. Experientia, 1964, 20, 559.
6. Deyl, Z.; Rosmus, J. J. Chromatog., 1965, 20, 514.
7. Morse, D.; Horecker, B. L. Anal. Biochem., 1966, 14, 429.
8. Mesrob, B.; Holeysovsky, V. J. Chromatog., 1966, 21, 135.
9. Frei, R. W.; Lawrence, J. F.; Hope, J.; Cassidy, R. M.
    J. Chromatogr. Sci., 1974, 12, 40.
10. Chen, R. F. Arch. Biochem. Biophys., 1967, 120, 609.
11. Froehlich, P. M.; Murphy, L. D. Anal. Chem., 1977, 49, 1606.
12. Lawrence, J. F.; Frei, R. W. Anal. Chem., 1972, 44, 2046.
13. Klimisch, H -J.; Ambrosius, D. J. Chromatog., 1976, 121, 93.
14. Moye, H. A.; Boning, A. J. Anal. Lett., 1979, 12(Bl), 25.
15. Moye, H. A.; Scherer, S. J.; St. John, P. A. Anal. Lett.,
    1977, 10(13), 1049.
16. Krause, R. T. J. Chromatogr. Sci., 1978, 16, 281.
17. Pesticide Analytical Manual, Food and Drug Administration,
    Washington, D. C., Pest. Reg. Sec. 180.364.

RECEIVED February 7, 1980.

Fluorescence and Ultraviolet Absorbance of Pesticides and Naturally Occurring Chemicals in Agricultural Products After HPLC Separation on a Bonded-CN Polar Phase

ROBERT J. ARGAUER

Analytical Chemistry Laboratory, Agricultural Environmental Quality Institute, Agricultural Research, Science and Education Administration, U.S. Department of Agriculture, Beltsville, MD 20705

Abstract

Some pesticides and many naturally occurring chemicals fluoresce sufficiently that direct monitoring of their natural fluorescence during HPLC is feasible. The fluorescence intensities of over thirty pesticides in hexane and methanol were measured at excitation wavelengths of both 254 nm and maximum absorbance. Carbaryl at 0.2 ppm was used as a model pesticide to contrast the relative merits of the fluorescence and absorbance modes for HPLC detection. Actual samples studied included rice, corn, green peas, potato, cucumber, lima beans, and orange. Pollen gathered by foraging honey bees proved the most challenging of the agricultural products studied because of the highly complex chromatograms obtained for methylene chloride extracts. Highly significant is the finding that the fluorescence efficiency of some pesticides varied dramatically with a change in polarity of the mobile phase.

Fluorescence spectrometry is used in both research and surveillance to help assure both the farmer and the consumer a continued bounty of high quality agricultural products while preserving the quality of the environment. My objective is to discuss some of our current research and to share with you several interesting observations where we have used fluorescence as a monitor in high performance liquid chromatography.

In 1970, we described in a book the research published by a large number of scientists from many disciplines who are contributing to the development of fluorescence as a useful and powerful analytical tool (1). Subsequently, in 1977, I published a chapter devoted specifically to the use of fluorescence as a practical technique for the analysis of certain pesticides (2). Figure 1 is taken from that chapter and illustrates what I consider to be the five fundamental approaches that we

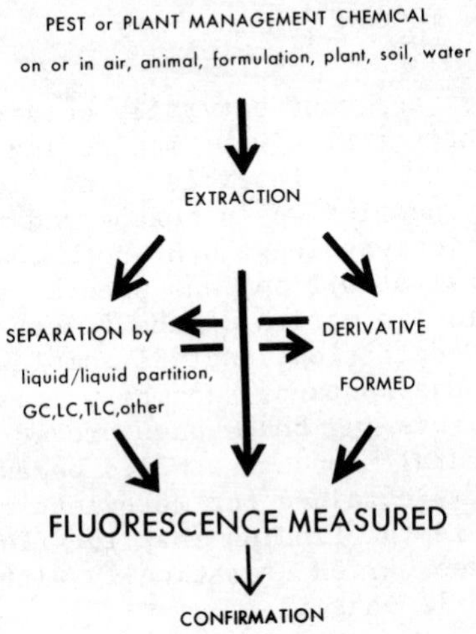

PEST or PLANT MANAGEMENT CHEMICAL

on or in air, animal, formulation, plant, soil, water

EXTRACTION

SEPARATION by   DERIVATIVE

liquid/liquid partition,   FORMED
GC,LC,TLC,other

FLUORESCENCE MEASURED

CONFIRMATION

*Figure 1. Fundamental approaches generally used in the analysis of pesticides
(or other chemicals) by fluorescence (or some other physical property)*

researchers generally use to quantitatively measure chemicals in
a particular substrate. One of the approaches is an idealized
analytical approach and includes just two steps---extraction
(including dilution or concentration of the extract) followed by
a direct determination of a physical property that is specific
for the pesticide in question. Because numerous coextractives
usually interfere with this idealized approach, you and I usually
resort to one of the remaining four approaches to minimize the
effects produced by the interferences. Indeed, one of the
speakers in this symposium has discussed one of these approaches,
i.e. the preparation and analysis of fluorescent derivatives.
Oftentimes though, when forming derivatives, interfering compounds also are formed. Co-extractants originally present as
non-interfering compounds might indeed now become highly fluorescent derivatized interfering compounds. The approach then must
include a separation of the pesticide from the interfering co-
extractants, either before or after formation of the derivative,
and prior to detection.

For purposes of this symposium, I have limited myself to
still another of these approaches. That approach includes
extraction, separation by HPLC, and direct measurement of the
relatively high natural fluorescence inherent in the molecular
structure of the pesticides themselves. You will find that
solvent polarity and instrumental parameters are important
variables when I attempt to contrast the chromatograms obtained
in the absorbance mode with those obtained in the fluorescence
mode.

Experimental Instrumentation

A Perkin-Elmer MPF-2A Fluorescence Spectrophotometer was
used to determine the excitation and emission wavelengths required for achieving maximum fluorescence intensity for the
pesticides studied. The MPF-2A contained a 150 watt xenon arc
and an excitation monochromator with a grating blazed at 300 nm
as the excitation unit; a Hamamatsu R 777 photomultiplier tube
(sensitivity range: 185 - 850 nm) and an emission monochromator
grating blazed at 300 nm as the emission detection unit. A
DuPont Model 848 Liquid Chromatograph was used for HPLC (Figure
2). The accessory injection device included a Rheodyne Model
70-10 six-port sample injection valve fitted with a 20 μ liter
sample loop. A Whatman HPLC column 4.6 mm x 25 cm that contained Partisil PXS 1025 PAC (a bonded cyano-amino polar phase
unspecified by the manufacturer) was used with various mobile
phases at ambient temperature and a flowrate of 1.25 ml/minute.
A dual beam 254 nm photometric absorbance detector with a cell
volume of 6.3 μl was used for absorbance measurement. The exit
port of the photometric absorbance detector cell was connected
in series to a 70 μ-liter flow cell in an Aminco Fluoromonitor.
The Fluoromonitor's excitation light source consisted of a

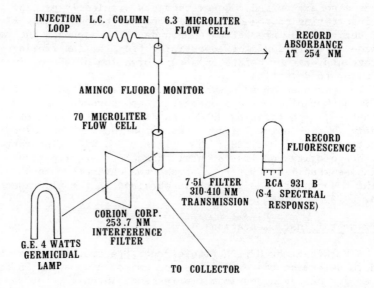

Figure 2.   Schematic of a fluorescence monitor in series with an absorbance detector in HPLC

General Electric 4 watt germicidal lamp and a Corion Corporation 2537 interference filter (15% transmission at 2537). The Fluoromonitor's emission detection unit consisted of a RCA 931B (S-4 spectral response) photomultiplier tube and a Corning 7-51 glass filter that transmits light of wavelengths between 310 and 410 nm.

Determination of Relative Fluorescence of Pesticides

We selected over sixty commercially important pesticides for this study. Some of the pesticides were known to fluoresce. Others were selected on the basis of most likely to fluoresce. The chemical designations of the pesticides mentioned in the text are listed in Table 1. Standard solutions of the pesticides were prepared in ethyl acetate at concentrations of 1 mg/ml. The standard solutions were diluted with hexane or methanol to prepare solutions that contained 2 µg/ml pesticide for the initial fluorescence measurements. Excitation and emission band widths on the spectrofluorometer were adjusted to 4 nm. A solution of quinine sulfate, 1 µg/ml in 0.1 N sulfuric acid, was used as a reference in determining the relative fluorescence intensity of the pesticides. The wavelengths for excitation and emission that would give the maximum fluorescence intensity in both hexane and methanol were obtained next by using 1 cm^2 quartz fluorometer cells. Finally the excitation monochromator was set at 254 nm, and the fluorescence intensity was again measured at wavelength of maximum emission in both hexane and methanol.

Determination of Carbaryl in Food Products

Extraction. One hundred grams of samples (corn, orange, potato, rice, cucumbers, lima beans, or green beans) were blended for five minutes with 300 ml of methylene chloride and 10 ml of 10% sulfuric acid. The filtered extract was dried over anhydrous sodium sulfate, and a 150 ml aliquot was concentrated on a Rinco evaporator to near dryness. The concentrate was dissolved into 10 ml of methylene chloride.

Pre-HPLC Column Chromatography. Five grams of 60-200 mesh silica gel powder (4.7% weight lost on ignition) supplied by J. T. Baker Chemical Company were poured into a 10 mm i.d. glass column. The concentrated extract redissolved in 10 ml methylene chloride was transferred to the column and allowed to soak into the silica gel. Then 65 ml of additional methylene chloride were added to the column. The first 10 ml of the eluate were discarded, and the next 50 ml collected in a 125 ml Erlenmeyer flask.

Chromatography by HPLC. The 50 ml fraction collected from the "dry-column" was concentrated to near dryness on a Rinco evaporator. Two ml of methylene chloride was next added to the

Table 1. Chemical Designations of Pesticides Mentioned in Text

Pesticide	Chemical Designation	Other Designation
Benomyl	methyl 1-(butylcarbamoyl)-2-benzimidazole-carbamate	Benlate; Tersan 1991
Bentazon	3-isopropyl-1H-2,1,3-benzothiadiazin-4(3H)-one 2,2-dioxide	Basagran
Carbaryl	1-naphthyl methylcarbamate	Sevin
Carbofuran	2,3-dihydro-2,2-dimethyl-7-benzofuranyl methylcarbamate	Niagara NIA-10242 Furadan
3-Hydroxycarbofuran	2,3-dihydro-3-hydroxy-2,2-dimethyl-7-benzofuranyl methylcarbamate	
3-Oxocarbofuran	2,3-dihydro-2,2-dimethyl-3-oxo-7-benzofuranyl methylcarbamate	
Coumaphos	O-(3-chloro-4-methyl-2-oxo-2H-1-benzopyran-7-yl)O,O-diethyl phosphorothioate	Co-Ral
Devrinol	2-(α-naphthoxy)-N,N-diethylpropionamide	Napropamide
Diphenyl	biphenyl	Biphenyl
Diphenylamine	diphenylamine or N-phenylbenzenamine	DPA; Scaldip
Ethoxyquin	6-ethoxy-1,2-dihydro-2,2,4-trimethylquinoline	Nix-Scald; Santoquin Stop-Scald
Guthion	O,O-dimethyl S-[(4-oxo-1,2,3-benzotriazin-3(3H)-yl)methyl] phosphorodithioate	Azinphosmethyl
Indolebutyric acid	indole-3-butyric acid	IBA; Hormodin
Maretin	N-hydroxynaphthalimide diethyl phosphate	Naphthalophos

Table 1. Chemical Designations of Pesticides Mentioned in Text (Continued)

Pesticide	Chemical Designation	Other Designation
Morestan	cyclic S,S-(6-methyl-2,3-quinoxalinediyl) dithiocarbonate	Bay 36205; Forstan Oxythioquinox Quinomethionate
Naphthalene	naphthalene	
1-Naphthaleneacetamide	1-naphthaleneacetamide	Rootone; Amid-Thin W
1-Naphthaleneacetic acid	1-naphthaleneacetic acid	NAA, Fruitone
1-Naphthol	1-naphthol	
2-Naphthol	2-naphthol	
1-Naphthylthiourea	1-naphthylthiourea	Antu
o-Phenylphenol	[1,1'-biphenyl]-2-ol	Dowicide 1
Phosalone	0,0-diethyl S-[(6-chloro-2-oxobenzoxazolin-3-yl)methyl] phosphorodithioate	Zolone Chipman RP-11974 Niagara NIA-9241
Piperonyl Butoxide	α-[2-(2-butoxyethoxy)ethoxy]-4,5-(methylenedioxy)-2-propyltoluene	Butacide
Pirimicarb	2-(dimethylamino)-5,6-dimethyl-4-pyrimidinyl dimethylcarbamate	Pirimor
Pyrazophos	ethyl 2-[(diethoxyphosphinothioyl)oxy]-5-methyl pyrazolo[1,5-a]pyrimidine-6-carboxylate	Hoe 2873; Afugan Curamil
Thiabendazole	2-(4-thiazolyl)benzimidazole	APL-Luster; Mertect TBZ; Tecto; Tobaz
Warfarin	3-(α-acetonylbenzyl)-4-hydroxycoumarin	Coumafene; Kypfarin

concentrate. Since the HPLC was fitted with a 20 µ-liter injection loop, the amount of extract injected into the HPLC was
equivalent to 500 mg of starting sample per injection as given
for the agricultural products in several of the figures.

Results and Discussion

Initial Measurements. The data in Table 2 for the fluorescence of pesticides in hexane and methanol were obtained with a
single-beam spectrofluorometer. No attempt was made to adjust
these values for either the intensity distribution of the
excitation source or the relative sensitivity of the emission
unit with wavelength. However, several observations can be
drawn from this data that can be useful when applied to a HPLC
fluorescence detector.

1. Pesticide-solvent interactions affect the fluorescence of
the pesticides.

2. Sensitivity and selectivity would be enhanced when one is
operating in the fluorescence mode if the light source selected
for excitation were to emit light concentrated solely in a
narrow band at wavelengths that correspond with the wavelengths
of maximum absorption (maximum excitation) for the particular
pesticide.

3. Tunable dye lasers could provide both high intensity and
some selectivity when used as excitation sources in a fluorescence monitor. However, mercury lamps provide a convenient and
inexpensive excitation source with emitted light largely concentrated at a wavelength of 254 nm. Many pesticides in Table 2
fluoresce sufficiently when excitated at 254 nm (xenon arc) so
that the strong 253.7 nm line in a mercury lamp can be advantageously used for excitation in a HPLC fluorescence monitor.
Indeed if one were to adjust the data in Table 2 for the quantum
distribution of the xenon arc source with wavelength, the values
tabulated for the fluorescence intensity with excitation at
254 nm would increase relative to the values given for the
fluorescence excited at wavelengths of maximum excitation/
absorption.

Pesticides Separated on a CN-Bonded Polar Phase. Figures 3
and 4 contain the chromatograms obtained for several of the
pesticides listed in Table 2. Various mobile phases were used
to facilitate separation on the CN-bonded polar phase. Detection is shown in both absorbance and fluorescence modes. It
would appear that fluorescence detection is more sensitive for
some pesticides, while absorbance detection is more sensitive
for others. However, comparisons of one manufacturer's absorbance monitor with another manufacturer's fluorescence monitor
could be misleading. (I will refer to this under Instrumental
Parameters.) Furthermore, the ultimate useful comparison is
obtained when practical samples are chromatographed since these

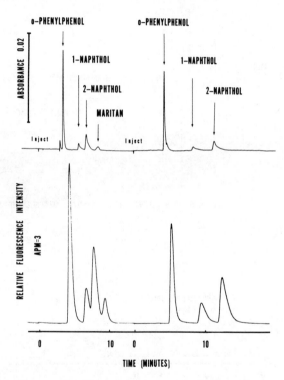

Figure 3.   HPLC of several pesticides (100 ng) on a CN-bonded polar phase with absorbance and fluorescence detection (APM = attenuation setting of photomultiplier signal)

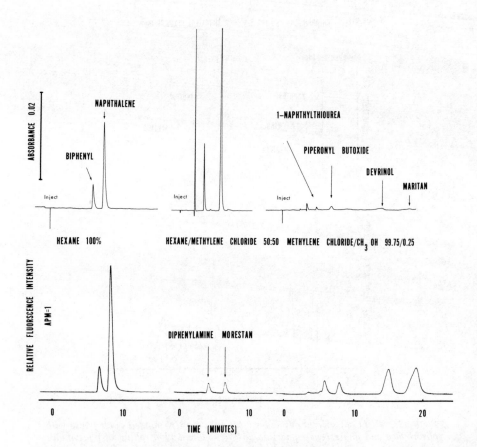

Figure 4.   HPLC of several pesticides (100 ng) on a CN-bonded polar phase with
absorbance and fluorescence detection

samples generally contain many naturally occurring chemicals that may absorb and fluoresce and be considered as interfering coextractives.

To illustrate the effects of coextractives, I fortified several agricultural products with carbaryl at 0.2 ppm. Carbaryl was chosen as a model pesticide to illustrate the effects of coextractives because its fluorescence intensity lies in the lower-middle range relative to the other pesticides in Table 2. In addition, we were familiar with its gas chromatographic and spectrofluorometric properties (2). Figure 5 contains the chromatograms obtained when carbaryl was injected into the HPLC at several instrument sensitivity settings. Amounts of carbaryl injected were between 1 μg and 4 ng.

The chromatograms show linearity in response with the amount of carbaryl injected in both the absorbance and fluorescence modes. Practical agricultural samples when analyzed for carbaryl, however, require a preliminary column cleanup before injection into the HPLC in order to preserve the integrity of the HPLC Column.

Function of Pre-HPLC Column. The schematic in Figure 6 for a HPLC chromatogram representative of extracts of agricultural products illustrates use of silica gel adsorption chromatography for the pre-HPLC cleanup step. The schematic shows that (A) a large part of the co-extractives can be removed in the first fraction from the precolumn, (B) the polarity of the mobile phase can be adjusted so the pesticide elutes in pre-HPLC column fraction B where the eluate can be collected and concentrated for injection into the HPLC, while (C) more polar compounds that would otherwise appear during HPLC have been eliminated by permanent adsorption on the pre-HPLC Column.

Practical Agricultural Samples. Chromatograms in Figures 7, 8, and 9 reflect HPLC injections of various crop extracts equivalent to 500 mg of crop and 100 ng of carbaryl. These chromatograms show that for carbaryl at the described operating conditions the fluorescence mode of detection is more sensitive and selective than is the absorbance mode of detection for many of these agricultural crops. There are several difficulties with these chromatograms, however, as the amount of organic coextractables increase relative to the amount of carbaryl present in the crops. Indeed, the chromatograms in Figure 8 show that carbaryl at 0.2 ppm in orange would be difficult to measure given the extraction, precolumn cleanup procedure, and the 500 mg crop injection described. However difficult, agricultural samples such as orange and pollen collected by foraging honey bees that contain a large amount of co-extracted interferences can still be analyzed by HPLC. Four modifications in the described procedure can be attempted.

Table 2. Fluorescence of Pesticides (2 μg/ml) in Hexane and Methanol

| Common/Chemical Name | Wavelength (nm) at Maximum[1] | | Relative Fluorescence Intensity[2] | | | |
| | Excitation | Emission | at maxima in | | at excitation set at 254 nm and emission at maximum in | |
			Hexane	Methanol	Hexane	Methanol
Benomyl	286	300	52	16	0.5	0.2
Bentazon	340(306)	450(370)	7	10	0.6	1.8
Carbaryl	280	332	22	41	3.7	8
Carbofuran	280	304	60	31	4	3
3-Hydroxycarbofuran	280	304	40	49	2	2.5
3-Ketocarbofuran	not detected	-	-	-	-	-
Coumaphos	320	380	4	48	0.1	3
Devrinol	293	335	85	71	15	5
Diphenyl	260	305	23	50	23	48
Diphenylamine	290	330	90	120	8	9
Ethoxyquin	360	440(420)	87	27	10	5
Guthion	290	340	not detected	6	not detected	2
Indolebutyric acid	290	340	67	107	17	16
Maretin	340(330)	375(365)	22	320	0.1	7
Morestan	360	380	6	22	0.3	1
Naphthalene	284	322	13	27	3	8
1-Naphthaleneacetamide	280	336	20	34	3	5
1-Naphthaleneacetic acid	290	324	20	35	3	5
1-Naphthol	295	340	100	160	7	7
2-Naphthol	285	355	59	90	17	27
1-Naphthylthiourea	325(332)	425(370)	7	6	0.6	0.5
o-Phenylphenol	290	335(320)	18	230	7	80
Phosalone	282	310	8	6	0.3	0.4
Piperonyl Butoxide	290	320	59	66	1.6	2

Table 2. (Continued)

| Common/Chemical Name | Wavelength (nm) at Maximum[1] | | Relative Fluorescence Intensity[2] | | | |
| | | | at maxima in | | at excitation set at 254 nm and emission at maximum in | |
	Excitation	Emission	Hexane	Methanol	Hexane	Methanol
Pirimicarb	310	380(355)	72	80	18	28
Pyrazophos	252	420	23	1	21	1
Thiabendazole	310	340(332)	210	183	9	12
Warfarin	310	390(340)	5	8	0.3	0.7

Reference:

Quinine Sulfate
Excitation λ 350
Emission λ 450
in 0.1N Sulfuric acid

Concentration (μg/ml)	Relative Fluorescence Intensity
1.0	78.0
0.1	7.8
0.01	0.77
0.001	0.10
0.000	0.033

(1) The wavelengths are given for pesticide in methanol and are the same in hexane except as indicated in parentheses.

(2) 50 μ-liters of 1 mg/ml stock solutions of pesticide in ethylacetate diluted with hexane or methanol to 25 ml and measured in a Perkin-Elmer Model MPF-2A Fluorescence Spectrophotometer.

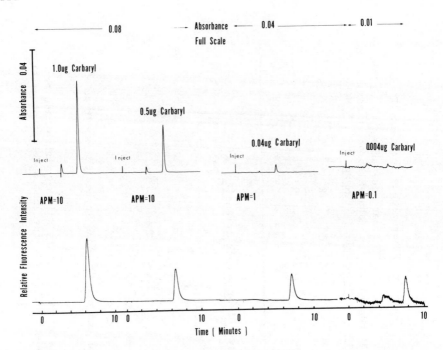

Figure 5.   Fluorescence and absorbance responses for carbaryl during HPLC on a
CN-bonded polar phase (mobile phase: methylene chloride/methanol 99.5/0.5)

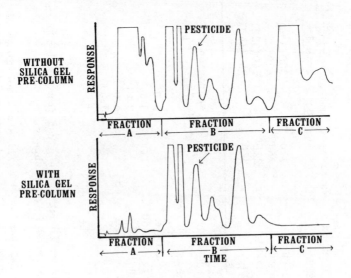

Figure 6.   Function of pre-HPLC column cleanup chromatography on HPLC
chromatograms

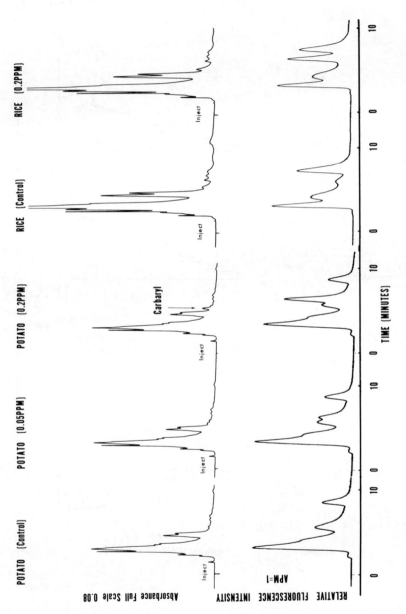

Figure 7. Potato and rice extracts: comparison of fluorescence and absorbance detection modes after HPLC on a CN-bonded polar phase (0.2 ppm = 100 ng carbaryl/500 mg crop equivalent injected; mobile phase: methylene chloride/methanol 99.5/0.5)

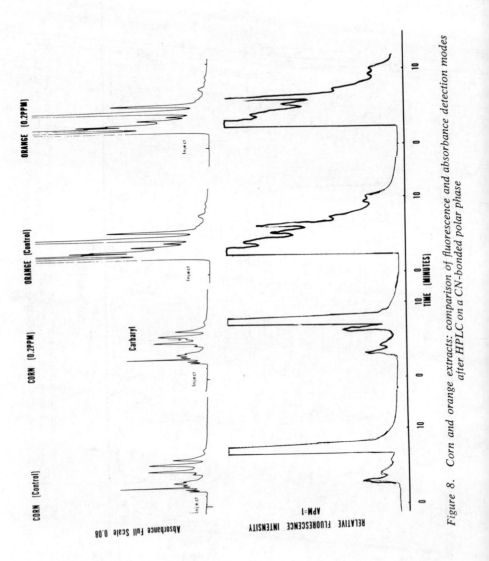

Figure 8. Corn and orange extracts: comparison of fluorescence and absorbance detection modes after HPLC on a CN-bonded polar phase

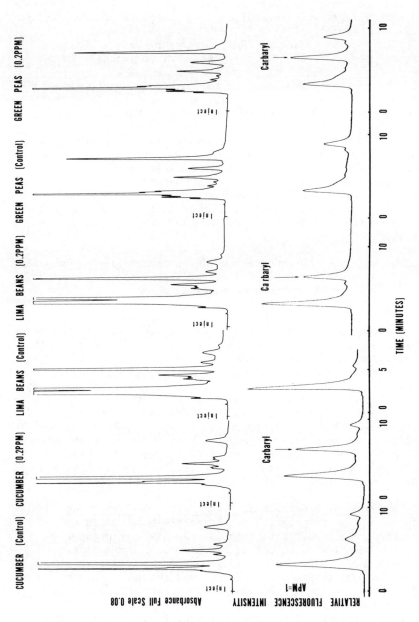

Figure 9. Extracts of cucumber, lima beans, and green peas: comparison of fluorescence and absorbance detection modes after HPLC on a CN-bonded polar phase

1. Additional pre-HPLC cleanup steps, their selection based on
the chemistry of the pesticide and impurities, are introduced to
help achieve the required separation.
2. The polarity of the mobile phase is reduced or modified to
facilitate separation.
3. The size of the HPLC column is increased to yield an
increased number of theoretical plates for an increase in the
power for separation.
4. The detectability level of the method is sacrificed simply
by reducing the amount of crop equivalent injected into the
chromatography, thereby reducing the size of the interfering
peaks.

The chromatograms we obtained for samples of pollen at a
detectability level for carbaryl of 5 ppm are given in Figure 10
and illustrate our use of the fourth modification.

Whereas chromatograms obtained as background controls for
many agricultural crops vary with the natural chemical composition and degree of ripeness, chromatograms for pollen can become
further complicated by the unpredictable foraging habits of honey
bees (Apis mellifera L.), and the question of what constitutes a
valid background control sample.

The beauty of absorbance and fluorescence modes as compli-
mentary detectors for HPLC lies in their nondestructive nature.
Fractions eluting from these detectors are readily available for
confirmation by mass spectrometry.

Decrease of Fluorescence Signal by Unresolved Co-Extract-
ables. The fluorescence peak for carbaryl in lima beans in the
chromatogram in Figure 9 is somewhat smaller than expected. Two
mechanisms can account for the decrease in the fluorescence
signal for carbaryl. I prefer the mechanism that suggests that
the co-elutants are absorbing some of the exciting light, thereby
decreasing the light quanta available for absorption by the
carbaryl molecules. A second mechanism suggests that the co-
elutants can quench the fluorescence by interacting with the
excited state of the carbaryl molecule to cause the absorbed
energy to be dissipated along a solute-solute interacting non-
fluorescent pathway.

Variation of Fluorescence Signal with Polarity of the Mobile
Phase. A change in polarity of the mobile phase often aids the
resolution of the pesticide from interfering co-elutants. However, the relative retention area for several chromatographic
peaks obtained in the fluorescence mode of detection was found to
vary as the mobile phase in the HPLC was changed. The data in
Table 3 show the effects of common chromatographic solvents on
the fluorescence of two pesticides. For maretin and o-phenylphenol, the relative fluorescence intensity, as measured in a
spectrofluorometer, increased substantially as the polarity of
the solvent increased. However, as shown for pyrazophos in

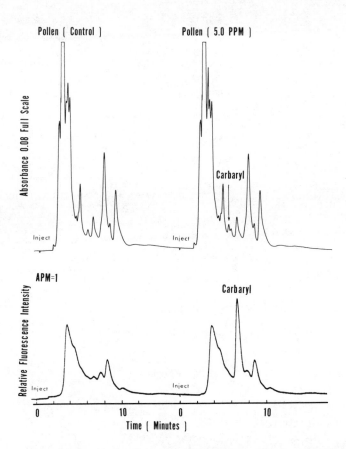

Figure 10. Extracts of pollen collected by foraging honey bees: comparison of fluorescence and absorbance detection modes after HPLC on a CN-bonded polar phase (0.1 μg carbaryl/20 mg pollen equivalent injected; mobile phase: methylene chloride/methanol 99.5/0.5)

TABLE 3. FLUORESCENCE OF MARETIN (2 μG/ML) AND O-PHENYLPHENOL
(2 μG/ML) IN COMMON CHROMATOGRAPHIC SOLVENTS

MARETIN   O-PHENYLPHENOL

SOLVENT[1]	EXλ	EMλ	RELATIVE FLUORESCENCE INTENSITY	EXλ	EXλ	RELATIVE FLUORESCENCE INTENSITY
HEXANE 100%	330	365	20	290	320	19
HEXANE/METHYLENE CHLORIDE 50:50	340	375	100	290	320	34
METHYLENE CHLORIDE 100%	340	375	230	290	320	41
METHYLENE CHLORIDE/ METHANOL 95:5	340	375	240	290	335	120
METHANOL 100%	340	375	320	290	335	230

(1) 1 MG/ML OF STOCK SOLUTIONS OF PESTICIDES IN ETHYLACETATE DILUTED
WITH SOLVENT.

TABLE 4. FLUORESCENCE OF PYRAZOPHOS (2 μG/ML) IN
COMMON CHROMATOGRAPHIC SOLVENTS.

PYRAZOPHOS

SOLVENT	RELATIVE (1,2) FLUORESCENCE INTENSITY
HEXANE 100%	21.0
HEXANE/METHYLENE CHLORIDE 50:50	11.0
METHYLENE CHLORIDE 100%	8.5
METHYLENE CHLORIDE/ETHYLACETATE 95:5	8.4
METHYLENE CHLORIDE/ETHYLACETATE 85:15	7.5
METHYLENE CHLORIDE/METHANOL 99.5:0.5	7.7
METHYLENE CHLORIDE/METHANOL 99:1	7.3
METHYLENE CHLORIDE/METHANOL 95:5	5.4
METHANOL 100%	1.0

(1) 1 MG/ML STOCK SOLUTION OF PYRAZOPHOS IN
ETHYLACETATE DILUTED WITH SOLVENT.

(2) EXCITATION WAVELENGTH SET AT 254 NM; EMISSION
WAVELENGTH AT 420 NM.

Table 4, the relative fluorescence intensity also can decrease substantially as the polarity of the solvent increases. Next the absorbance for o-phenylphenol, maretin, and pyrazophos was compared in both hexane and methanol. A Beckman DG-GT Spectrophotometer was used to obtain the absorbance values.

Table 5. Absorbance of Pesticides in Hexane and Methanol

Pesticide	Concentration	Wavelength (nm)	Absorbance in hexane	in methanol
o-phenylphenol	20 μg/ml	286	0.59	0.60
maretin	20 μg/ml	343	0.74	0.72
pyrazophos	10 μg/ml	248	1.32	1.26

No appreciable difference in absorbance values was observed between solutions in hexane or methanol (Table 5). We therefore attribute the change in fluorescence to solute-solvent interactions between the solvent and the pesticide in its excited state.

Several Instrumental Parameters. In a dynamic flow system such as HPLC, the signal generated by the fluorescence monitor is proportional to several variables in addition to the two (pesticide concentration and solvent used as the mobile phase) already discussed. These variables include:

1. Size of the flow cell.

2. The geometry or design of the detector, which also includes selection of optical filters and types of light detectors with their differing relative responses to light of various wavelengths.

3. The intensity of the excitation source.

4. The spectral (absorbance and fluorescence) characteristics of the solute (pesticide).

The 70 μ-liter size of the flow cell contributed to the broadening of the chromatographic peaks observed in the fluorescence mode and could account for some loss in selectivity. The second variable, detector design, could account for the unexpected ratio in peak heights (compared with data in Table 2) obtained in Figure 4 for diphenyl and naphthalene: the glass envelop that houses the photomultipler tube may absorb some of the fluorescence of biphenyl (emission maximum 305 nm) while the fluorescence of naphthalene (emission maximum 322 nm) passes through.

Variables 3 and 4 are found in the familiar relationship, i.e., that fluorescence in dilute solutions is directly proportional to the absorbance of the solute $A\lambda$ at the wavelength selected for excitation, times the quanta of light $Io\lambda$ available at the wavelength selected for excitation, times the fluorescence quantum efficiency $\phi\lambda$ for the solute. Oftentimes the wavelengths available for maximum illumination in an excitation source do not

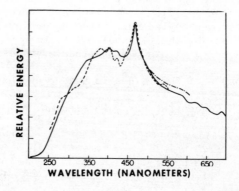

Analytical Chemistry

Figure 11.   Relative energy distribution of a xenon-arc source (3)

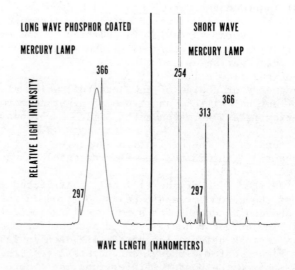

*Figure 12.   Relative spectral distribution of a low-pressure mercury vapor lamp
with visible light cut-off filter as measured through the emission monochromator of
a spectrofluorometer (2 nm slit width). (Typical of lamps used to view fluorescence/
fluorescence quenching on thin-layer chromatographic plates.)*

correspond with wavelengths of maximum light absorption for the solution. In Figures 11 and 12 the relative energy distribution of a xenon arc source and two mercury lamps is compared. The 254 nm mercury line emitted by a germicidal mercury lamp was selected for excitation in our HPLC fluorescence monitor. The product of $I0_{254}$ (a relatively high energy source for irradiation) and A_{254} (the absorbance of the pesticide at 254 nm) was expected to produce a sufficiently large fluorescence for many of the pesticides. Furthermore any 254 nm excitation radiation that is scattered by the flow cell and reflected toward the detector is filtered out by the glass envelope surrounding the RCA 931B photomultiplier tube and redundantly by an additional glass filter placed in the entrance window to the detector. It was anticipated however that there would be some loss of selectivity due to fluorescing co-elutants that absorb at 254 nm.

Conclusions

Many pesticides fluoresce sufficiently when excited at 254 nm to permit the advantageous use of the intense 253.7 nm emission from a mercury light source for excitation in a HPLC fluorescence monitor.

A fluorescence monitor can conveniently confirm and support data obtained with an absorbance detector. However, any comparison of the relative sensitivity/selectivity of the absorbance mode vs the fluorescence mode depends on the spectrochemical nature of both the pesticide itself, and the co-extracted co-elutants found in the agricultural product extracted. A judicious selection of mobile phase is required to optimize separation of the pesticide and co-extractives on a CN-bonded polar stationary phase.

Chromatographic peak areas in the fluorescence mode can be affected by the choice of the mobile phase because pesticide – solvent interactions affect the fluorescence. We have identified this unusual phenomena for both the pesticides themselves and for unidentified co-extractives from several agricultural crops. Quenching of the fluorescence emitted by pesticides can occur when co-elutants are present in large amounts.

Acknowledgments

The author appreciates the assistance of Ernest J. Miles, Physical Science Technician, of the Analytical Chemistry Laboratory, AEQI, AR, SEA, USDA, Beltsville, Md. in obtaining the data for this contribution.

Literature Cited

1. White, C. E., Argauer, R. J., "Fluorescence Analysis, A Practical Approach," Marcel Dekker, New York, N.Y., 1970 (contains numerous literature references)
2. Robert J. Argauer, "Fluorescence Methods for Pesticides." Chapter 4 in Analytical Methods for Pesticides and Plant Growth Regulators, Volume IX, "Spectroscopic Methods of Analysis," Academic Press, Inc., New York, N.Y., 1977. (contains over 100 literature references to fluorescence of pesticides)
3. Argauer, R. J. and White, C. E. Anal. Chem. 1964, 36, 368.

RECEIVED February 7, 1980.

Quantitative Thin-Layer Chromatography of Pesticides by In Situ Fluorometry

VICTORIN N. MALLET

Chemistry Department, Université de Moncton,
Moncton, New Brunswick, Canada, E1A 3E9

Thin-layer chromatography is well established as a clean-up technique for environmental samples and it has been used successfully for many years to separate pesticide residues from interfering co-extractives. In the late sixties instrumentation for measuring the absorbance, transmission and reflectance of compounds separated as spots on thin-layer chromatograms (tlc) was developed but detection limits were not very good and reproducibility was a deterrent. Then fluorometers and spectrofluorometers having better sensitivity and selectivity became available for measuring the fluorescence of organic compounds on tlc. At the time gas liquid chromatography (glc) was rapidly becoming accepted as a selective analytical technique for the quantitative determination of pesticides but there were some drawbacks. Many pesticides were thermally unstable or non-volatile which made their quantitative determination very difficult. There was also the need for alternative methods to glc for confirmation purposes. This incited many scientists to develop fluorometric methods suitable for direct quantitative measurements on tlc.

The main object of this review is to look at some of the early developments of quantitative tlc in terms of the fluorometric detection methods that were developed for the evaluation of pesticides. It is also intended to discuss the present status of the technique and consider its future.

Theory of Fluorescence

The subject of fluorescence, the theory and practical applications, has been treated very thoroughly in many textbooks. Some of the most important have been written or edited by Guilbault (1), Winefordner

0-8412-0581-7/80/47-136-127$07.75/0

et al (2), Hercules (3), and White and Argauer (4).
The literature on fluorescence analysis has also been
reviewed periodically since 1949, originally by White
(5) and lately by O'Donnell and Solie (6). Only a
very brief introduction to the subject as it relates
to fluorescence measurements on thin-layer chromatograms will be given in this paper.

Molecular fluorescence can be described as the
immediate emission of energy from a molecule after it
has been irradiated at a particular wavelength or by
a band of light. The lifetime of the excited state
is of the order of 10^{-9}-10^{-7}sec (3) and the emission is
usually at a longer wavelength than that of the excitation light. Fluorescence is said to occur from a
singlet state of a molecule in contrast to phosphorescence which is emitted from a triplet state after
"intersystem crossing". Some compounds do not fluoresce, that is, the energy emitted is dissipated by
the medium or emitted as phosphorescence which has a
longer excited lifetime. The fluorescence (quantum)
yield may be defined as the ratio of molecules that
fluoresce to the number of molecules excited. Those
compounds that are highly fluorescent have a fluorescence efficiency that approaches 100%.

Fluorescence spectroscopy is characterized by a
greater selectivity when compared with other spectrophotometric techniques because there is an excitation
and an emission spectra, with maxima usually quite
characteristic of a particular compound. It is also
selective because of the limited number of organic
compounds that fluoresce. It has greater sensitivity
than spectrophotometric methods; in solution 10^{-9}-
10^{-12} M can usually be measured while on a thin-layer
chromatogram some naturally fluorescent compounds may
be detected instrumentally in sub-nanogram amounts.

Fluorescence is greatly affected by the structure
of a molecule. Usually only aromatic compounds fluoresce although some aliphatic and alicyclic molecules
are known to fluoresce. Electron-donating groups
such as -OH and -OCH$_3$, that can increase the electron
flow of an aromatic system usually increase the fluorescence while other groups that contain hetero atoms
with n-electrons that can absorb the emitted energy,
will usually quench the fluorescence. However, it is
always difficult to predict whether or not, or to what
extent, a compound will fluoresce.

Fluorescence is affected by the medium, such as
the solvent or surface. Polar media in general tend
to enhance fluorescence and pH has a drastic effect
creating hypsochromic or bathochromic shifts as well as

an increase or decrease in fluorescence. Concentration is also very important. If it is too high self-quenching usually occurs and calibration curves become non-linear, thus care must be taken to assure that measurements are made in the linear portion of a curve. As it will be revealed later on, the nature of the adsorbent used as thin-layer may have a profound effect on the nature and intensity of fluorescence.

Instruments for Measuring Fluorescence on Thin-Layer Chromatograms

Various instruments are available for the quantitative measurement of fluorescence on a thin-layer chromatogram and some have been described in earlier reviews (7, 8). They are usually of two types, one that enables only the quantitative scanning of a plate and the other that can be used to record the excitation and emission spectra of a given fluorescent compound. Since they more or less all perform on the same principle, only two will be described in this article.

Usually, simple fluorometers are equipped only with filters and the cost is relatively low. One example is the Turner Model III equipped with a tlc scanner (Figure 1). The incident light from a mercury vapour lamp passes through a primary filter and irradiates the chromatogram (Figure 2). The emitted fluorescence passes through a secondary filter before being picked-up by a photomultiplier tube and the signal is amplified. The sensitivity is controlled by admitting more light to pass through a control slit. The instrument offers very good detection limits (sub-nanograms per spot) and reproducibility is no deterrent, both factors depending on the nature of the fluorescent species which in turn may depend on the fluorometric detection technique.

A more sophisticated instrument is the Farrand VIS-UV Chromatogram Analyser (Figure 3) that can also be used to measure the fluorescence spectra as well as to obtain reliable quantitative data. This particular instrument is equipped with a Xenon lamp that emits a continuous spectrum of wavelengths. It has two monochromators and two filters, for both the excitation and emission modes and can be operated either in double-beam ratio or single beam.

An optical diagram of the Farrand instrument is shown in Figure 4. The excitation light goes successively through the optical system, the first mono-

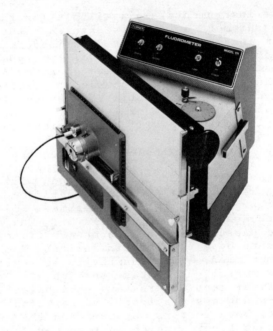

G. K. Turner Associates

Figure 1.   Turner fluorometer Model III

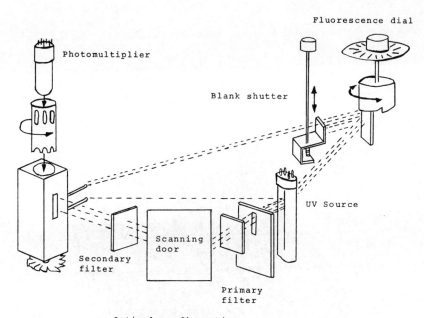

Fluorescence dial

Photomultiplier

Blank shutter

UV Source

Scanning door

Secondary filter

Primary filter

Optical configuration

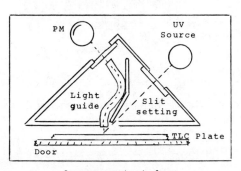

Arrangement at door

Figure 2. *Optical configuration of the Camag–Turner fluorometer*

Farrand Export Sales, Inc.

Figure 3.   Farrand VIS–UV chromatogram analyzer

chromator and the primary filter before it hits the
chromatogram. The emitted fluorescence goes through
a secondary filter, an emission monochromator and the
optical system before it is picked-up by the photomultiplier. The double-beam ratio system offers an
added advantage when there is background fluorescence.
     There are many more instruments on the market
including newer models but those mentioned above have
been used on a regular basis by the author and have
been found to be sensitive and reliable in terms of
reproducibility.

The Fluorescence of Pesticides on Thin-Layer Chromatograms

     As mentioned earlier fluorescence measurements
made directly on tlc are usually more advantageous
than UV-VISIBLE densitometry in terms of sensitivity
and selectivity for organic compounds in general.
The main drawback with pesticides is that only a limited number are naturally fluorescent. The procedure
is usually straightforward for fluorescent pesticides
and measurements can be made immediately after separation. But for those that are not fluorescent, physical, chemical or biochemical treatments are required.
     There are various alternatives to obtain fluores-
cence with a pesticide. For instance, fluorescence
may be derived from the pesticide by treatment with
heat, acid, base, inorganic salts or a combination of
these. Another approach is to prepare a derivative
(fluorogenic labelling) in solution; the derivative
is then extracted and applied directly on a chromatogram for separation. A fluorogenic reagent that
will become fluorescent on contact with the pesticide
may also be used. These alternatives are summarized
in Figure 5.

     ### Pesticides Possessing Native Fluorescence. The
number of pesticides that possess enough natural fluorescence to enable their quantitative determination
directly on thin-layer chromatograms is rather limited.
Typical examples are given in Table I which includes
their spectral characteristics (9). With the exception of diphacinone, most fluoresce in the blue region
of the spectrum. The limit of detection for each
compound is in the low nanogram per spot range.
     A residue analytical method has been developed
for quinomethionate (MORESTAN) in tap water (10) and
in various crop produce (11). In tap water, as low
as 0.1 ppb was determined with over 90% recovery. The

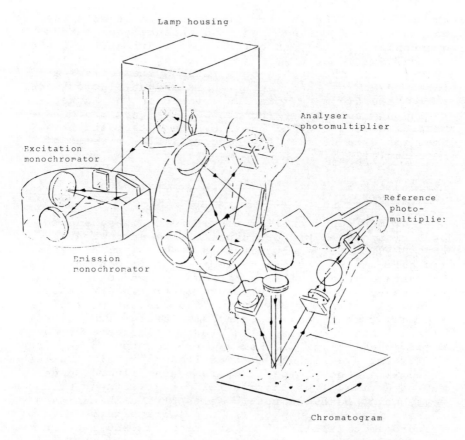

Figure 4.   Optical scheme of the Farrand spectrofluorometer

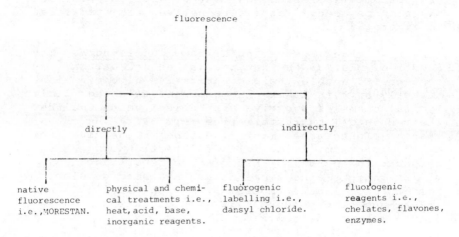

Figure 5.   Alternatives to the fluorescence of pesticides

Table I. Some Naturally Fluorescent Pesticides (9).

Common Name	Structure	λEx (nm)	λEm
Benomyl	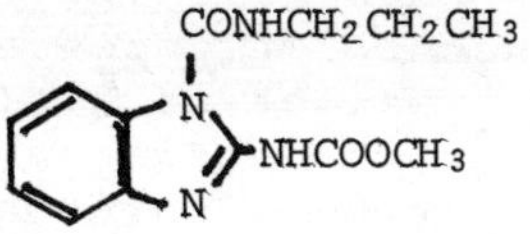	292	422
Coumatetralyl	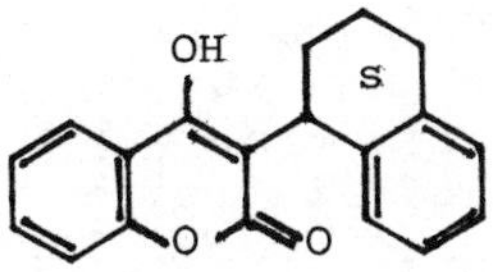	330	415
Diphacinone	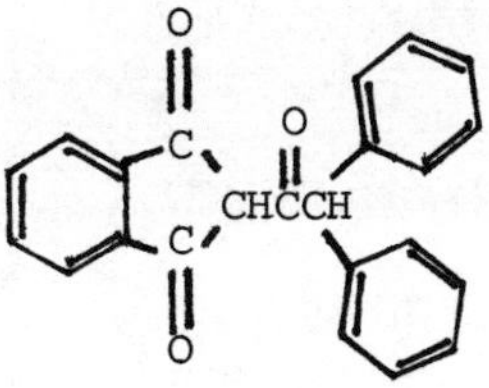	330 (365)	518
Fuberidazole	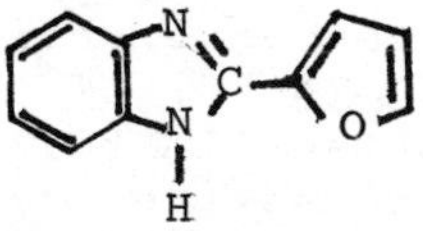	328	402
Propyl isome	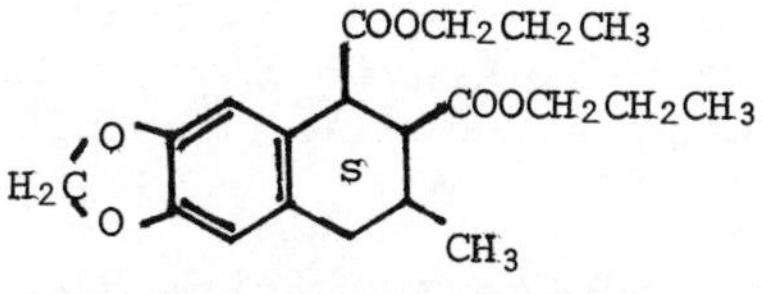	343	460
Quinomethionate	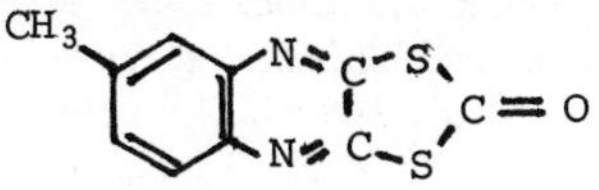	363	418

s = saturated

Journal of Chromatography

average relative standard deviation for the detection
technique at that concentration (0.1 μg per spot) is
2.5%. In apples, peaches, pears and tomatoes 0.05 ppm
has been determined with recoveries over 90% (11).
The method can be used to confirm glc data or as an
independent analytical technique.

Maretin (N-hydroxynaphthalimide diethyl phosphate)
(I) possesses enough natural fluorescence to be determined directly on tlc. Mallet et al (12) proposed a
method for its quantitative determination in water and
milk by measuring the native fluorescence directly on
a silica-gel tlc.

The relative standard deviation was 3% for 1 μg
spots and the fluorescence was linear up to 3 μg.
Recoveries were greater than 90% for 0.1 ppb in water
and 10 ppb in milk. The method was both rapid and
simple but since the natural fluorescence of Maretin
is unstable on silica-gel layers, instrumental measurements had to be done quickly and chromatograms
could not be kept for future reference.

(I) $(C_2H_5O)_2$ — P — O — N

It is advantageous in terms of an analytical
technique for a pesticide to be naturally fluorescent.
There is no background fluorescence on the plate,
except that occuring from co-extractives. This allows
using maximum sensitivity (gain) of an instrument,
thus giving better detection limits. Since no reagents are added or sprayed on the chromatogram, the
spots are usually well defined and the reproducibility
is very good. The procedure is shorter and a complete
determination can be achieved faster, as compared with
a detection technique requiring special treatment of
the chromatogram.

Fluorescence of Pesticides by the Indirect
Approach. This is the traditional approach. The
thin-layer chromatogram is usually sprayed with a
fluorogenic reagent which becomes fluorescent by reacting with the pesticide; the latter may have been pre-
treated to allow the reaction. The important aspect
of this approach is that the fluorescence originates
from the spray reagent and is almost always the same.
With fluorogenic labelling compounds, the fluorescence
normally will vary slightly with different pesticides
but for all practical purposes quantitative measurements can usually be made at fixed wavelengths.

In some of the early experiments, organochlorine
compounds were detected by fluorescence when the chromatogram was sprayed with N-methylcarbazole or rhoda-
mine B (13). Also, fluorescence whiteners (14) known
as Calcofluors, were used to visualize carbamates,
uracils, and ureas. Both of these techniques, however,
were only of a semi-quantitative nature.

   1. **Chelate spray reagents.** A procedure for the
fluorometric detection of phosphate esters on paper
chromatograms was suggested by Gordon et al. (15) in
1964. It was later applied by Ragab (16) to organothiophosphorus pesticides on silica-gel layers. The
method required treatment of the chromatogram with
bromine vapours, spraying with a solution of ferric
chloride followed by 2-(o-hydroxyphenyl) benzoxazole.
The technique was modified by Belliveau et al. (17)
who combined the fluorogenic reagent and the salt solution to form a single spray.

   The presence of the metal cation quenches the
fluorescence of the ligand. Treatment of an organophosphorus pesticide containing sulfur atoms with
bromine vapours produces HBr, a strong acid. Spraying
the fluorogenic reagent onto the chromatogram liberates the ligand and a blue fluorescent spot appears in
presence of a pesticide (7) whilst the background
remains essentially non-fluorescent. The reaction is
summarized below:

$$M^n + nHR \rightleftharpoons MRn + nH^+$$

where nHR represents the ligand and MRn the metal-
ligand complex which is not fluorescent.

   One such chelate spray reagent prepared by
mixing SAQH (II) and manganous chloride was used successfully to detect organothiophosphorus pesticides
on tlc with a detection limit of 0.02 µg/spot (18).
Linear calibration curves were obtained up to 6 µg
per spot and the relative standard deviation for 0.1µg

(II)

Salicyl Aldehyde Quinolyl Hydrazone

spots was 10.2% (Table II). Although quantitative measurements could be done at a lower concentration the reproducibility decreased sharply.

Table II.   Reproducibility as a Function of Concentration of Trithion (SAQH-Mn Spray) (18).

Concentration (µg)	Average (n=5) Relative Standard Deviation (%)
4.0	2.6
1.0	2.5
0.5	5.5
0.1	10.2

International Journal of Environmental Analytical Chemistry

Nevertheless, the method was used to determine Guthion (an organothiophosphorus compound) in water at the 0.5 ppb level. This particular insecticide is difficult to determine by glc and at the time the method offered a reasonable alternative. Guthion was also determined in blueberries with 85% recovery at 0.5 ppm using the SAQH-Mn spray reagent after separation on tlc (19). The major difficulty with the chelate spray method is that bromine vapours will also react with a variety of compounds. Thus, a rigorous clean-up may be necessary for samples with many co-extractives.

Bidleman et al. (20) used a palladium chloride-calcein chelate spray for organothiophosphorus compounds and detected 10 ng per spot. The mechanism is based on the release of the ligand because of the affinity of palladium ions for sulfur atoms. The technique does not necessitate bromine vapours and is therefore an improvement in terms of selectivity. Unfortunately, the reaction is slow and the fluorescence increases with time.

2.   Flavones (fluorescence enhancement). Flavones have been used to detect pesticides on cellulose thin-layer chromatograms (21). Fluorogenic reagents such as Fisetin (III) are relatively non-fluorescent on a non-polar background like cellulose. The presence of

(III)

a polar impurity is immediately detected as a yellow-wish-green fluorescent spot. Figure 6, shows that the weak fluorescence of the background is "enhanced" by the presence of Baygon, a carbamate insecticide.

By this method a variety of pesticides including carbamates and organophosphates can be detected (22) at 0.01 µg per spot. A quantitative method using flavones was developed for Proban (an organothiophosphate), at 1 ppb in water, with a relative standard deviation of 12.2% for 0.1 µg spots (23). The drawback with this type of spray reagent is that the thin-layer must be free from impurities otherwise high background fluorescence is recorded hindering the measurement of the fluorescence of the spot. Also, cellulose layers are not that good for the separation of pesticides from co-extractives and since the spray is almost universal, this becomes a serious problem.

   3. pH-Sensitive spray reagents. Flavones have also been used on silica-gel layers (24). Since the latter is a relatively polar surface the background is fluorescent but heat-treatment of the chromatogram for a short period quenches the fluorescence. In this way an organothiosphosphorus compound previously treated with bromine vapours to give HBr, appears as yellowish-green fluorescent spot against a relatively non-fluorescent background. It is thought that the flavone forms a fluorescent salt-like species (IV) with HBr.

(IV)

Since the separation is done on a silica-gel thin-layer the method is preferred to that using flavones on cellulose. However, the technique does not offer any added advantage over chelate spray reagents and in general the reproducibility is not as good because there is always some residual background fluorescence, but it may be used for quantitative purposes (24).

   Another spray reagent used by Belliveau and Frei (25) after treatment of the chromatogram with bromine vapours, is 1,2-dichloro-4,5-dicyanobenzoquinone (DDQ). This compound was used to detect a large number of sulfur-containing pesticides. Unfortunately the reagent is sensitive to light which causes

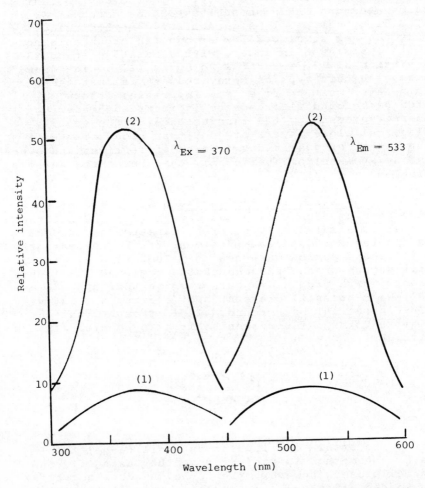

Figure 6.   Emission and excitation spectra of fisetin on cellulose layers: (1) background at meter multiplier = 0.03; (2) Baygon spot at meter multiplier = 0.01 (21)

problems with reproducibility if the chromatogram is
not scanned immediately after development (26). Other
pH-sensitive fluorogenic spray reagents were discovered by Belliveau (7) but were never studied in view
of developing an analytical technique for pesticide
residues.

4. <u>Fluorogenic labelling of pesticides</u>. The
subject has been reviewed earlier by Lawrence and
Frei (27). Labelling consists in replacing a proton
or other atom of a pesticide with a so-called labelling compound such as dansyl chloride (V) or fluorescamine (VI). The former reacts with primary and
secondary amines, phenols, some thiols and aliphatic
alcohols. The latter reacts very selectively with
primary amines.

(V)   N(CH₃)₂   (VI)

N-Methylcarbamate insecticides have been labelled
with dansyl chloride (28). The procedure involves
hydrolysis with aqueous base to form a phenol and
methylamine. The reagent reacts rapidly with both
compounds and the newly formed fluorescent derivatives
can then be applied and separated by tlc (29, 30). A
typical reaction scheme is shown in Figure 7. Detection limits are good (1 ng per spot) and dansylation
of pesticides can be accomplished with extracts from
water and soil samples (31).
Dansylation has also been used to detect N-phenyl-
carbamate and urea herbicides by adding excess reagent
on the applied spot directly on the tlc (32). This
<u>in situ</u> reaction proved to be cleaner and more prac-
tical than labelling in solution and was applied to
the detection of Linuron (33) in agricultural crops
at the 0.05 ppm level. Other applications involving
hydrolysis and dansylation in solution prior to tlc
have been reported with organophosphorus insecticides
(34).
The disadvantage with dansyl chloride is the lack
of selectivity and many hydrolysis or degradation
products of pesticides will give fluorescence, such
that it is sometimes difficult to get a good separation

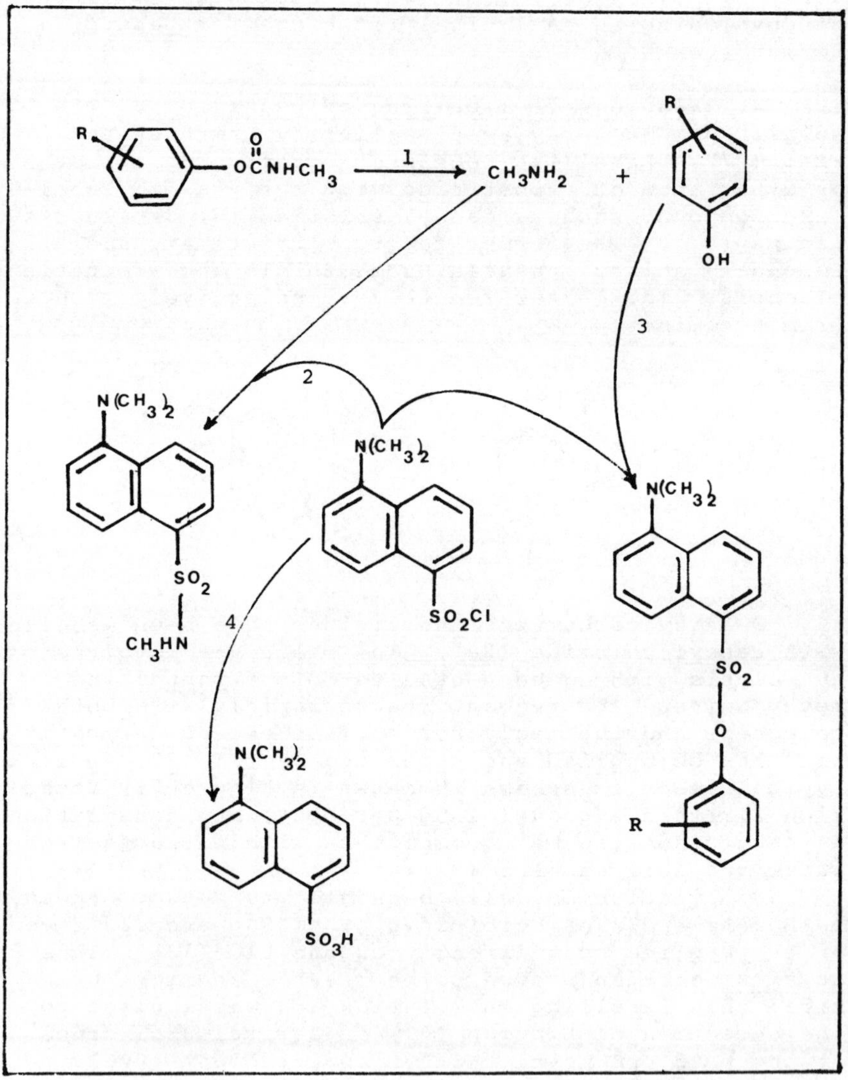

Journal of Chromatography

Figure 7. Reaction scheme for the dansyl labeling of N-*methylcarbamate insecticide: (1) hydrolysis of the carbamate; (2 and 3) labeling of the amine and phenol hydrolysis products; (4) hydrolysis of the reagent by carbamate (29)*

for quantitative purposes. However, the technique is
very sensitive (nanogram per spot) and the reproducibility is good.

Another labelling compound, NBD-chloride (VII)
has been used for the analysis of alkylamine-genera-
ting pesticides (35). Sub-nanogram quantities can

(VII)

sometimes be detected but as is the case with dansyl
chloride many derivatives are formed during the
reaction.

Fluorescamine (VI) is more advantageous because
of its specificity. The reaction can be performed
with any compound that has a primary amine group or
that can be chemically treated to yield a primary
amine; this is the case with a pesticide such as
fenitrothion, that has a nitro group which can be
reduced to an amino group. This property of fenitrothion has been used to advantage and a very reliable
quantitative method has been developed (36).

The procedure involves reduction of the nitro
group with stannous chloride in HCl and is applicable
to all derivatives of fenitrothion that have a nitro
group, such as fenitrooxon, and nitrocresol. (Table
III). With aminofenitrothion, a degradation product
commonly found in anaerobic water, the detection is
instantaneous.

Thus, by the technique of in situ fluorometry on
tlc it is possible to analyze the parent compound and
all of the common degradation products simultaneously.
The detection limit is 0.01 μg per spot with an
average standard deviation of 7.25% for 0.4 μg spots.
The method has been used on a routine basis in our
laboratory to determine fenitrothion, aminofenitrothion,
aminocresol and fenitrooxon simultaneously in water
(37, 38, 39) and in biological samples (unpublished
results).

The reproducibility is not as good at low concentrations as that obtained with chelate spray
reagents and this is usually attributed to the multiple steps involved in the detection technique.
However, it has the very distinct advantage of being
selective towards primary amino groups. A typical
chromatogram showing the separation and detection of

Table III.　Fenitrothion and Some of its Most
　　　　　　　Important Derivatives (<u>36</u>).

Compound	Structure
Aminofenitrothion	(CH$_3$O)$_2$—P(=S)—O—⟨C$_6$H$_3$(CH$_3$)⟩—NH$_2$
4-Amino-3-methylphenol	HO—⟨C$_6$H$_3$(CH$_3$)⟩—NH$_2$
Fenitrooxon	(CH$_3$O)$_2$—P(=O)—O—⟨C$_6$H$_3$(CH$_3$)⟩—NO$_2$
Fenitrothion	(CH$_3$O)$_2$—P(=S)—O—⟨C$_6$H$_3$(CH$_3$)⟩—NO$_2$
3-Methyl-4-nitrophenol	HO—⟨C$_6$H$_3$(CH$_3$)⟩—NO$_2$

fenitrothion and three important degradation products on a silica-gel tlc is shown in Figure 8.

Thus, fluorogenic labelling offers an interesting approach to the analysis of pesticides on tlc. It is preferable to perform the reaction on the chromatogram, rather than is solution since it is less time consuming and more practical.

Fluorescence of Pesticides by the Direct Approach. The fluorescence obtained by the methods just mentioned comes from the reagents, i.e., fluorogenic sprays and labelling compounds. There is sometimes a lack of selectivity since the fluorescence is nearly always the same for all the compounds on a particular chromatogram whether they are pesticides or impurities. In some cases there is also background fluorescence in addition to interferences from co-extractives which fluoresce under the experimental conditions. But for many pesticides there is no other choice.

Another approach is to make the pesticide fluoresce. This is advantageous since the fluorescence is particular to the original compound, and different pesticides give different fluorescence spectral characteristics.

One earlier example is the analysis of carbaryl (Sevin) and α-naphthol on tlc (40). Carbaryl (VIII) is hydrolyzed with 1.0 N NaOH to α-naphthol, a highly fluorescent species. The instrumental limit of detection is 1 ng/spot, which is comparable to expected glc data.

(VIII)

1. Heat Treatment, Acids and Bases. Many pesticides are non-fluorescent or only barely fluorescent because they contain fluorescence quenching groups such as nitro, carboxyl or phosphate. The removal of these groups liberates the aromatic moiety which may be fluorescent depending on the structure. One obvious approach to remove these groups is to use an acid or a base, but sometimes heat is sufficient to break down the molecule into new fluorescent species.

One very successful method (41) was developed for coumaphos (IX) which gives a very fluorescent spot

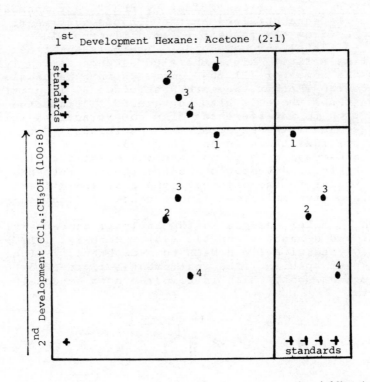

Figure 8.   Two-dimensional thin-layer chromatographic separation of fenithrothion and possible breakdown products (spots: (1) fenithrothion; (2) fenitrooxon; (3) aminofenitrothion; (4) 3-methyl-4-nitrophenol (36)

on silica-gel after heat treatment of the chromatogram
(detection limit 1 ng per spot). The technique also
gives fluorescence for coroxon, the oxygen-analog
(X), potasan an impurity in the formulation (XI)
(same as coumaphos without the chlorine atom) and the
corresponding hydrolysis products of coumaphos (XII)
and potasan. The basic scheme for the degradation of
coumaphos upon heat treatment or in water is summarized in Figure 9. This technique has led to the development of a routine procedure to verify for the presence of potasan in formulations (42) and the degradation of coumaphos as a function of pH has also been
studied (43) which demonstrates its applicability for
routine purposes. Thus a practical quantitative tlc
technique was developed for coumaphos which has been
determined at the 0.01 ppb level in lake and sewage
water (44). Linear calibration curves were obtained
up to 10 µg per spot. Coumaphos and its oxygen analog
coroxon, were also determined in eggs (45).

Another organophosphorus insecticide, Bayrusil
(XIII) becomes fluorescent upon heat treatment but the
fluorescence yield is less than with coumaphos (around
0.01 µg/spot) (44). This pesticide has also been
analysed in foodstuffs at the 0.02 ppm level (46).

(XIII) $(C_2H_5O) \overset{\overset{\text{S}}{\underset{||}{}}}{P} O$ —

Heat treatment not only gives fluorescence in
many cases, but addition of acid or base to the chromatogram causes either a bathochromic or hypsochromic
shift of the excitation and emission maxima (Table IV).
For instance, coumaphos fluoresces at λEx 344, and
λEm 440 when heat-treated only, but the wavelengths
shift to 372 and 474 nm, respectively if the chromatogram is sprayed with NaOH (47). This particular
aspect is very important because the data may be utilized for identification purposes. In most cases sub-
microgram amounts of pesticides can be detected.

This technique of heat treatment in combination
with acid or base has been used by Mallet and Surette
(48) with a variety of other types of pesticides
(Table V). Good detection limits (less or equal to
0.01 µg/spot) were obtained with coumatetralyl, fuberi-
dazole, methabenzthiazuron and quinomethionate.

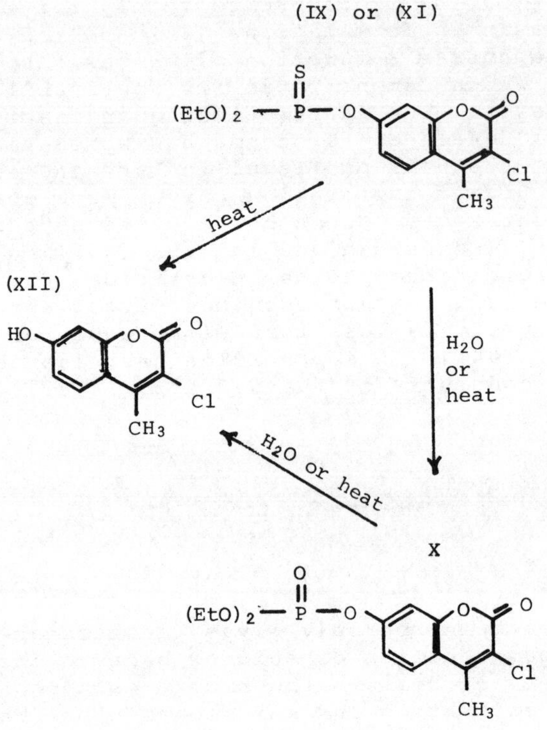

Analytica Chimica Acta

Figure 9. Scheme for the degradation of coumaphos (42)

Table IV. Fluorescence Spectral Data of Some Organophosphorus Pesticides upon Heat, Acid
and/or Base Treatment (<u>47</u>).

Pesticide	Spray Reagent	Optimum Heating Conditions		Wavelength of Maximum (nm)	
		Temp (°C)	Time (min)	Ex	Em
Bayrusil					
	-	100	30	353	441
	NaOH	200	30	356	440
	H₂SO₄	200	30	373	510
Coumaphos	-	200	20	344	440
	NaOH	200	30	372	474
Maretin	-	200	20	352	435
	NaOH	200	20	370	500
	H₂SO₄	75	20	355	440

Journal of Chromatography

Table V. Fluorometric Response of Some Pesticides (<u>48</u>).

Pesticide	Filter Combination	Wavelength (nm)		Optimum Conditions	I.L.D.
		Ex	Em		
Coumatetralyl	B	358	450	a,c	0.01
Diphacinone	A	330	514	a,b	2.0
Fuberidazole	B	333	410	a,d	0.006
Methabenzthiazuron	B	353	439	a	0.002
Naptalam	A	361	455	a	0.08
	A	298	455	b,f	0.04
Propyl isome	A	352	472	a,d	0.008
Quinomethionate	B	337	458	a,e	0.04
	B	335	455	a,c	0.01
	B	362	417	a,d	0.06
Rotenone	A	362	453	F	0.8
	A	370	440	a	0.6
Thioquinox	B	329	441	a	0.04
	B	329	435	a,c	0.08
Warfarin	A	363	456	a	0.06

F=Fluorescent naturally; a heated at 200° for 45 min;
b=sprayed with 2.5 N KOH; c=sprayed with 0.1 N NH₄OH;
d=sprayed with 0.1 N HCl; e=sprayed with 0.1 N H₂SO₄;
f=heated at 220° for 30 min.
I.L.D.=Instrumental limit of detection.   Journal of Chromatography

2. <u>Inorganic Reagents</u>. Another way to obtain
fluorescence from pesticides is with inorganic salts.
On the one hand, pesticides may act as ligands capable of combining with metal ions to yield fluorescence,

and on the other, the presence of inorganic ions may
catalyze the degradation of pesticides to fluorescent
species.

A large number of pesticides were tested by
Surette and Mallet (49) who sprayed the chromatograms
with inorganic salts such as $AlCl_3$ (which may also be
incorporated in the layer). Good detection limits
(around 0.01 μg/spot) were obtained with at least ten
(Table VI). An attempt to identify the fluorescent
species responsible was made by Surette (50) without
too much success; for instance, Morestan (XIV) gives
5 fluorescent degradation products while naptalam (XV)
gives over 60.

Table VI.   Spectral Data and Detection Limits of
             Pesticides (49).

Abbreviations: C=plate covered; R.T.=room temperature;
UV=irradiated with UV 366 nm; NF=natural fluorescence;
SG-H=silica gel H; SG-60 silica gel 60; (I)=incorpora-
ted reagent; (S)=sprayed reagent.

Pesti-cide	Layer	Reagent	Temp (°C)	Time (min)	Wave-length (nm) Ex	Em	I.L.D. (μg)
Captan	SG-H	(1)$AlCl_3$ (2)$NaClO_3$	100	45	360	465	0.02
Devrinol	SG-60	$AlCl_3$	R.T.	30	355	428	0.008
Difola-tan	SG-H	(1)$AlCl_3$ (2)$NaClO_3$	100	45	360	465	0.02
Diquat	SG-60	$AlCl_3$	R.T.	30	375	472	0.02
Maretin (NF)	(a)SG-H				358	412	0.008
	(b)SG-H	$AlCl_3$	R.T.	20	358	412	0.002
Menazon	SG-H	H_2SeO_3	R.T.	5	366	466	0.01
Napta-lam	SG-60	$AlCl_3$	200	45(C)	312	482	0.01
Para-quat	SG-H	$Na_2B_4O_7$	100	45	420	510	0.04
Propyl isome	(a)SG-H	$AlCl_3$	R.T.		359	476	0.01
	(b)SG-H	$AlCl_3$	R.T.		359	483	0.004
Rote-none	SG-H	$AlCl_3$	100	45	362	450	0.1-0.06

3. **Type of Layer.** The influence of the type of thin-layer was investigated by Caissie and Mallet (51) and it was shown to have an important effect on the fluorescence of pesticides. For instance, the spectral data vary from one layer to another (Table VII); some pesticides are fluorescent on basic aluminium oxide layers but they are not fluorescent on acidic layers. All these experiments demonstrate the flexibility of using fluorometric procedures and the selectivity that could be achieved if desired.

Table VII.   Comparison between Silica Gel and Alumina (51).

Pesticide & Conditions	Silica gel		Aluminium oxide			
			Acidic		Basic	
	Ex	Em	Ex	Em	Ex	Em
Coumaphos (R.T.)	325	434	340	412	340	418
(H)	344	440	360	435	363	447
Diphacinone (R.T.)	330	518	368	490	370	495
Guthion (H)	342	442	347	421	350	431
Warfarin (H)	363	456	375	462	374	456
Captan (I.R.)						
(H)			372	460	372	465
Difolatan (I.R.)	360	465				
			372	467	375	462
Diquat (I.R.)	375	472				
(H)					372	452
Paraquat (I.R.)	420	510				
(H)					367	460

R.T.=room temperature; H=chromatogram heated at 200° for 45 min; I.R.=inorganic reagent (see 49).

Journal of Chromatography

Residue Methods

    Residue analytical methods other than those already mentioned and based on the aforementioned results were only developed with a few additionnal pesticides. For instance a method has been described for Captan and Captafol in apples and potatoes by Francoeur and Mallet (52). Basically the procedure

requires that the silica-gel layer be impregnated
with aluminium chloride. After separation the chromatogram is heat-treated for 45 min at $100°C$ and a
yellowish-green fluorescence is obtained for both
Captan and Captafol (XVI and XVII). Greater that 90%
recoveries were obtained at the 0.2 ppm level (Table
VIII) which compares favorably well with a glc method
using the electron capture detector (53).

(XVI) [structure] NSCCl$_3$ (XVII) [structure] NSCCL$_2$CHCl$_2$

Table VIII. Recovery^a(%) Data for Captan and Captafol
             from Fortified Crops at 0.2 ppm (52).

Crop	Captan	Captafol
Apples^b	100	96
Apples^c	97	103
Potatoes	93	94
Potatoes	90	96

a Average of 3 samples
b Captan and Captafol determined in different samples
c Captan and Captafol determined in the same sample

Journal of the Association of Official Analytical Chemists

Another method has been developed for Maretin (I)
in milk and eggs (54). This procedure requires hydrolysis of the pesticide in solution and extraction of
the fluorescent hydrolysis product which is then chromatographed on a thin-layer. A method for Maretin had
previously been developed by measuring its native fluorescence (12). However the fluorescence was not
stable which is undesirable. The fluorescence of the
hydrolysis product of Maretin, namely, naphthostyril
(XVIII) is very stable. By this way Maretin was determined at the 0.01 ppm level with over 90% recoveries
(Table IX).

(XVIII)   [structure]

Table IX.   Recovery of Maretin from Eggs and Milk (54).

Sample	Quantity added µg	Conc. ppm	Quantity recovered µg	Recovery %
Milk	20.0	0.10	19.9	99
	2.0	0.01	2.0	100
Eggs	5.0	0.10	4.5	90
	0.50	0.01	0.48	96

Journal of Agricultural and Food Chemistry

Conclusion

The technique of quantitative thin-layer chromatography by in situ fluorometry can be used successfully for the quantitative determination of pesticides.
When the pesticide is naturally fluorescent the procedure is straightforward and good sensitivity and
reproducibility may be achieved. Otherwise fluorogenic reagents that become fluorescent in presence of
a pesticide (which may have been chemically altered),
as well as a variety of treatments such as heat that
may convert the pesticide into a fluorescent species,
can be used. Fluorogenic reagents sometimes give
fluorescence of the background or they react with co-
extractives on the chromatogram. This renders the
quantification step more difficult. Chemical transformation of the pesticide into a fluorescent species
does not have these drawbacks and the sensitivity and
selectivity approaches that obtained with naturally
fluorescent compounds. Under optimum conditions low
nanograms per spot can be detected with 2-4% relative
error.

Unfortunately the technique is used only on a
limited scale for the determination of pesticides on
a routine basis. Most pesticide analysts prefer gas
liquid chromatography and more recently high pressure
liquid chromatography to the more cumbersome handling
of thin-layer chromatograms.

Nevertheless thin-layer chromatography and in
situ fluorometry still offer a challenging approach
to the quantitative analysis of pesticides particularly in the determination of multi-residues, such as
the parent compound and degradation products. This
was demonstrated in this article by the determination
of coumaphos and fenitrothion, with some of their
respective degradation products. The technique is
also very adaptable to a limited budget requiring only
simple chemicals and labware for the thin-layer chromatography and a good photofluorometer with tlc attachment is still relatively inexpensive.

Literature Cited

1. Guilbault, G.G., Ed. "Fluorescence: Theory,
    Instrumentation and Practice"; Marcel Dekker,
    New York, 1967.

2. Winefordner, J.D.; Shulman, S.G.; O'Haver, T.C.
    "Luminescence Spectrometry in Analytical Chemis-
    try", Interscience, New York, 1972.

3. Hercules, D.M. "Fluorescence and Phosphorescence
    Analysis: Principles and Applications", Inter-
    science, New York, 1966.

4. White, C.E.; Argauer, R.J. "Fluorescence Analysis:
    a Practical Approach", Marcel Dekker, New York,
    1970.

5. White, C.E. Anal. Chem., 1949, 21, 104.

6. O'Donnell,C.M.; Solie, T.N. Anal. Chem., 1978,
    50, 189 R.

7. Mallet, V.N.; Belliveau, P.E.; Frei, R.W. Residue
    Revs, 1975, 59, 51.

8. Guilbault, G.G. Photochem. and Photobiol., 1977,
    25, 403.

9. Mallet, V.N.; Surette, D.; and Brun, G.L. J.
    Chromatogr., 1973, 79, 217.

10. Mallet, V.N.; Lebel, C.; Surette, D.P. ANALUSIS,
    1973-1974, 2, 643.

11. Francoeur, Y.; Mallet, V.N. J. Assoc. Offic. Anal.
    Chem., 1976, 59, 172.

12. Mallet, V.N.; Zakrevsky, J.-G.; Brun, G.L. Bull.
    Soc. Chim. Fr., 1974, 9-10, 1755.

13. Ballschmitter, K.H.; Tölg, G. Z. Anal. Chem.,
    1966, 215, 305.

14. Abbott, D.C.; Blake, K.W.; Tarrant, K.R.; Thomson,
    J. J. Chromatogr., 1967, 30, 136.

15. Gordon, H.T.; Werum, L.N.; Thornburg, W.W.; J.
    Chromatogr., 1964, 13, 272.

16. Ragab, M.T.H. J. Assoc. Offic. Anal. Chem.,
    1967, 50, 1088.

17. Belliveau, P.E.; Mallet, V.; Frei, R.W. J. Chro-
    matogr., 1970, 48, 478.

18. Frei, R.W.; Mallet, V.N. Intern. J. Environ. Anal.
    Chem., 1971, 1, 99.

19. Frei, R.W.; Mallet, V.N.; Thébaud, M. Intern, J.
    Environ. Anal. Chem., 1971, 1, 141.

20. Bidleman, T.; Nowlan, P.; Frei, R.W. Anal. Chim.
    Acta, 1972, 60, 13.

21. Mallet, V.; Frei, R.W. J. Chromatogr.,1971, 54,
    251.

22. Mallet, V.; Frei, R.W. J. Chromatogr.,1971, 56, 69.

23. Mallet, V.; Frei, R.W. J. Chromatogr.,1971, 60,
    213.

24. Frei, R.W.; Mallet, V.N.; Pothier, C. J. Chroma-
    togr. 1971, 59, 135.

25. Belliveau, P.E.; Frei, R.W. Chromatographia, 1971,
    4, 189.

26. Frei, R.W.; Belliveau, P.E. Chromatographia, 1972,
    5, 296.

27. Lawrence, J.F.; Frei, R.W. J. Chromatogr.,1974,
    98, 253.

28. Lawrence, J.F.; Frei, R.W. Intern. J. Environ.
    Anal. Chem., 1972, 1, 317.

29. Frei, R.W.; Lawrence, J.F. J. Chromatogr.,1971,
    61, 174.

30. Lawrence, J.F.; LeGay, D.S.; Frei, R.W. J. Chro-
    matogr., 1972, 66, 295.

31. Frei, R.W.; Lawrence, J.F. J. Chromatogr., 1972,
    67, 87.

32. Frei, R.W.; Lawrence, J.F.; LeGay, D.S. Analyst
    (London), 1973, 98, 9.

33. Lawrence, J.F.; Laver, G.W. J. Assoc. Offic.
    Anal. Chem., 1974, 57, 1022.

34. Lawrence, J.F.; Renault, C.; Frei, R.W. J. Chroma-
    togr., 1976, 121, 343.

35. Lawrence, J.F.; Frei, R.W. Anal. Chem., 1972, 44,
    2046.

36. Zakrevsky, J.-G.; Mallet, V.N. J. Chromatogr.,
    1977, 132, 315.

37. Berkane, K.; Caissie, G.E.; Mallet, V.N., J. Chro-
    matogr., 1977, 139, 386.

38. Mallet, V.N.; Brun, G.L.; MacDonald, R.N.; Berkane,
    K. J. Chromatogr., 1978, 160, 81.

39. Mallet, V.N., Francoeur, J.M.; Volpé, G. J. Chro-
    matogr., 1979, 172, 388.

40. Frei, R.W.; Lawrence, J.F.; Belliveau, P.E. Anal.
    Chem., 1971, 254, 271.

41. Brun, G.L.; Mallet, V.N. J. Chromatogr., 1973, 80,
    117.

42. Volpé, Y.; Mallet, V.N. Anal. Chim. Acta, 1976, 81,
    111.

43. Mallet, V.N.; Volpé, Y. Anal. Chim. Acta, 1978,
    97, 415.

44. Mallet, V.N.; Brun, G.L. Bull. Environ. Contamin.
    and Toxicol., 1974, 12, 739.

45. Zakrevsky, J.-G.; Mallet, V.N. J. Assoc. Offic.
    Anal. Chem., 1975, 58, 554.

46. Zakrevsky, J.-G.; Mallet, V.N. Bull. Environ.
    Contamin. Toxicol., 1975, 13, 633.

47. Brun, G.L.; Mallet, V.N. J. Chromatogr., 1973, 80,
    117.

48. Mallet, V.N.; Surette, D.P. J. Chromatogr., 1974,
    95, 243.

49. Surette, D.P.; Mallet, V.N. J. Chromatogr., 1975,
    107, 141

50. Surette, D.P., M.Sc. thesis "Quelques Réactifs Inorganiques pour la Détection Fluorimétrique de Pesticides", Université de Moncton, Moncton, N.-B. 1977.

51. Caissie, G.E.; Mallet, V.N. J. Chromatogr., 1976, 117, 129.

52. Francoeur, Y., Mallet, V.N. J. Assoc. Offic. Anal. Chem., 1977, 60, 1328.

53. Baker, P.B.; Flaherty, B. Analyst, 1972, 97, 713.

54. Zakrevsky, J.-G.; Mallet, V.N. J. Agr. Food Chem., 1975, 23, 754.

RECEIVED December 28, 1979.

Recent Developments in High-Performance Thin-Layer Chromatography and Application to Pesticide Analysis

H. E. HAUCK and E. AMADORI

E. Merck, D-6100 Darmstadt, Federal Republic of Germany

Summary

The HPTLC pre-coated layers examined in this paper, namely silica gel 60, RP, cellulose and silica gel 60 with concentrating zone, are new developments or modifications of existing TLC pre-coated layers offering new potential for pesticide analysis by thin-layer chromatography.
　　A number of separations (of BHC isomers, of bromophos-ethyl and dimethoate, and of the active constituents of Trevespan 6038) were performed in order to provide a qualitative and quantitative comparison between TLC and HPTLC pre-coated plates, or between HPTLC plates coated with various sorbents, and in order to demonstrate their use in pesticide chemistry.

1. Introduction

Silica gel 60, the most versatile and most frequently used TLC sorbent, was taken as a basis. The mean particle size of this sorbent was optimized; simultaneously, the particle size distribution was brought to within as narrow limits as possible (1-6).
　　The sorbent material thus obtained was used to prepare HPTLC pre-coated plates silica gel 60, being followed subsequently by the development of other sorbents for processing into HPTLC pre-coated layers. The materials chosen were largely "reversed phase" sorbents, that is to say, chemically modified silica gels with a non-polar surface (7, 8), as well as microcrystalline cellulose (9).
　　In the case of HPTLC pre-coated layers, in particular, optimum use of the available separating performance is dependent on the compactness of the

starting spots. These narrowly delimited starting
spots are obtained with the aid of suitable,
relatively involved application techniques and by the
application of small volumes. In order to make better
use of the potential performance of HPTLC pre-coated
plates silica gel 60, and to permit application of
larger quantities with a rapid and non-problematical
technique, special HPTLC pre-coated plates silica
gel 60 with a so-called "concentrating zone" were
developed (10, 11, 12).

2. Experimental

2.1 Reagents and solvents. All the reagents and
solvents used in this work were of analytical grade
from E. Merck, Darmstadt.

2.2 Application of the samples. The volumes of
0.75 μl were applied as a dot using appropriate glass
capillaries (E. Merck, Darmstadt).

2.3 Development. All the plates investigated
here were each kept overnight at a relative humidity
of 20%. The plates were then developed to heights of
5 cm (HPTLC) and 7 cm (TLC) in normal chambers,
without chamber saturation, using the appropriate
solvent systems.

2.4 Evaluation. The evaluation was carried out
by reflectance using a Zeiss PMQ II chromatogram
spectrophotometer (Zeiss, Oberkochen), with a direct
link of the analogue output to an IBM 1800 process
control computer.

3. HPTLC pre-coated plates silica gel 60

Optimization of the mean particle size of silica gel
60 and the consequent reduction of the particle size
distribution produced a number of advantages in HPTLC
pre-coated plates silica gel 60 prepared from this
improved material which the traditional TLC silica
gel 60 pre-coated layers do not possess. These
advantages are to be found not only in the already
mentioned optimized mean particle diameter and the
narrower particle size distribution, but also in the
smoother and more homogeneous surface of the plates,
which leads to an increase in chromatographic
performance.
     The improvements made in the HPTLC pre-coated
plate silica gel 60 as compared with the TLC silica

gel 60 plates meant that HPTLC pre-coated plates could also be used advantageously in pesticide analysis (13, 14, 15).

In this paper the advantages of HPTLC pre-coated plates silica gel 60 (E. Merck, Darmstadt) as compared with TLC pre-coated plates silica gel 60 (E. Merck, Darmstadt), are presented in the separation of benzene hexachloride (BHC) isomers.

Several publications (16, 17, 18) were taken as a basis for determining the optimum conditions for the separation and detection of BHC isomers, as regards both eluant and detection mode.

Both for TLC and for HPTLC pre-coated plates silica gel 60 the best results were obtained with a solvent mixture consisting of cyclohexane and chloroform in a ratio of 65/35 by volume, with a 5 cm migration distance for HPTLC plates and a 7 cm migration distance for the TLC plates. In both cases the plates were adjusted prior to use to 20% relative humidity and were developed in normal chambers without chamber saturation. For both types of plates, 0.75 µl quantities of a test mixture of α-, β-, γ-, δ- and ϵ-BHC were applied with glass capillaries (E. Merck, Darmstadt) at a concentration of 10 mg/ 10 ml dissolved in methanol. Once separation had been completed, the best visualization of BHC isomers was achieved by spraying the plates with 1% o-toluidine solution in ethanol followed by 15 minutes' irradiation under UV light (254 nm). This led to the formation of greyish-green spots on a greyish-brown background; even after 3 days the spots did not turn pale. The plates were evaluated with a Zeiss chromatogram spectrophotometer PMQ II in reflectance mode, with the analogue output directly coupled to an IBM 1800 process computer.

Table I shows that retention of the individual BHC isomers is virtually identical on the TLC and HPTLC pre-coated silica gel 60 plates. The hRf values of the BHC isomers on the HPTLC plates lie only slightly above those achieved on the TLC pre-coated plates.

Table II compares a number of additional chromatographic data achieved for BHC isomer separations on TLC and HPTLC silica gel 60 plates.

These values show that the optimum separation characteristics on HPTLC silica gel 60 plates are achieved after a migration distance of only 5 cm, whereas on the TLC plates optimum separation is achieved after a migration distance of 7 cm. This means that in spite of a roughly 35% smaller flow

Table I. Retention data (hRf values) of BHC
isomers on TLC and HPTLC pre-coated
plates silica gel 60; solvent system :
cyclohexane/chloroform 65/35

types of plates	hRf values of BHC isomers				
	α	γ	ϵ	β	δ
TLC silica gel 60	59,6	51,0	47,5	39,1	27,3
HPTLC silica gel 60	62,6	53,4	48,2	40,6	28,3

Table II. Chromatographic data on the separation
of BHC isomers on TLC and HPTLC pre-
coated plates silica gel 60; solvent
system: cyclohexane/chloroform 65/35

types of plates	velocity coefficient κ [mm^2/s]	opt. migration distance [mm]	development time [s]	resolution R_s of α-and γ-BHC
TLC silica gel 60	3,67	70	1540	1,34
HPTLC silica gel 60	2,38	50	1306	1,62

rate the required development time for the HPTLC
plate is about 240 seconds or 15% shorter than on
the TLC plate. In spite of this shorter migration
distance and development time, the chromatographic
separation efficiency, expressed by the resolution R_s
between the two neighbouring α- and γ-isomers of BHC,
is about 21% higher on the HPTLC pre-coated plate
silica gel 60 than on the corresponding TLC pre-coated
plate.

As a means of comparing the potential for
quantitative determinations on the TLC and HPTLC
pre-coated plates silica gel 60, the calibration
curves of δ-BHC were recorded.

In Figure 1 the peak heights (expressed in mV,
after evaluation by reflectance using a Zeiss PMQ II
chromatogram spectrophotometer) are plotted against
the applied quantity (expressed in ng) of δ-BHC on
the TLC and HPTLC silica gel 60 plates.

It can be seen that in both cases a well
correlating calibration curve in the shape of a
straight line is obtained. It is also evident that
for equal applied quantities the peak heights on the
HPTLC plate are slightly more than twice as high as
on the TLC plate. For this reason and on account of
the fact that the baseline noise is markedly less on
the HPTLC plate evaluated in reflectance mode as a
result of its more homogeneous and smoother surface,
the detection limit for δ-BHC under the given
conditions on the HPTLC pre-coated plate silica gel
60 is considerably less at 18 ng than on the TLC
pre-coated plate silica gel 60 at 40 ng.

4. HPTLC pre-coated plates RP

Now that reversed phase column packing materials have
been used successfully in column liquid chromatography
for a number of years, success has also been achieved
in thin-layer chromatography in that it is now
possible to prepare HPTLC pre-coated plates with the
same types of surface-modified sorbents and to use
them for separations (7, 8, 19, 20).

The basic skeleton of the modified sorbents
consists of a surface-active silica gel. Modification
is in the form of a surface reaction with specially
active silanes at the silanol groups of the silica
gel. Following elimination of these silanol groups,
new siloxane groupings are formed, on which aliphatic
hydrocarbon groups are chemically bound by silicon-
carbon bonds to the silica gel skeleton. Pre-coated
plates with these modified silica gels are designated:

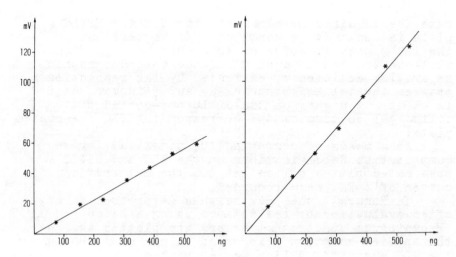

Figure 1. Calibration curves for δ-BHC on a TLC pre-coated plate silica gel 60 (left) and on an HPTLC pre-coated plate silica gel 60 (right) (abscissae: applied quantity (ng); ordinates: peak height (mV))

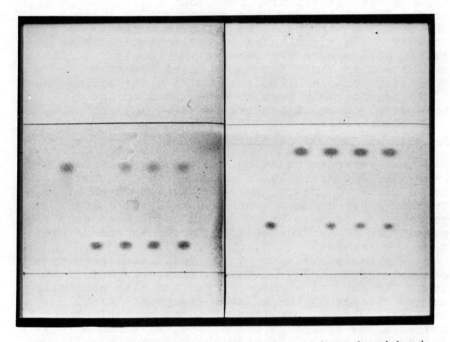

Figure 2. Linear chromatograms for the separation of bromophos-ethyl and dimethoate on an HPTLC pre-coated plate silica gel 60 (left) and an HPTLC pre-coated plate RP-18 (right) (solvent system on silica gel 60: n-heptane/acetone 65/35; on RP-18: acetone/water 80/20; application from left to right—(1st band) bromophosethyl; (2nd band) dimethoate; (3rd–5th band) mixture; migration distance: 5 cm; applied quantities: 0.75 μL = 750 ng; detection: spraying with PdCl₂ in ethanol heating to 120°C)

HPTLC pre-coated plate RP-2, RP-8 and RP-18. The
figure following the RP abbreviation refers to the
chain length of the aliphatic hydrocarbon.

One example for the use of an HPTLC pre-coated
plate with a reversed phase material in the field of
pesticide analysis is the separation of bromophos-
ethyl and dimethoate on an HPTLC pre-coated plate
RP-18 F 254 s (E. Merck, Darmstadt). This is described
and compared with the separation of this substance
mixture on an HPTLC pre-coated plate silica gel 60
F 254 (E. Merck, Darmstadt).

The solvent system most suitable for separating
bromophos-ethyl and dimethoate is an 80/20 mixture
of acetone/water for the HPTLC pre-coated plate RP-18,
and a 65/35 mixture of n-heptane/acetone in the case
of the HPTLC pre-coated plate silica gel 60. The
optimum migration distance was 5 cm in both cases.
The applied substance quantity was 0.75 μl on both
plates, at a concentration of bromophos-ethyl and
dimethoate of 10 mg/10 ml in methanol in both cases.

Detection of substance spots was performed by
spraying with a solution of 0.5 g palladium(II)-
chloride in 100 ml of ethanol with subsequent heating
to 120°C. The spots appearing are brown on a grey-
pink background. A Zeiss PMQ II chromatogram
spectrophotometer coupled to an IBM 1800 process
computer was used for evaluations.

Figure 2 shows the separation of bromophos-ethyl
and dimethoate on an HPTLC pre-coated plate silica
gel 60 (left) and on an HPTLC pre-coated plate RP-18
(right). The application scheme was the same in both
cases, namely from left to right :
1st band   : bromophos-ethyl
2nd band   : dimethoate
3rd-5th band : mixture of the two substances.
The two chromatograms show that in changing from the
silica gel 60 plate to the RP-18 plate there is a
reversal of substance sequence. This is also apparent
in Table III with the hRf values of the two sample
substances.

Table IV compares a few chromatographic data
relating to the separation of bromophos-ethyl and
dimethoate on the HPTLC pre-coated plate silica gel 60
and the HPTLC pre-coated plate RP-18.

These figures show that resolution between
bromophos-ethyl and dimethoate on the HPTLC RP-18
plate is roughly 60% greater than on the silica gel 60
plate. Though the optimum migration distance of 5 cm
is the same in both cases, the development time on
the RP-18 plate is roughly twice as long as on the

Table III Retention data (hRf values) for bromophos-ethyl and dimethoate on an HPTLC
pre-coated plate silica gel 60 (solvent
system: n-heptane/acetone 65/35) and on
an HPTLC pre-coated plate RP-18 (solvent
system: acetone/water 80/20)

| types of plates | hRf values of | |
	Bromophos-Ethyl	Dimethoate
HPTLC silica gel 60	70,7	30,7
HPTLC RP-18	27,3	79,8

Table IV Chromatographic data for the separation
of bromophos-ethyl and dimethoate on
HPTLC pre-coated plates silica gel 60
and HPTLC pre-coated plates RP-18

types of plates	velocity coefficient κ [mm^2/s]	opt. migration distance [mm]	development time [s]	resolution R_s of Bromophos-Ethyl and Dimethoate
HPTLC silica Gel 60	3,73	50	814	3,43
HPTLC RP - 18	1,69	50	1841	5,53

silica gel 60 plate. This is attributed to the much
lower velocity coefficient κ of the solvent system
used on the RP-18 plate as compared to that on the
silica gel 60 plate.

Quantitative in-situ evaluation of the bromophos-
ethyl and dimethoate separation in this case of
visualization has no real value on account of the
rather non-homogeneous background colouration
produced by spraying with the detection reagent used
here. Therefore the routine, quantitative in-situ
evaluation of the unstained bromophos-ethyl and
dimethoate in our laboratory is carried out by UV
measurement in reflectance mode.

5. HPTLC pre-coated plates cellulose

Apart from the sorbents already mentioned in
connection with HPTLC pre-coated layers, a microcrystalline cellulose has also been produced in an
average particle size and a narrow particle size
distribution suitable for HPTLC (9). HPTLC pre-coated
plates cellulose F 254 s (E. Merck, Darmstadt),
composed of this microcrystalline cellulose, were
used to separate Trevespan 6038. This is a mixture of
the substances ioxynil (3,5-diiodo-4-hydroxy-benzo-
nitrile), flurenol (9-hydroxyfluorenecarboxylic acid)
and MCPA (2-methyl-4-chlorophenoxyacetic acid). For
comparison purposes Trevespan 6038 was also separated
on HPTLC pre-coated plates silica gel 60 F 254
(E. Merck, Darmstadt).

The solvent system most suitable for separating
Trevespan on HPTLC pre-coated plates cellulose
consists of tert. butanol/water/25% ammonia in a
ratio of 80/20/1 by volume, whilst for HPTLC pre-
coated plates silica gel 60 the best mixture is
n-hexane/ethyl acetate/formic acid in the ratio
40/60/0,5. Prior to the separation proper, the HPTLC
pre-coated plates were adjusted to a relative humidity
of 20%. Quantities of o.75 µl of a test solution in
methanol, composed as follows, were applied to the
plates and migrated over 5 cm :
Ioxynil : 0.6 µg
Flurenol : 0.8 µg
MCPA   : 2.8 µg.
The plates, once developed, were checked visually
under UV light at 254 nm, while qualitative and
quantitative evaluation ensued with the aid of a
Zeiss chromatogram spectrophotometer PMQ II in
reflectance mode at a wavelength of 280 nm.

Owing to the differing activities of the sorbents

used and the fact that different solvent systems were
employed for the different types of plates, the hRf
values for the three constituents of Trevespan 6038
differ quite considerably, as is shown in Table V.
The substance sequence is identical in both cases,
however.

Table VI provides a further comparison of
chromatographic data obtained for the separation of
Trevespan 6038 on HPTLC pre-coated plates silica gel
60 and cellulose.

Although the migration distance was the same in
both cases, namely 5 cm, the development time for the
HPTLC pre-coated plate silica gel 60 is considerably
shorter than for the HPTLC pre-coated plate cellulose
on account of the higher flow rate. In both cases the
three substance components were completely separated,
though the R_s values achieved on the HPTLC silica gel
plate were considerably higher for this particular
separation than the values obtained on the HPTLC
cellulose plate.

For quantitative determination of the individual
Trevespan 6038 components, calibration curves were
drawn up for the HPTLC pre-coated plate silica gel 60
and for the HPTLC pre-coated plate cellulose.

Figure 3 shows the peak heights for ioxynil,
MCPA and flurenol (expressed in mV, after evaluation
by reflectance using a Zeiss PMQ II chromatogram
spectrophotometer) plotted against the respective
applied quantities (expressed in ng) for both the
HPTLC silica gel 60 plate and the HPTLC cellulose
plate.

Well correlating calibration curves in form of
straight lines are obtained on both types of HPTLC
plate for all three sample substances in the range of
quantities examined. Although the applied quantities
are identical, the peak heights on the HPTLC pre-
coated plate cellulose are somewhat higher than those
obtained on the HPTLC pre-coated plate silica gel 60.
The detection limits determined by the peak heights
and by the baseline noise for the three substances
examined are listed in Table VII.

These values show that for the separation
performed the detection limits of all three sample
substances on the HPTLC cellulose plate are somewhat
lower than on the HPTLC silica gel 60 plate.

6. HPTLC pre-coated plates silica gel 60 with concentrating zone

A further new development from the field of HPTLC

Table V. Retention data (hRf values) for ioxynil,
MCPA and flurenol on HPTLC pre-coated
plates silica gel 60 (solvent system:
n-hexane/ethyl acetate/formic acid
40/60/0.5) and on HPTLC pre-coated
plates cellulose (solvent system: tert.
butanol/water/25% ammonia 80/20/1)

types of plates	hRf values of		
	Ioxynil	MCPA	Flurenol
HPTLC silica gel 60	49,4	31,6	16,2
HPTLC cellulose	89,3	72,7	60,2

Table VI. Chromatographic data for the separation
of ioxynil, MCPA and flurenol on HPTLC
pre-coated plates silica gel 60 and on
HPTLC pre-coated plates cellulose

types of plates	velocity coefficient κ [mm^2/s]	opt. migration distance z_f [mm]	development time [s]	resolution R_s of	
				Ioxynil/MCPA	MCPA/Flurenol
HPTLC silica gel 60	3,67	50	858	2,45	2,24
HPTLC cellulose	0,67	50	4615	1,49	1,18

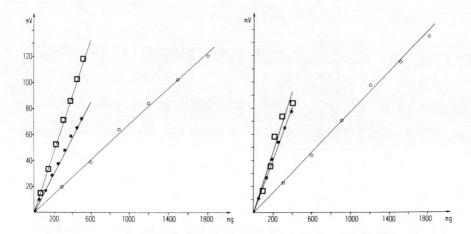

*Figure 3.   Calibration curves for ioxynil (*), MCPA (○), and flurenol (□) on HPTLC pre-coated plates silica gel 60 (*left) *and on HPTLC pre-coated plates cellulose (*right) *(abscissae: applied quantity (ng); ordinates: peak height (mV))*

Table VII. Detection limits (in ng) for ioxynil, MCPA and flurenol on HPTLC pre-coated plates silica gel 60 and on HPTLC pre-coated plates cellulose

types of plates	limit of detection [ng]		
	Ioxynil	MCPA	Flurenol
HPTLC silica gel 60	10	15	6
HPTLC cellulose	7	12	5

pre-coated layers represents the HPTLC pre-coated plate silica gel 60 with concentrating zone. The size of the spot of the substance mixture applied has a considerable bearing on the separating performance attainable on a thin-layer plate. The more compact the application of the spot of substances to be separated, the more compact and thus the more favourable, the spot will also be after chromatography has taken place. Such small starting spots can be obtained with the aid of suitable, sometimes laborious application techniques combined with small application volumes.

In order to improve the separating performance of HPTLC pre-coated plates silica gel 60 even at larger applied volumes, as may be necessary at low sample concentrations, and with a rapid and simple technique of application, HPTLC pre-coated plates silica gel 60 with so-called "concentrating zones" were developed (10, 11, 12). This type of plate consists of two distinct layer sections, namely the separating layer proper consisting of silica gel 60 and a concentrating zone composed of an inert, porous silicon dioxide. These two sorbent materials pass into one another at a clearly defined boundary-line in such a way that the eluant is offered no resistance as it passes through. What happens when an HPTLC pre-coated plate silica gel 60 with concentrating zone is used is as follows : the substance mixture is applied in spots or streaks at any point and in particular at any height on the concentrating zone and the spots or streaks are then carried by the eluant to the border between the two sorbent materials. At this boundary-line the substance spot is automatically concentrated into a very narrow band. From the boundary-line then begins the chromatographic separation on the silica gel 60 layer, with the concentrating phase passing over into the separating phase without interruption.

Improvements in the separating performance achieved with HPTLC pre-coated plates silica gel 60 with concentrating zone (E. Merck, Darmstadt) compared with HPTLC pre-coated plates silica gel 60 without concentrating zone, taking as an example the application of a relatively large sample volume of 0.75 µl (in HPTLC sample volumes of between 0.02 µl and 0.1 µl are common practice!), is demonstrated by the separation of Trevespan 6038, as already described.

The solvent system used for both types of plates was n-hexane/ethyl acetate/formic acid in a ratio of 40/60/0.5 by volume. The chromatographic conditions,

applied quantities and sample concentrations are the
same as described under point 4 for the separation of
Trevespan 6038 on HPTLC pre-coated plates silica gel
60.
       By comparing the chromatographic data given in
Table VIII it can be seen that for the optimum
migration distance of 5 cm and with a development
time shortened by 6%, the resolution R_s on the HPTLC
pre-coated plate silica gel 60 with concentrating
zone is on average about 26% better than on the
corresponding plate without concentrating zone.
       The calibration curves for the substances
ioxynil, MCPA and flurenol on the two HPTLC pre-
coated plates silica gel 60, one with and one without
concentrating zone, show, as in Figure 4, that the
peak heights for identical applied quantities are
higher for all the substances examined on plates with
a concentrating zone than on corresponding plates
without a concentrating zone. This is attributed to
the already mentioned more compact spot formation on
the HPTLC pre-coated plate silica gel 60 with
concentrating zone.
       Because of the more highly concentrated spots
obtained on the HPTLC pre-coated plates silica gel 60
with concentrating zone, the detection limits shown
in Table IX for this type of plate are lower than in
the case of the HPTLC pre-coated plate silica gel 60
without concentrating zone.
       The improvements achieved in HPTLC pre-coated
plates silica gel 60 with concentrating zone as
compared with HPTLC pre-coated plates silica gel 60
are all the more pronounced the higher the application
volume is. There is no point in using HPTLC pre-coated
plates without concentrating zone at an application
volume of more than 0.75 µl. However, on HPTLC pre-
coated plates with concentrating zone it is possible
to chromatograph even larger quantities with very
good separation. Apart from these improvements in
separation performance, particularly at large applied
quantities, the HPTLC pre-coated plates silica gel 60
with concentrating zone offer the following additional
advantages over corresponding pre-coated plates
without the concentrating zone :
1. Following spot-wise or streak-wise application of
   a substance mixture, the spot or streak is automatically concentrated into a compact starting
   line during the actual development process, separation of this starting line then continuing
   without interruption.
2. No special skills are required during application,

Table VIII. Chromatographic data for the separation
of ioxynil, MCPA and flurenol on HPTLC
pre-coated plates silica gel 60 and on
HPTLC pre-coated plates silica gel 60
with concentrating zone; solvent system:
n-hexane/ethyl acetate/formic acid
40/60/0.5; migration distance: 5 cm

types of plates	velocity coefficient κ [mm^2/s]	development time [s]	resolution R_s of	
			Ioxynil/MCPA	MCPA/Flurenol
HPTLC silica gel 60	3,67	858	2,45	2,24
HPTLC silica gel 60 with conc. zone	3,86	806	3,27	2,64

Table IX. Detection limits (in ng) for ioxynil,
MCPA and flurenol on HPTLC pre-coated
plates silica gel 60 and on HPTLC pre-
coated plates silica gel 60 with concentrating zone

types of plates	limit of detection [ng]		
	Ioxynil	MCPA	Flurenol
HPTLC silica gel 60	10	15	6
HPTLC silica gel 60 with concentrating zone	7	13	5

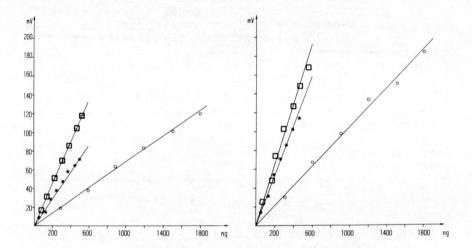

Figure 4.   Calibration curves for ioxynil (*), MCPA (○), and flurenol (□) on HPTLC pre-coated plates silica gel 60 (left) and on HPTLC pre-coated plates silica gel 60 with concentrating zone (right) (abscissae: applied quantity (ng); ordinates: peak height (mV))

in respect of geometry, positioning and spreading of the applied spots in the concentrating zone. This simplification of application techniques leads in itself to considerable saving of time.

3. Apart from the concentrating effect, there is also a certain "clean-up" process which takes place in the concentrating zone in certain cases; this can replace the extraction of substances to be chromatographed, which may otherwise have been necessary.

4. Possible decomposition or irreversible adsorption of substances on to the active silica gel of the separating layer during drying of the applied spot is avoided in the inert concentrating zone. The substances come into contact with the active sorbent only after passing through the concentrating zone and then only in dissolved form. In this respect, it is possible to speak of wet dosing, as in column liquid chromatography.

5. It is possible to apply sample substances by immersing the concentrating zone of the plate into a dilute sample solution. This method can replace a multiple, streak-wise application of very dilute sample solutions using specialised apparatus.

Literature Cited

1 Ripphahn, J.; Halpaap, H. J. Chromatogr., 1975, 112, 81.

2 Halpaap, H.; Ripphahn, J. Kontakte (Merck), 1976, 3, 16.

3 Hezel, U.; Blome, J.; Halpaap, H.; Jänchen, D.; Kaiser, R. E.; Ripphahn, J. "Einführung in die Hochleistungs-Dünnschicht-Chromatographie, HPDC"; Institute of Chromatography; Bad Dürkheim, 1976.

4 Zlatkis, A.; Kaiser, R. E. "HPTLC-High Performance Thin-Layer Chromatography"; J. Chrom. Library 9, Elsevier Publishing Company; Amsterdam-New York, 1977.

5 Halpaap, H.; Ripphahn, J. Chromatographia, 1977, 10, (10), 613.

6 Halpaap, H.; Ripphahn, J. Chromatographia, 1977, 10, (11), 643.

7 Kaiser, R. E.; Rieder, R. J. Chromatogr., 1977, 142, 411.

8 Halpaap, H.; Ripphahn, J. Lecture, 13. Symposium on Advances in Chromatography, 1978, St. Louis.

9 Hauck, H. E. Lecture, 2. Danube Symposium on Progress in Chromatography, 1979, Carlsbad.

10 Halpaap, H.; Krebs, K.-F. J. Chromatogr., 1977, 142, 823.

11 Gerke, H. E. GIT Fachz. Lab., 1977, 21, 685.
12 Halpaap, H. Kontakte (Merck), 1978, 1, 32.
13 Jork, H.; Roth, B. J. Chromatogr., 1977, 144, 39.
14 Davies, R. D.; Pretorius, V. J. Chromatogr., 1978,
   155, 229.
15 Davies, R. D. J. Chromatogr., 1979, 170, 453.
16 Granger, C.; Zwilling, J. P. Bull. Soc. Chim.
   France, 1950, 873.
17 Thielemann, H. Z. Chem., 1976, 16, 155.
18 Tewari, S. N.; Sharma, I. C. J. Chromatogr., 1977,
   131, 275.
19 Brinkman, U. A. Th.; de Vries, G. J. HRC & CC,
   1979, 2, 79.
20 Becker, H.; Exner, J.; Bingler, T. J. Chromatogr.,
   1979, 172, 420.

RECEIVED March 3, 1980.

Analysis of Pesticidal Residues in the Air Near Agricultural Treatment Sites

JAMES N. SEIBER, GERALDO A. FERREIRA[1], BRUCE HERMANN[2], and JAMES E. WOODROW

Department of Environmental Toxicology, University of California, Davis, CA 95616

Substantial quantities of pesticides become airborne during and following spraying operations. This loss route has economic implications; at the least, airborne material leaves the intended use area and is no longer efficacious and, if the chemical is phytotoxic, drift damage to non-target foliage may result. Furthermore, airborne residues present a potential exposure route for farm workers and other individuals dwelling near agricultural sites, and atmospheric transport may be a major pathway for widespread distribution of pesticides--particularly the more persistent ones--in the environment. Evidence to support these contentions comes from several sources: (1) Assessment of drift during spraying (1, 2, 3). These studies typically employ fallout collectors, particulate air samplers, and, sometimes, sensitive plants or animals placed at distances from a spraying operation (4). The emphasis has been on particulates-- their nature, concentration, and size--in relation to spray variables and meteorological conditions. (2) Assessment of evaporative losses following application (5, 6). Experimental data come from air samplers placed within a pre-treated plot, preferably at 2 or more heights above the soil or crop surface (7). Sampling is continued for several days following treatment to allow determination of the quantity lost by vaporization in relation to other loss routes. A variation involves the use of model chambers rather than open plots, but with the same goals in mind (8). (3) Ambient air sampling in a defined geographical area (9, 10, 11). The objective in these studies is to determine the nature, frequency of occurrence, and levels of airborne residues in relation to use patterns within a network of sampling sites. Extensions involve regional (12) or national networks (13, 14, 15) which assess atmospheric residues on a

Current address: [1]Department of Chemistry, University of Brasilia, Brasilia DF 70000, Brazil; [2]Department of Analytical Chemistry, Stanford Research Institute International, Menlo Park, California 94025.

0-8412-0581-7/80/47-136-177$08.00/0
© 1980 American Chemical Society

larger scale. (4) Long-range transport studies. Pesticides in the air passing remote sites--frequently islands or ships stationed at sea (15, 16, 17, 18)--are analyzed for relatively long periods to determine global circulation pathways (19). Extensions include the analysis of rainwater (20) or snow-melt (21) from remote areas, and the collection of particulate matter from dust storms of a known or suspected origin (22). (5) Atmospheric degradation studies. Recent efforts have been aimed at determining the mechanisms and rates of pesticide residue degradation in the atmosphere (23, 24). The proportions of products and parent are related to residence times following release at the source (25). Parallel experiments are carried out in irradiated vapor-phase laboratory chambers in which the levels of reactants and exposure times may be controlled (26).

These considerations are summarized in a general way in Figure 1. That is, the relative amounts of a given pesticide entering the air and surviving to some downwind site depend on the nature of the source (e.g., the type of application in agricultural operations), its physical properties and chemical reactivity, its form once within the air, and the meteorological conditions (wind, temperature, sunlight, etc.) which hold throughout the process. As might be expected, it is those compounds possessing a reasonable degree of volatility and stability, and which enjoy a fairly sizeable use which are most likely to survive to the atmospheres of remote sampling sites.

Some aspects of the sources, occurrence, and dispersion of airborne pesticide residues (6, 27) and methods for their sampling and analysis (28, 29, 30, 31) have been reviewed elsewhere. In this paper, the focus will be on sampling methodology, experimental design, and some results from recent field tests aimed at determining the entry and proximate fate of airborne residues in relation to specific agricultural treatments.

Methods of Sampling and Analysis

A variety of methods have been developed for sampling pesticides in air. Suitable procedures must deal with difficulties posed by the uncertainty regarding the physical state (aerosols, solid particles, vapors) of airborne residues, their relatively low concentrations (less than 1 mg/m^3, ca 80 ppb), fluctuations in pesticide concentrations and the levels of potential interferences with time, potential reactivity during the sampling process, and limited availability of sampling devices. These are in addition to the problems of cleanup, recovery, quantitation, and confirmation which are common to trace analytical processes once the sample has been collected and brought to the laboratory for determination.

Particulates

Fairly straightforward methods may be used to assess particulate drift within 100 m or so downwind of the treatment
site (32). These include droplet counts and sizing from de-
posits on plastic sheets, glass plates, or other surfaces.
Addition of colored, fluorescent (4), or radiolabeled tags
(33) to the spray formulation are useful refinements. Biological
indicators, particularly sensitive plants in the case of herbicides, are also employed in many experiments (3,4).

For measurement of particulates, continuous filtration of
air through glass fiber, paper, or cloth filters is frequently
practiced, providing a sample of total particulate matter for
analysis. Alternatively, Anderson or cascade impactors may be
used to fractionate particulates in size categories during sampling (30). For pesticides of negligible volatility, whose
airborne residues are completely associated with particles,
filtration or impaction may work quite well providing that
collected residues are stable in the airstream (34). We have
developed a procedure for determining low concentrations of
paraquat downwind from treatment sites by combining accumulative
filtration for sampling with N-selective gas chromatographic
analysis of reduced paraquat derivative. This method, outlined
in Figure 2, involves extraction of paraquat from the filter by
sonication with 6N HCl, concentration and centrifugation to remove
insoluble materials, and dissolution of soluble matter in saturated ammonium biocarbonate. The latter solution was taken to
dryness, including evaporation of the volatile ammonium bi-
carbonate, in a sand bath; the residue was then dissolved in
aqueous sodium hydroxide and derivatized by reduction with
sodium borohydride. The reduced product, a mixture of mono-
and diunsaturated tertiary amines, was extracted in hexane
and quantitated by gas chromatography using NP-selective
thermionic detection. Recoveries of paraquat dichloride
spiked to glass fiber filters were 95, 92, 80, and 74% for
1.0, 0.5, 0.1, and 0.05 µg fortifications. Considering these
recovery data and the background from an air sample lacking
paraquat (Figure 3), the limit of detection of the method was
estimated to be 0.1 to 1.0 ng/m^3.

This method is considerably more sensitive than the colorimetric assay applied to fall-out or high volume filter samples.
We have applied it to a determination of paraquat residues in
air collected over 200 m from a spraying operation, both
during and for several hours following treatment. Additionally,
the method may be used to measure the quantity of paraquat
associated with various particulate size fractions collected
by cascade impaction.

Filtration and impaction devices have little or no affinity
for vapors and, furthermore, have the distinct disadvantage of
releasing residues initially associated with the particulate

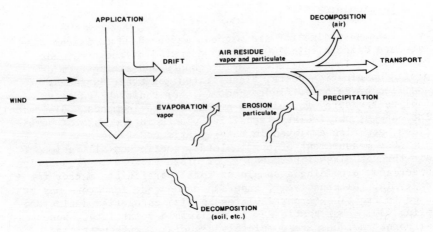

Figure 1. *The source and fate of airborne residues related to pesticide applications*

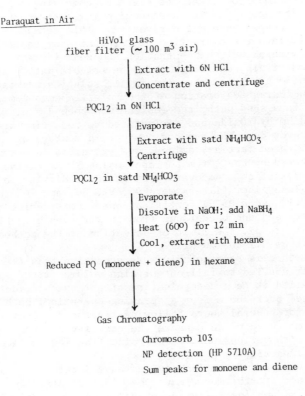

Paraquat in Air

HiVol glass
fiber filter (∼ 100 m³ air)

| Extract with 6N HCl
| Concentrate and centrifuge

PQCl₂ in 6N HCl

| Evaporate
| Extract with satd NH₄HCO₃
| Centrifuge

PQCl₂ in satd NH₄HCO₃

| Evaporate
| Dissolve in NaOH; add NaBH₄
| Heat (60°) for 12 min
| Cool, extract with hexane

Reduced PQ (monoene + diene) in hexane

Gas Chromatography

Chromosorb 103
NP detection (HP 5710A)
Sum peaks for monoene and diene

Figure 2. *Schematic for the determination of airborne residues of paraquat*

fraction through volatilization to the airstream during sampling.
This was clearly demonstrated by Lewis (31), who found 0-5%
retention after 24 hours of sampling through particulate-coated
glass fiber filters spiked with 19 common pesticides. It appears
that the only way to overcome this problem would be to use a
sampler in which particles are dropped out of the air stream
following collection. Cyclone separators accomplish this, but
are inefficient for particles less than ca 10 μm in diameter--
just the fraction most likely to remain airborne for significant
times following release (35). Thus, while much has been written
on the need for developing samplers which discriminate between
vaporized and particulate residues for covalent pesticides (13,
29, 31, 36), no useful techniques have yet been forthcoming.

Vapors

For determination of vapor concentrations of pesticides,
some form of dynamic accumulative sampling followed by solvent
extraction of trapped material has predominated in most
applications. Principal variables are the rate of sampling--
arbitrarily divisible into "low volume" (ca 1-50 L/min) or
"high volume" (ca 0.2-2 m^3/min) categories; the configuration
of the sampler collector; and, particularly, the nature of
the collection medium. Of several devices used to collect
airborne pesticide residues, glycol-filled Greenburg-Smith
impingers have been the most thoroughly tested and widely
used (13, 36). Enos et al. (37) employed ethylene glycol in
the larger impinger in methodology used in the Community
Studies air monitoring program in 1970-72 (14). The method
allows for determination of a fairly wide range of pesticides
at levels as low as 0.1 ng/m^3, with efficiencies considering
both the sampling and post-sampling workup of approximately
40-70% (28). Method recoveries from extraction of the glycol
through the modified Mills cleanup were greater than 85% for
the pesticides tested by Thompson (38), while an alternative
extraction-cleanup scheme based on silica gel fractionation
gave method recoveries of 75-100% for 25 pesticides (39).

Impinger methods have the advantages of using commercially
available glassware and a sequential sampler specially designed
for monitoring, in addition to having been relatively thoroughly
evaluated. Disadvantages include the possibility of degradation
of labile compounds on the filter or in the impinger fluid,
restrictions of air flow to ca 30 L/min, and the expense associated with the glassware and sequential sampler.

There have been several attempts to develop alternate sampling methods, with primary attention devoted to substitution of
solids or coated solids as the trapping media. Inorganic adsorbents examined include alumina (13), charcoal (40), Florisil
(41), molecular sieves (42), and silica gel (9, 28, 33, 43).
Surface-catalyzed degradation of some labile compounds and the

difficulty in extracting strongly sorbed pesticides limits
the utility of the inorganic adsorbents.

Liquid phase coated solids tested as sampling media for
pesticides include polyethylene on silica gel (44), polyethylene
glycol on stainless steel nets (45), cottonseed oil on glass
beads (12, 46), paraffin oil on Chromosorb A (28, 47), and OV-17
on ceramic chips (48), among the coatings, while ODS-
Chromosorb (49) and GC Durapak-Carbowax 400-Porosil F (50) ex-
emplify the bonded phase approach. The major difficulty with
the coated phases lies in the difficulty in separating them from
post-sampling extracts containing the pesticides of interest (31).
Bonded phases, on the other hand, provide cleaner extracts but
suffer the severe limitation at present of being quite expensive
and available in too small particle sizes for use at flow rates
higher than a few L/min. Their utility would seem to lie more
with personnel monitors in work-place air than with ambient monitoring out of doors.

Much of the attention in recent years has been directed
toward organic polymers--particularly the polystyrene-based
macroreticular resins and polyurethane foam--as sampling media
for airborne pesticide residues. Among the macroreticular
resins Chromosorb 102 was found by Thomas and Seiber (51) to
satisfactorily trap ca 0.1-10 µg quantities of several pesticides
ranging in volatility from lindane, trifluralin, and diazinon,
to DDT and methoxychlor in ca 10 m^3 air volumes. The air
volume to medium weight ratio giving > 50% sampling efficiency
was 10 m^3/g for Chromosorb 102 vs only 0.2 m^3/g for ethylene
glycol in the Greenburg-Smith impinger (29). Several reports
have subsequently appeared dealing with the applications of
Chromosorb 101 (52), Chromosorb 102 (53), XAD-2 (54), and
XAD-4 (25, 55, 56, 57)--all of which are closely related resins
having, particulary for the latter three, very similar composition, surface area, pore size, and polarity--to sampling pesticides and related compounds in air. These materials are available as free flowing beads in a variety of particle sizes which
conform to virtually any sampling configuration. For example,
commercial high volume air samplers may be readily adapted to
hold up to 60 g of 20/50 mesh XAD-4 in the sampler intake, with
little restriction to air flow (Figure 4, (28). The procedure
gave > 67% recovery for 15 of the 16 organochlorine,
organophosphorus, and dinitroaniline pesticides tested, aldrin
(22%) being the exception (Table I, 58). Alternatively, the
resin may be packed in tubes for low-volume sampling (Figure
5, 59) or, through a manifold to a high volume sampler capable
of powering several such packed tubes (Figure 6, 57). The
latter arrangement, patterned after that of Turner and
Glotfelty (60), is particularly useful when sampling air at
several heights above ground. It is worth noting that the
mean sampling efficiency for four organophosphorus pesticides
(DEF, malathion, mevinphos, and parathion) was 84% for XAD-4

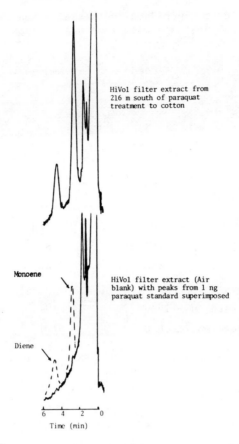

HiVol filter extract from
216 m south of paraquat
treatment to cotton

Monoene

HiVol filter extract (Air
blank) with peaks from 1 ng
paraquat standard superimposed

Diene

6   4   2   0
Time (min)

*Figure 3. Gas chromatograms for air blank (no paraquat) and field air sample
collected downwind from a paraquat treatment*

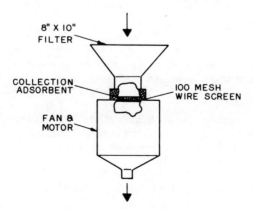

8" X 10"
FILTER

COLLECTION
ADSORBENT

100 MESH
WIRE SCREEN

FAN &
MOTOR

Plenum Press

*Figure 4. High volume air sampler modified for holding solid sampling media
for adsorbing pesticide vapors (28)*

Table I. Pesticide Vapor Sampling Efficiencies of XAD-4 Resin
          and Polyurethane Foam in a High Volume Air Sampler[a,b]

Pesticide	Amount Added	Sampling Efficiency (%)	
		XAD-4	Polyurethane Foam
Organochlorines			
α-BHC	80 µg	79	60
Lindane	80	71	69
Aldrin	100	22	43
DDT	800	101	93
DDE	200	96	90
Heptachlor	180	81	82
Dieldrin	400	86	78
Organophosphates			
Methyl parathion	1620	101	80
Ethyl parathion	1620	101	63
	0.2 - 5	85	95
Diazinon	544	67	50
Mevinphos	0.2 - 5	86	80
Malathion	0.2 - 5	92	78
DEF	0.2 - 5	82	93
Dimethoate	0.2 - 5	86	-[c]
Carbophenothion	0.2 - 5	72	93
Dinitroaniline			
Trifluralin	0.2 - 5	93	99

[a] Chemicals were introduced as vapors, by evaporation from a
filter or glass wool surface upstream from the collection
media. Air was sampled for 2 hr at a flow rate of ca 1 m^3/min.

[b] Recoveries using the analytical method were determined by
spiking XAD-4 or foam extracting solvent prior to silica gel
clean-up and GLC analysis; recoveries were >80% for all compounds except mevinphos (36%).

[c] Interferences from polyurethane foam precluded quantitation.

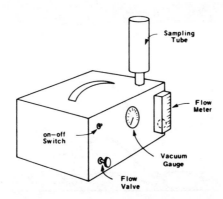

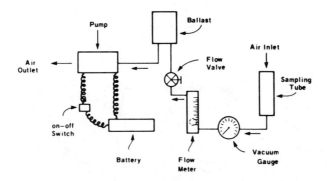

Analytical Chemistry

Figure 5. External appearance and schematic diagram of portable air sampler for pesticide vapors (59)

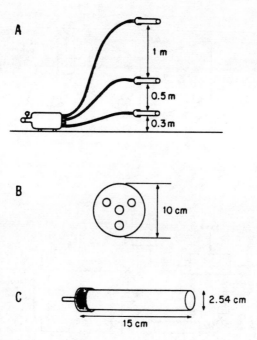

Figure 6. Arrangement for sampling pesticide vapors at three heights above a treated surface via a manifold interface (B) between XAD-packed tubes (C) and high volume air sampler motor (A) (57)

<u>vs</u> 64% for ethylene glycol when the two were evaluated at
nearly identical flow rates (~30 L/min) and sampling times
(59). The latter figure is close to the reported mean for
seven organophosphorus compounds (68%) evaluated in the
glycol impinger (28). Besides its advantageous efficiency,
applicability, and adaptability, XAD-4 is readily available
in bulk quantities, has a generally low gas chromatographic
background after cleanup by Soxhlet extraction, is stable in
the airstream, and readily releases trapped chemicals during
post-sampling extraction, often by a simple solvent wash.
Additionally, the washing may be chosen to provide some
fractionation, as was shown by Farwell <u>et</u> <u>al</u>. (54) in the
elution from XAD-2 of 2,4-D esters with hexane followed by
elution of the free acid and amine salt with KOH in aqueous
methanol.

Porous polyurethane foams possess many of the same attributes as the macroreticular resins and, in fact, the range of
applicability is quite similar (16, 18, 60, 62, 63, 64, 65).
Plugs of the foam may also be deployed in low-volume (63),
high volume (61), or multiple air samplers (60). Foam preparation is similar to that of the XAD's, and its commercial
availability through local sources is a distinct advantage.
Some consider the wide variation in resistance to air from
one foam plug to another (58) and the lengthy extraction (54)
as disadvantages.

In one study where polyurethane foam and XAD-4 were compared, Hermann et al (58) found generally equivalent sampling ef-
fiencies (> 50%) for 14 of 16 organochlorine, organophosphorus,
and dinitroaniline pesticides (Table I). Foam gave higher recovery for aldrin (43% <u>vs</u> 22%) while XAD-4 gave acceptable
results for one compound (dimethoate) which was subject to
severe foam-derived interference. Lewis et al (64) found
that foam/sorbent combinations, including foam/chromosorb
102 and foam/XAD-2, gave better sampling efficiencies for the
more volatile PCB isomers, aldrin, and diazinon than foam
alone.

No firm conclusions can be reached concerning detection
limits with the resins or foam, since these vary so much with
the rigor of cleanup, the chemicals tested, and the particular
air environment. It appears, however, that both materials can
achieve detection limits of 0.01 ng/m^3 under the most favorable
conditions when employed in high-volume samplers, while the
limit for the glycol impingers is on the order of 0.1 ng/m^3
(28).

The macroreticular resins, porous polyurethane foam, and
glycol impingers all exhibit characteristics of gas/matrix partitioning agents with retention volumes varying with analyte
polarity and volatility. Sydor and Pietrzyk (66) determined
the retention of low molecular weight organic compounds varying
widely in polarity using a number of porous copolymers. Overall

capacity increased in the order Tenax GC < XAD-1 < Porapak R
<< XAD-2 < Poropak Q < XAD-7 < XAD-4. For XAD-2 and XAD-4,
the capacity factor increased roughly with increasing lipo-
philicity of the analyte, being lowest for water and ethanol
and highest for carbon tetrachloride, MEK, and benzene.
Concomitant elution of water or MEK did not markedly affect
retentions on XAD-2, XAD-4, or XAD-7--consistent with a partitioning mechanism.

Working with porous polyurethane foam, Simon and Bidleman
(67) determined that penetration of PCB isomers through 1 cm
segments of foam at high volume flow rates was correlated with
GC retention times. PCB components with the highest volatility
(i.e., shortest GC retention) penetrated furthest in the foam
sampling bed. Retention volumes were 1250 and 2700 m^3 for
3,3'-DCB and 2,4',5'-TCB, respectively, for a 7.6 cm diameter x
15 cm length polyurethane foam column. In a related study,
Erickson et al (63) found that monochloronaphthalene vapor passed
through two 13 x 5 cm diameter foam plugs, 78% of the dichloro-
naphthalene was on the second (downstream) plug and 100% of
trichloronaphthalene was on the 1st plug for 90 m^3 of air sampled
at 60 L/min. Taken collectively, these studies indicate that the
vapor pressure cut-off for efficient retention of chemicals on
polyurethane foam is on the order of 10^{-3} mm Hg, but obviously is
affected by foam dimensions and the volume of air processed. No
comparable study has been reported for the macroreticular resins,
although Thomas and Seiber (51) found no significant penetration
of trifluralin, lindane, or diazinon (vapor pressures 10^{-5} to
10^{-4} mm Hg) occurred beyond a 1-g segment of Chromosorb 102 with
10 m^3 air volumes, and Woodrow and Seiber (59) found nearly
complete retention of the polar insecticide mevinphos (2.2 x 10^{-3}
mm Hg) by 18 g of XAD-4 when spiked to 1.7-1.8 m^3 volumes of air.
In recent tests of the applicability of XAD-4 resin in a high
volume air sampler to trapping volatile formulation impurities
of DEF[R] and merphos, dibutyldisulfide (b.p. 226°C) was quantitatively trapped from 120 m^3 of air using 100 ml of resin, but
butyl mercaptan (b.p. 98.2°C) was essentially not retained
under the same conditions (68). For the latter compound, a
mercuric acetate-impregnated silica gel compartment was placed
downstream from the resin in the high volume air sampler. For
ethylene glycol, volatility affected retention of chlordane
components as the two most rapidly-eluting major GC peaks of
technical chlordane were sampled with much lower efficiencies
(38-39%) then the six later-eluting major peaks (67-84%) during
a 12-hour vapor spiking run (69).

Summarizing, it appears that the macroreticular resins
such as XAD-2, XAD-4, and Chromosorb 102 or porous polyurethane foam can be considered as alternatives to the glycol
impinger for sampling many of the common pesticides and related
chemicals in air. The glycol impinger has been more thoroughly
tested, while the resins and foams are clearly to be preferred

for high-volume sampling. Such parameters as sampling efficiency, background, and analyte stability should be evaluated
thoroughly before a final choice is made for specific applications.

Field Applications

It has only been in the past 10 years or so that serious
attempts have been made to quantify the source of airborne
residues of pesticides or their atmospheric fate. The examples
which follow are meant to illustrate experimental design and
some results from which quantitative conclusions regarding
source and fate may be drawn for agricultural applications.
The emphasis in these examples is on volatilization processes
and the fate of vapors, although equally impressive advances
have been made in understanding particulate drift phenomena.
The first direct field measurements of pesticide concentrations in air were obtained by Willis et al. above endrin (70) and
DDT-treated plots (71). In the endrin study, air was sampled
at 1 L/min through ethylene glycol using a boom support suspend-
ed 4 feet above ground level within a sugar cane canopy. Granular endrin was applied at 2 lb/acre on a 24 x 30 foot plot. The
mean atmospheric endrin concentration reached a maximum of 540
ng/m^3 during the first 3-day sampling period after treatment,
and then decreased to 123 ng/m^3 at 21 days. The cumulative recovery for endrin approached 12 μg and reflected a decrease in
volatilization with time. Assuming a mean lateral air movement
of 0.1 mph through a 24 x 30 x 10 foot volume above the plot,
the cumulative total was estimated to translate to 5% volatilization of the total endrin applied.
In the DDT study, DDT (100 lb/acre) and DDD (32 lb/acre)
were simultaneously incorporated to a 6 inch depth in commerce
silt loam and also applied to the soil surface after incorporation. Two sampling heights (10 and 30 cm) and two moisture
regimes (flooded and unflooded) were employed. Maximum air
levels were ca 2 μg/m^3 at 10 cm for the 1st day after treatment.
Levels at 30 cm were generally about 1/2 those at 10 cm, and
vaporization was most pronounced from the non-flooded plot over
the duration of the experiment. The cumulative recovery of DDD
from the non-flooded plot was similar to that of DDT. Given
similar volatilities for the two pesticides, it is possible
that some reduction of DDT to DDD, or breakdown of DDT to DDE
(which was not measured) may have occurred. Subsequent experiments by Spencer and Cliath (72, 73) indicated that DDE volatili-
zes much more rapidly than either DDT or DDD from soil or inert
surfaces. Combining the results of these experiments with
those of other studies aimed at characterizing the mechanism of
pesticide vaporization from soils (5), it appears that substantial loss of pesticides may occur through volatilization of the

parent chemical or degradation products formed in the soil
layer, even when vapor pressures are relatively low.

Knowledge of the relative role of vaporization in pesticide field dissipation has been markedly advanced by development
of methods for measuring the vertical flux of vapors from a soil
or soil-crop field plot. The basis and techniques for this
method have been presented by Caro et al. (74) and Parmele et al.
(7), and summarized recently by Taylor (6). The equation

$$F \uparrow = K_z \, (dc/dz)$$

describes the pesticide vapor vertical flux ($F\uparrow$) through a
horizontal plane at height z above the soil or crop, where dc/dz
is the gradient of vapor concentration with height and K_z is the
vertical eddy diffusivity coefficient at height z. In the
commonly employed "aerodynamic method" the eddy diffusivity is
measured from the gradient of the wind profile over the surface
and the thermal stability of the atmosphere, and the equation for
pesticide flux is:

$$F \uparrow = k_2 \frac{(C_1 - C_2)(u_2 - u_1)}{Ln \, (z_2/z_1)^2 \phi^2}$$

where C_1 and C_2 are pesticide concentrations and u_1 and u_2 are
windspeeds at heights z_1 and z_2 above the surface; k is
von Karman's constant (= 0.4) and ϕ is a thermal stability term.
The field method, then, employs air sampling and wind speed
measurements at 2 heights or more above the surface.

A detailed study for dieldrin and heptachlor surface-applied
to a grass pasture at 5.6 kg/ha illustrates the method (75). The
rectangular plot (3.34 ha) had a site near the center of the plot
for vapor measurements at 10, 20, 30, 50, and 100 cm above the
grass surface, sites for grass and soil sampling, and wind stations for recording anemometers at 10, 50, 100, 150, 200, and
250 cm above the grass surface. Vapor samples were collected at
8.3 L/min through hexylene glycol scrubbers. Raw data, selected
for the 20 and 50 cm heights on the day of spraying in Table II,
were used along with corresponding wind data in the flux equation
to generate hourly flux intensities. Most notably, the data
showed high initial vapor losses (12% for dieldrin, 46% for
heptachlor) in the first 12 hours after application. This
contrasts sharply with much lower volatilization losses when the
same chemicals were incorporated to 7.5 cm prior to planting corn
in an earlier study (76). With soil incorporation 1st day losses
were less than 1 g/ha/day, and maximum flux was measured ca 50
days after treatment. In both types of treatment, however, a
marked diurnal variation in flux was obtained, the maxima occurring between 8 am and 4 pm and roughly centered at noon.

Table II. Dieldrin and Heptachlor Concentrations in Air and Hourly Flux Intensities from Orchard Grass Surface on Day of Application.[a,b]

Time	Dieldrin			Heptachlor		
	Concn in μg/m³		Flux[c] (g/ha/hr)	Concn in μg/m³		Flux[c] (g/ha/hr)
	20cm	50cm		20cm	50cm	
1100–1300 EDT	60.0	33.0	169	277	171	822
1300–1500	42.2	32.0	80.4	124	92.7	296
1500–1700	31.6	20.9	60.6	57.8	34.8	128
1700–1900	14.3	11.3	16.2	23.4	18.1	29.4
1900–2100	7.6	5.6	0.6	14.4	8.9	1.4
2100–2300	3.9	1.8	0.2	10.4	5.0	0.6

[a] Data selected from Ref 75.

[b] Rate was 5.6 kg/ha for both chemicals; application was made between 0930 and 1030 EDT.

[c] For 35 cm above grass surface.

The dissipation of the residues of these two chemicals on grass and soil could be entirely accounted for by volatilization. No run off or degradation losses were found.

A somewhat different approach was taken by Seiber et al. (56) to reach similar conclusions for the multicomponent insecticide toxaphene. Residues of toxaphene were analyzed in leaf, air, and topsoil samples taken for up to 58 days following an application at 9 kg/ha to a cotton field. Analyses were conducted using packed column GLC for total toxaphene residues and by open-tubular (capillary) column GLC for enhanced component resolution. Leaf residue analyses to the 50th day following treatment (135 ppm) showed a regular trend toward greater loss of components of higher volatility when compared with the day 0 sample (661 ppm). Peak by peak inspection of capillary chromatograms (Figure 7) showed that no dramatic qualitative changes occurred since all of the peaks present at day 0 were still there after 50 days. These data were converted into quantitative terms by computing heights of individual peaks relative to an arbitrarily chosen reference peak (peak 6) in the capillary GLC profile of the day 0 and day 50 cotton leaf extracts. The percentage change in the relative peak height for each numbered peak is presented in Figure 8 for the day 0 and day 50 leaf samples. The graphic data show that for most of the numbered peaks the trend was quite regular in that the peaks showing shorter retention times decreased in relative abundance on weathering.

To determine whether the loss from leaf surfaces was matched by a corresponding enrichment of the volatile components in the air, air samples were collected at several points within and toward the edges of the field using Hivol samplers containing XAD-4 resin.

Comparison of a capillary chromatogram from an early air sample (Figure 9) with those from leaves indeed showed a higher proportion of early-eluting peaks in the air samples than in the foliage samples. Using the peak ratio technique for a 2-day air sample and 0-day leaf sample, the trend was once again regular but the inverse of that observed for weathered versus fresh foliage residues (Figure 8). That is, comparison of the two graphs showed that those components most rapidly lost from the leaf surface were proportionately enriched in the air.

Taken collectively, the data clearly indicated differential vaporization as the primary mode of toxaphene loss from leaf surfaces and gave no indication that chemical reactivity played even a minor role. If toxaphene had been degraded either on surfaces or during its brief residence time in the air prior to sampling, changes in the chromatographic profile would have been erratic with new peaks observed in the capillary chromatograms such as occur in samples of anaerobic soil and ditch sediment where microbial decomposition is extensive.

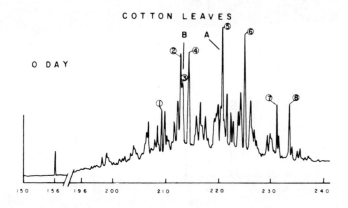

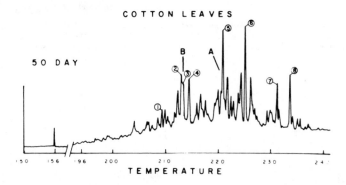

Journal of Agricultural and Food Chemistry

Figure 7. Capillary gas chromatograms of toxaphene residue in 0- and 50-day cotton leaf extracts (56)

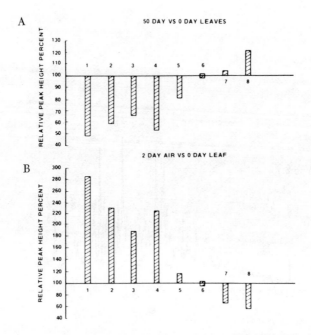

Journal of Agricultural and Food Chemistry

Figure 8.   Relative peak height percents for the major toxaphene GLC peaks in the residue from sample pairs. For example, the top graph (A) shows the heights (relative to Peak 6) of peaks from the 50-day leaf extract expressed as percents of corresponding peaks from the 0-day leaf extract (56).

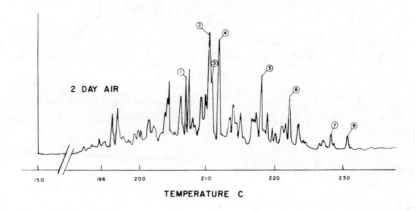

Journal of Agricultural and Food Chemistry

Figure 9.   Capillary gas chromatogram of toxaphene residue in air sample taken 2 days after treatment to cotton (56)

The heptachlor, dieldrin, and toxaphene examples show that foliage and soil residue dissipation data can be used to estimate the amounts of residues volatilized when no significant degradation or runoff losses are incurred. For toxaphene, residue analyses indicated that 80% of the foliage residue and 51% of the top soil residue was lost by volatilization within ca 50 days. This is considerably more than the 24% vaporization loss reported for toxaphene within 90 days in a model chamber (8), but is comparable to foliage-applied heptachlor and dieldrin (75) both of which overlap in volatility with components of the toxaphene mixture.

Evidence was obtained recently that pesticide vapors may enter the air by still another mechanism, involving plant circulation and water loss (57). Rice plants were found to efficiently transport root-zone applied systemic carbamate insecticides via xylem flow to the leaves, eventually to the leaf surface by the processes of guttation and/or stomatal transpiration, and finally to the air by surface volatilization. Results from a model chamber showed that 4.2, 5.8, and 5.7% of the residues of carbaryl, carbofuran, and aldicarb, respectively, present in rice plants after root soaking vaporized within 10 days after treatment. The major process was evaporation of surface residues deposited by guttation fluid.

To find out whether evaporative losses are important in the field, air was sampled above rice plants systemically treated with carbofuran. Two kinds of air samplers were used; one was a high volume (623 L/min) sampler placed 30 cm above the soil and the other a multiple air sampler which processed air at 30 L/min at 3 heights above the soil (Figure 6). XAD-4 was the sampling medium in all cases. Young rice plants were transported to an experimental plot after soaking for 24 hours in carbofuran solution. The air samplers were on for 8 hours/day for 8 consecutive days after treatment. The same plots were then treated by the gelatin capsule root zone technique (77), and four days later the same schedule of sampling was repeated. Four days delay in the latter case was to give time for the plants to absorb material from the capsules. Two other subplots were treated by conventional methods--foliage spray or foliage spray followed by granular broadcast to the paddy water (77).

Measureable air concentrations of parent carbofuran were present from both systemic treatments (Figure 10). The root soak treatment gave a maximum vapor concentration 1 day after treatment (i.e., transplanting), and then a slow decrease thereafter. This is because the root soaking is essentially a one-shot treatment and evaporation from the leaf surface is apparently the slow step in the distribution/loss process. By contrast, the gelatin capsule treatment boosted air concentrations continually over the period of sampling. This arises through an apparently continual supply of insecticide to the

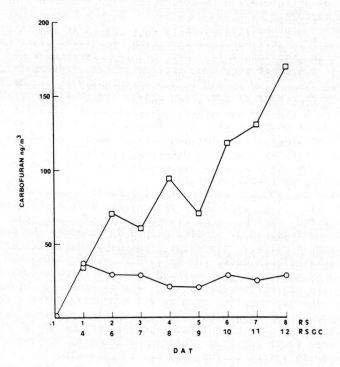

Figure 10.   Carbofuran residue in high volume air samples taken above rice paddies treated by root soaking seedlings (○) or by root soaking followed by gelatin capsule root zone treatment (□) (57) (DAT = days after treatment; each point = 8 hr of sampling at 623 L/min)

leaf surface, allowing for a buildup of surface residue, and hence vaporization, with time. Apparently the sampling was not carried out long enough to see the leveling out and eventual decline of vapor concentration which inevitably must occur.

For the conventional treatments (Figure 11), foliage spray produced a maximum vapor concentration at day 1 with a fairly rapid decrease thereafter--consistent with data for foliage-spray treatments of other insecticides mentioned previously (75). The granular broadcast produced vapor concentrations which were similar to the root-zone treatment, except that the magnitude is only ca 1/5th-1/10th that in the gelatin capsule experiment. Gelatin capsule treatment has previously been shown to produce up to 200 times the leaf residue observed with granular broadcast (77). [An apparent increase in carbofuran air concentration at days 8 (FS) and 12 (FSB) reflects a subsequent application not connected with these experiments.]

Perhaps most remarkably, the maximum vapor concentration observed for the systemic root soak followed by gel capsule treatments was over 150 ng/m^3 (day 12), while for foliage spray it was only 50 ng/m^3 (day 1). This in itself confirms the postulated evaporative loss route for systemic insecticides, and suggests that it is far more pronounced than indicated by the chamber study. Calculation of vapor fluxes from multiple air sample data--and thus quantitation of vapor losses--has not yet been completed.

Another type of experiment has been used to assess the chemical reactivity of pesticides in the air. This principally employs downwind sampling from a treatment site during application (for measuring conversion in the spray drift) and for several days following application (for conversions involving volatilized residues) (24). The principal data are in the form of product(s)/parent ratios with increasing downwind distance, from which estimates of the rate of conversion can be made knowing the air residence time calculated from windspeed measurements.

An early study along this line was with a parathion-treated plum orchard, in which the paraoxon/parathion ratio was the index of atmospheric conversion (55). By conducting air sampling during daylight and after sunset, the role of sunlight in the conversion was determined. Sampling was done principally using XAD-4 resin in high volume air samplers, with some low volume glycol impingers employed to confirm results.

Residues of airborne parathion within the orchard declined regularly during the 21 days following treatment. Significant levels of paraoxon were also present in the orchard air. The ratio paraoxon/parathion increased initially and then leveled off from day 2 through the end of sampling. The PO/PS ratio in the air was higher during the early sampling dates than for soil, leaf surfaces, or whole leaf samples--an indication that conversion did in fact occur in the air. At later sampling dates the leaf surface residue PO/PS ratio increased to greater

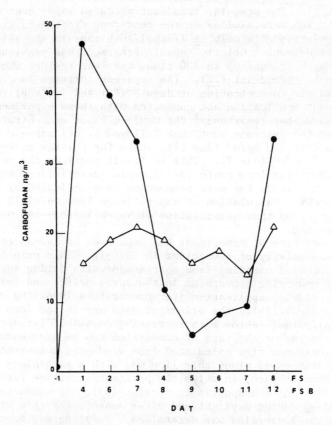

Figure 11. Carbofuran residue in high volume air samples taken above rice paddies treated by foliage spray (●) or by foliar spray followed by granular broadcast treatment (△) (57) (each point = 8 hr of sampling at 623 L/min)

than 1, in accord with the findings of others (78) for oxon buildup on orchard foliage.

Most significantly, downwind sampling during daylight always gave higher PO/PS ratios than occurred within the orchard--a clear indication of conversion in the air (Table III). Air samples collected after sunset showed the same trend, although conversion was not nearly as extensive as in sunlight. Samples of airborne dust collected at some downwind sites, gave very low residues of parathion and paraoxon, and in fact a lower PO/PS ratio than existed in the total air sample collected through XAD resin. It was concluded that parathion conversion to para-oxon occurred rapidly in the air, was promoted by sunlight, and took place largely in the vapor-phase.

To get a better handle on rates of airborne conversion free of complications from surface-formed products, we have used a technique in which the pesticide is released directly to the air

Table III. Afternoon Air Residues (ng/m^3) and Paraoxon (PO): Parathion (PS) Ratios from Hivol XAD-4 Sampling within a Plum Orchard and 100 m Downwind Following Treatment with Parathion -- 1975[a]

Day following treatment	In orchard			100 m Downwind		
	PS	PO	PO:PS	PS	PO	PO:PS
1	4100[b]	611[b]	0.146[b]	-	-	-
2	394	128	0.324	35.0	28.8	0.82
3	397	127	0.321	8.69	9.27	1.06
5	149	70.4	0.472	-	-	-
6	55.6	30.4	0.547	1.55[b]	1.69[b]	1.07[b]
14	21.4[b]	15.3[b]	0.711[b]	-	-	-
21	16.1	6.4	0.395	0.86	1.19	1.39

[a] Data selected from Ref 55.

[b] Averages of two samples; all others represent single determinations.

along a broad front (25). Samples were collected downwind from the spraying operation. Samples from each station were analyzed for both parent compound and products. Knowing the transit time from point of release to downwind stations, and assuming first-order kinetics, the rate expression in Table IV can be used to compute an approximate rate constant. R_t is simply the ratio of moles of parent to the sum of the moles of parent plus products. R_t at t = 0 is unity, and decreases with residence time.

This "spray-drift" technique assumes that both parent and products undergo the same dilution with downwind distance. Limitations in the method are that (1) rates are computed from data points taken early in the reaction, usually well before one half-life occurs, (2) the rates are subject to conditions of sunlight, temperature, and oxidant level which exist during the experiment, and (3) the airborne residue is a mixture of vapors and aerosols. Regarding the latter, an attempt was made to maximize vapor formation by using fine sprays, and conducting the experiments when the relative humidity was low (~30%) and the air temperature was fairly high (~30°C). Such conditions are common in California during the summer months.

Using this "spray-drift" technique, half-lives for trifluralin conversion were measured (Table IV, 25). Under optimum sunlight conditions (August, 12:00 noon) the half-life for trifluralin dealkylation was 21 minutes. The rate fell off later in the summer, but was still on the order of 3 hours in October. By comparison, laboratory photolysis in air containing ozone was 47 minutes, in reasonable agreement with the field. The rate was slower when no ozone was present in the reactor, and little or no reaction took place without light.

Parathion gave a field half-life of 5 minutes for conversion to paraoxon (Table IV). The reaction was considerably slower when the experiment was repeated after sunset. Laboratory experiments gave a half-life of 23 min when irradiation was carried out with ozone present, and was only slightly slower without ozone. Without light very little or no reaction occurred.

From these experiments it may be concluded that sunlight is indeed the predominate factor in the airborne conversions of both trifluralin and parathion, but that oxidant enhances the reaction rate in both cases. Vapor-phase photodecomposition may be a primary dissipation process for the significant fractions of trifluralin (47, 79) and parathion (5) known or suspected to be volatilized following application. It may also explain why surprisingly small residues of potentially photoreactive pesticides such as trifluralin, parathion, and related chemicals are found in surveys of atmospheric levels of pesticides near heavy use areas (9, 10, 11, 13, 14, 41).

By far the fastest reaction observed in the spray-drift tests was with Folex[R] (merphos). This organophosphorus defoliant was essentially all oxidized to DEF by the time the spray drift had reached the 1st sampling station ($t_{\frac{1}{2}}$ ~ seconds) (80). The P(III) to P(V) oxidation is known to be a rapid one which requires only oxygen (81, 82). DEF on the other hand is fairly stable; we have observed slow breakdown in the laboratory

Table IV. Half-lives and Rate Constants for Trifluralin and
Parathion Transformations in Field Air[a,b]

Compound	Half-life (min)	Rate Constant (min^{-1})	Date
trifluralin	21	0.0325	8/05/76
	63	0.0110	8/12/76
	193	0.0036	10/05/76
	182^c	0.0038^c	
parathion	5	0.1430	6/04/76
	131^d	0.0053^d	7/23/75

[a] Data selected from Ref 25.

[b] Calculated from

$$R_t = R_o e^{-kt}$$

where: $\quad R = \dfrac{\text{parent (mol)}}{\text{product (mol) + parent (mol)}}$

except for lindane tracer where

$$R = \frac{\text{trifluralin (mol)}}{\text{lindane (mol)}}$$

[c] Computed using the lindane tracer rather than trifluralin.

[d] In the absence of sunlight.

photoreactor (to P-containing acids) when ozone was present, but the reaction was too slow to be measured by the spray-drift protocol. This was supported by recent field measurements surrounding DEF-treated cotton in which there was no evidence of vapor-phase decomposition to butyl mercaptan or dibutyl-disulfide--known DEF formulation impurities and suspected conversion products (68).

Conclusions

Virtually every step within the generalized scheme (Figure 1) for pesticide atmospheric entry, behavior, and fate has been demonstrated with at least a few examples. Developments in sampling-analysis methodology, and its application to selected field treatments, have provided the data upon which present knowledge is primarily based. The best data probably lie in the areas of drift losses during spraying and volatilization losses from soil and plant surfaces following treatment. Information on atmospheric reactivity and precipitation fate processes has been developed with just a few chemicals presently, and certainly many more examples must be studied before firm conclusions can be reached. Very little is known about residue distribution between particulate and vapor phases largely due to a lack of sampling methods able to discriminate between these two forms.

A more complete knowledge of pesticide ambient atmospheric levels would be immensely helpful in determining residue transport and residence times within the air. Particularly lacking at present are data from a systematic network of sampling sites within a major agricultural basin for which pesticide use and weather data are available. It appears that the time is ripe for such a study, so that the pieces of the puzzle which are now at hand can be fitted into a more comprehensive "model" of the role of atmospheric processes in overall pesticide environmental fate.

Acknowledgements

Aspects of the authors' work cited in this chapter were funded by the following sources: NIEHS Training Grant ES07059; NIEHS Grant ES00054; Grant DEB 76-22390; and the Western Regional Pesticide Impact Assessment Program.

Literature Cited

1. Akesson, N.B.; Yates, W.E. Problems relating to application of agricultural chemicals and resulting drift residues. Ann. Rev. Entomol., 1964, 9, 285.

2. Yates, W.E.; Akesson, N.B.; Coutts, H.H. Drift hazards related to ultra-low-volume and diluted sprays applied by agricultural aircraft. Trans. Am. Soc. Agric. Engrs., 1967, 10, 628.

3. Yates, W.E.; Akesson, N.B.; Bayer, D.E. Effects of spray adjuvants on drift hazards. Trans. Am. Soc. Agric. Engrs., 1976, 19, 41.

4. Byass, J.B.; Lake, J.R. Spray drift from a tractor-powered field sprayer. Pestic. Sci., 1977, 8, 117.

5. Spencer, W.F.; Farmer, W.T.; Cliath, M.M. Pesticide volatilization. Residue Revs., 1973, 49, 1.

6. Taylor, A.W. Post-application volatilization of pesticides under field conditions. J. Air Pollut. Contr. Assoc., 1978, 28, 922.

7. Parmele, L.H.; Lemon, E.R.; Taylor, A.W. Micrometeorological measurement of pesticide vapor flux from bare soil and corn under field conditions. Water, Air, Soil Pollut., 1972, 1, 433.

8. Nash, R.G.; Beal, M.L., Jr.; Harris, W.G. Toxaphene and
    1,1,1-trichloro-2,2-bis(p-chlorophenyl)ethane (DDT) losses
    from cotton in an agroecosystem. J. Agric. Food Chem.,
    1977, 25, 336.
9. Que Hee, S.S.; Sutherland, R.G.; Vetter, M. GLC analysis of
    2,4-D concentrations in air samplers from central
    Saskatchewan in 1972. Environ. Sci. Technol., 1975, 9, 62.
10. Barquet, A.; Morgade, C.; Shafik, T.M.; Davies, J.E.;
    Danauskas, J.X. Pesticides in Air: Air monitoring studies
    in South Florida. Presented at 170th National Meeting of
    the American Chemical Society (PEST 089), Chicago, Aug. 24-9,
    1975.
11. Arthur, R.D.; Cain, J.D.; Barrentine, B.F. Atmospheric
    levels of pesticides in the Mississippi delta. Bull.
    Environ. Contam. Toxicol., 1976, 15, 129.
12. Compton, B.; Bazydlo, P.P.; Zweig, G. Field evaluation of
    methods of collection and analysis of airborne pesticides.
    Contract CPA 70-145 for the U.S. EPA, Research Triangle
    Park, N.C., May 1972 (NTIS No. PB-214008/5, 154 pp; PB-
    214-009/3, 219 pp).
13. Stanley, C.W.; Barney, J.E.; Helton, M.R.; Yobs, A.R. Mea-
    surement of atmospheric levels of pesticides. Environ. Sci.
    Technol., 1971, 5, 430.
14. Kutz, F.W.; Yobs, A.R.; Yang, H.S.C. National pesticide
    monitoring programs. In "Air Pollution from Pesticides and
    Agricultural Processes." R.E. Lee, Jr. (Ed.), CRC Press,
    Cleveland, 1976; pp. 95-136.
15. Risebrough, R.W.; Huggett, R.J; Griffin, J.J.; Goldberg,
    E.D. Pesticides: Transatlantic movements in the northeast
    trades. Science, 1968, 159, 1233.
16. Bidleman, T.F.; Olney, C.E. Chlorinated hydrocarbons in the
    Sargasso sea atmosphere and surface water. Science, 1974,
    183, 516.
17. Seba, D.B.; Prospero, J.M. Pesticides in the lower atmos-
    phere of the northern equatorial Atlantic Ocean. Atmos.
    Environ., 1971, 5, 1043.
18. Bidleman, T.F.; Olney, C.E. Long range transport of toxa-
    phene insecticide in the atmosphere of the Western North
    Altantic. Nature, 1975, 257, 475.
19. Woodwell, G.M.; Craig, P.P.: Johnson, H.A. DDT in the
    biosphere: Where does it go? Science, 1971, 174, 1101.
20. Tarrant, K.R.; Tatton, J.O'G. Organochlorine pesticides in
    rainwater in the British Isles. Nature, 1968, 219, 725.
21. Peterle, T.J. DDT in Antarctic snow. Nature, 1969, 224,
    620.
22. Cohen, J.M.; Pinkerton, C. Widespread translocation of
    pesticides by air transport and rain-out. In "Organic
    Pesticides in the Environment." R.F. Gould (Ed.), Adv.
    Chem. Ser. No. 60, ACS, Washington, D.C., 1966; pp. 163-176.

23. Moilanen, K.W.; Crosby, D.G.; Soderquist, C.J.; Wong, A.S.
    Dynamic aspects of pesticide photodecomposition. In
    "Environmental Dynamics of Pesticides." R. Haque and
    V.H. Freed (Eds.), Plenum Press, New York, 1975; pp. 45-60.
24. Crosby, D.G.; Moilanen, K.W.; Seiber, J.N.; Woodrow, J.E.
    Chemical reactions of pesticides in air. Presented at the
    American Chemical Society/Chemical Society of Japan Chemical
    Congress (PEST 053), Honolulu, April 1-6, 1979.
25. Woodrow, J.E.; Crosby, D.G.; Mast, T.; Moilanen, K.W.;
    Seiber, J.N. Rates of transformation of trifluralin and
    parathion vapors in air. J. Agr. Food Chem., 1978, 26,
    1312.
26. Crosby, D.G.; Moilanen, K.W. Vapor-phase photodecomposition
    of aldrin and dieldrin. Arch. Environ. Contamin. Toxicol.,
    1974, 2, 62.
27. Lewis, R.G.; Lee, Jr., R.E. Air pollution from pesticides:
    Sources, occurrence, and dispersion. In "Air Pollution
    From Pesticides and Agricultural Processes." R.E. Lee, Jr.
    (Ed.), CRC Press, Cleveland, 1976; pp. 5-50.
28. Seiber, J.N.; Woodrow, J.E.; Shafik, T.M.; Enos, H.F. De-
    termination of pesticides and their transformation products
    in air. In "Environmental Dynamics of Pesticides."
    R. Haque and V.H. Freed (Eds.), Plenum Press, New York,
    1975; pp. 17-43.
29. Seiber, J.N.; Woodrow, J.E. The determination of pesticide
    residues in air. Proceedings of International Conference
    on Environmental Sensing and Assessment. Las Vegas, Nevada.
    Institute of Electrical and Electronics Engineers, 1976,
    1, 7-2.
30. Van Dyk, L.P.; Visweswariah, K. Pesticides in air: Sam-
    pling methods. Residue Revs., 1975, 55, 91.
31. Lewis, R.G. Sampling and analysis of airborne pesticides.
    In "Air Pollution from Pesticides and Agricultural
    Processes." R.E. Lee, Jr. (Ed.), CRC Press, Cleveland,
    1976; pp. 51-94.
32. Akesson, N.B.; Yates, W.E.; Cowden, R.E. Procedures for
    evaluating the potential losses during and following
    pesticide application. Presented at the 1977 Winter
    Meeting of the American Society of Agricultural Engineers,
    Chicago, Ill. Dec. 13-16, 1977.
33. Grover, R.; Maybank, J.; Yoshida, K. Droplet and vapor
    drift from butyl ester and dimethylamine salt of 2,4-D.
    Weed Science, 1972, 20, 320.
34. Pitts, J.N., Jr.; Van Cauwenberghe, K.A.; Grosjean, D.;
    Schmid, J.P.; Fitz, D.R.; Belser, Jr., W.L.; Knudson, G.B.;
    Hynds, P.M. Atmospheric reactions of polycyclic aromatic
    hydrocarbons: Facile formation of mutagenic nitro deriva-
    tives. Science, 1978, 202, 515.
35. Furmidge, G.G.L. The application of flying-spot scanning
    to particle size analysis in the formulation of pesticides.
    Analyst, 1963, 88, 686.

36. Miles, J.W.; Fetzer, L.E.; Pearce, G.W. Collection and determination of trace quantities of pesticides in air. Environ. Sci. Technol., 1970, 4, 420.
37. Enos, H.F.; Thompson, J.F.; Mann, J.B.; Moseman, R.F. Determination of pesticide residues in air. Presented at the 163rd National Meeting of the American Chemical Society (PEST 022), Boston, Mass., April, 1972.
38. Thompson, J.F. (Ed.). Analysis of pesticide residues in human and environmental samples. Primate and Pesticides Effects Laboratory, U.S. Environmental Protection Agency, Perrine, Florida, 1972.
39. Sherma, J.; Shafik, T.M. A multiclass, multiresidue analytical method for determining pesticide residues in air. Arch. Environ. Contamin. Toxicol., 1975, 3, 55.
40. Batora, V.; Kovác, J. The determination of small quantities of some insecticidal dialkyl phosphorodithioates in factory air. Chem. Tech. (Berlin), 1964, 16, 230.
41. Yule, W.M.; Cole, A.F.W.; Hoffman, I. A survey for atmospheric contamination following forest spraying with fenitrothion. Bull. Environ. Contamin. Toxicol., 1971, 6, 289.
42. Caro, J.H.; Bierl, B.A.; Freeman, H.P.; Sonnet, P.E. A method for trapping disparlure from air and its determination by electron-capture gas chromatography. J. Agric. Food Chem., 1978, 26, 461.
43. Grover, R.; Kerr, L.A. Evaluation of silica gel and XAD-4 as adsorbents for herbicides in air. J. Environ. Sci. Health B, 1978, 13(3), 311.
44. Herzel, F.; Lahmann, E. Polyethylene-coated silica gel as a sorbent for organic pollutants in air. Z. Anal. Chem., 1973, 264, 304.
45. Beyermann, K.; Eckrich, W. Trennung des Insecticidgehaltes der Luft in den aerosol-gebundenen und den gasförmigen Anteil. Z. Anal. Chem., 1974, 269, 279.
46. Compton, B.; Bjorkland, J. Design of a high-volume sampler for airborne pesticide collection. Presented at the 163rd National Meeting of the American Chemical Society (PEST 021), Boston, Mass. April, 1972.
47. Soderquist, C.J.; Crosby, D.G.; Moilanen, K.W.; Seiber, J.N.; Woodrow, J.E. Occurrence of trifluralin and its photoproducts in air. J. Agri. Food Chem., 1975, 23, 304.
48. Harvey, G.R.; Steinhaver, W.G. Atmospheric transport of polychlorobiphenyls to the North Atlantic. Atmos. Environ., 1974, 8, 777.
49. Aue, W.A.; Teli, P.M. Sampling of air pollutants with support bonded chromatographic phases. J. Chromatogr., 1971, 62, 15.
50. Melcher, R.G.; Garner, W.L.; Severs, L.W.; Vaccaro, J.R. Collection of chlorpyrifos and other pesticides in air on chemically bonded sorbents. Anal. Chem., 1978, 50, 251.

51. Thomas, T.C.; Seiber, J.N. Chromosorb 102, an efficient medium for trapping pesticides from air. Bull. Environ. Contamin. Toxicol., 1974, 12, 17.

52. Mann, J.B., Enos, H.F.; Gonzalez, J.; Thompson, J.F. Development of sampling and analytical procedure for determining hexachlorobenzene and hexachloro-1,3-butadiene in air. Environ. Sci. Technol., 1974, 8, 584.

53. Caro, J.H.; Glotfelty, D.E.; Freeman, H.P. (Z)-9-Tetradecen-1-ol formate: Distribution and dissipation in the air within a corn crop after emission from a controlled-release formulation. Presented at the American Chemical Society/Chemical Society of Japan Chemical Congress (PEST 081), Honolulu, April 1-6, 1979.

54. Farwell, S.O.; Bowes, F.W.; Adams, D.F. Evaluation of XAD-2 as a collection sorbent for 2,4-D herbicides in air. J. Environ. Sci. Health (B), 1977, 12(1), 71.

55. Woodrow, J.E.; Seiber, J.N.; Crosby, D.G.; Moilanen, K.W.; Soderquist, C.J.; Mourer, C. Airborne and surface residues of parathion and its conversion products in a treated plum orchard environment. Arch. Environ. Contamin. Toxicol., 1977, 6, 175.

56. Seiber, J.N.; Madden, S.C.; McChesney, M.M.; Winterlin, W.L. Toxaphene dissipation from treated cotton field environments: Component residual behavior on leaves and in air, soil, and sediments determined by capillary gas chromatography. J. Agric. Food Chem., 1979, 27, 284.

57. Ferreira, G.A.L.; Seiber, J.N. Volatilization of three N-methylcarbamate insecticides from rice plants following root-soak systemic and foliage spray treatments. Presented at the American Chemical Society/Chemical Society of Japan Chemical Congress (PEST 067), Honolulu, April 1-6, 1979.

58. Hermann, B.; Woodrow, J.E.; Seiber, J.N. A comparison of XAD-4 resin with polyurethane foams for use in air sampling of pesticides. Presented at the 175th National Meeting of the American Chemical Society (PEST 015), Anaheim, March 13-17, 1978.

59. Woodrow, J.E.; Seiber, J.N. Portable device with XAD-4 resin trap for sampling airborne residues of some organophosphorus pesticides. Anal. Chem., 1978, 50, 1229.

60. Turner, B.C.; Glotfelty, D.E. Field air sampling of pesticide vapors with polyurethane foam. Anal. Chem., 1977, 49, 7.

61. Bidleman, T.F.; Olney, C.E. High-volume collection of atmospheric polychlorinated biphenyls. Bull. Environ. Contamin. Toxicol., 1974, 11, 442.

62. Lewis, R.G.; Brown, A.R.; Jackson, M.D. Evaluation of polyurethane foam for sampling of pesticides, polychlorinated biphenyls and polychlorinated naphthalenes in ambient air. Anal. Chem., 1977, 49, 1668.

63. Erickson, M.D.; Michael, L.C.; Zweidinger, R.A.; Pellizzari, E.D. Development of methods for sampling and analysis of polychlorinated naphthalenes in ambient air. Anal. Chem., 1978, 12, 927.

64. Lewis, R.G.; Macleod, K.E.; Jackson, M.D. Sampling methodologies for airborne pesticides and polychlorinated biphenyls. Presented at the American Chemical Society/ Chemical Society of Japan Chemical Congress (PEST 065), Honolulu, April 1-6, 1979.

65. Cliath, M.M.; Spencer, W.F.; Shoup, T.; Grover, R.; Farmer, W.J. Volatilization of EPTC from water during flood irrigation of an alfalfa field. Presented at the American Chemical Society/Chemical Society of Japan Chemical Congress (PEST 056), Honolulu, April 1-6, 1979.

66. Sydor, R.; Pietrzyk, D.J. Comparison of porous copolymers and related adsorbents for the stripping of low molecular weight compounds from a flowing air stream. Anal. Chem., 1978, 50, 1842.

67. Simon, C.G.; Bidleman, T.F. Sampling airborne poly-chlorinated biphenyls with polyurethane foam-- Chromatographic approach to determining retention efficiencies. Anal. Chem., 1979, 51, 1110.

68. Hermann, B.; Seiber, J.N. Analysis of atmospheric residues from applications of DEF[R] and Folex[R] cotton defoliants. Presented at the 177th National Meeting of the American Chemical Society (PEST 010), Washington, D.C. Sept. 9-14, 1979.

69. Seiber, J.N.; Woodrow, J.E. Unpublished results.

70. Willis, G.H.; Parr, J.F.; Papendick, R.I., Smith, S. A system for monitoring atmospheric concentrations of field-applied pesticides. Pest. Monit. J., 1969, 3, 172.

71. Willis, G.H.; Parr, J.F.; Smith, S. Volatilization of soil-applied DDT and DDD from flooded and nonflooded plots. Pest. Monit. J., 1971, 4, 204.

72. Spencer, W.F.; Cliath, M.M. Volatility of DDT and related compounds. J. Agr. Food Chem., 1972, 20, 645.

73. Cliath, M.M.; Spencer, W.F. Dissipation of pesticides from soil by volatilization of degradation products. Environ. Sci. Technol., 1972, 6, 910.

74. Caro, J.H.; Taylor, A.W.; Lemon, E.R. Measurement of pesticide concentrations in air overlying a treated field. In "Proceedings of International Symposium on Measurement of Environmental Pollutants." National Research Council of Canada, Ottawa, 1971; pp. 72-77.

75. Taylor, A.W.; Glotfelty, D.E.; Turner, B.C.; Silver, R.E.; Freeman, H.P.; Weiss, A. Volatilization of dieldrin and heptachlor residues from field vegetation. J. Agr. Food Chem., 1977, 25, 542.

76. Taylor, A.W.; Glotfelty; D.E.; Glass, B.L.; Freeman, H.P.;
     Edwards, W.M. Volatilization of dieldrin and heptachlor
     from a maize field. J. Agr. Food Chem., 1976, 24, 625.
77. Aquino, G.B.; Pathak, M.D. Enhanced absorption and persis-
     tence of carbofuran and chlordimeform in rice plants on
     root zone application. J. Econ. Entomol., 1976, 69, 686.
78. Spear, R.C.; Lee, Y.S.; Leffingwell, J.T.; Jenkins, D. Con-
     version of parathion to paraoxon in foliar residues:
     Effects of dust level and ozone concentration. J. Agr.
     Food Chem., 1978, 26, 434.
79. Harper, L.A.; White, Jr., A.W.; Bruce, R.R.; Thomas, A.W.;
     Leonard, R.A. Soil and microclimate effects on trifluralin
     volatilization. J. Environ. Qual., 1976, 5, 236.
80. Woodrow, J.E.; Mast, T.; Seiber, J.N.; Crosby, D.G.;
     Moilanen, K.W. Rates of photochemical conversion of pesti-
     cides in the atmosphere. Presented at the 173rd National
     Meeting of the American Chemical Society (PEST 003),
     New Orleans, LA., March 21, 1977.
81. Kosolapoff, G.A. Organophosphorus chemistry. Wiley, London,
     1950; p. 198.
82. Teasley, J.I. Identification of cholinesterase-inhibiting
     compounds from an industrial effluent. Environ. Sci.
     Technol., 1967, 1, 411.

RECEIVED December 28, 1979.

New Technology for Pesticide Residue Cleanup Procedures

M. E. GETZ and K. R. HILL

Analytical Chemistry Laboratory, Agricultural Environmental Quality Institute,
Agricultural Research, Science and Education Administration, U.S. Department of
Agriculture, Beltsville, MD 20705

Pesticide residue analysis are conducted for three main
purposes: [1] Agriculturists and entomologists are interested in
how much residual pesticide is needed to control a pest problem.
[2] Regulatory agencies and monitoring stations screen for
environmental residues and enforce tolerance regulations.   [3]
Academic researchers and toxicologists are concerned with metabolites and degradated products.

Widely used techniques are: GLC, HPLC, and TLC for general
identification and quantitation, with mass spectrometry and infrared spectroscopy for specific identification.

Before any of these analyses can be performed, the residue
has to be extracted from the containing matrix and isolated in
a pure enough state for the particular analytical method. This
isolation procedure is called cleanup. As this paper deals
primarily with cleanup, we shall assume that the solvent or solvents used for extraction are in general use and that they
quantitatively extract the residues.

Residues and extractives from different environmental
matrices exhibit similar physical and chemical properties and
can be classified as lipophilic, hydrophilic, or amphiphilic
(exhibiting a dual nature).

The Food and Drug Administration [FDA] Pesticide Analytical
Methods Manual (1) and the Environmental Protection Agency [EPA]
Manual for Environmental Analysis (2) describe procedures that
have been used for many years. Two of the commonly applied
techniques are liquid-liquid partitioning and column adsorption
chromatography. These approaches are used to isolate lipohilic
and moderately polar residues for primary identification and
quantitation with GLC. An evaluation of the number of pesticide
residues that were satisfactorily analyzed by this approach was
published by McMahon and Burke (3). When one looks at the data
it can be seen that the highly polar and water soluble residues
do not fit into the analytical scheme very well. In an attempt
to rectify this problem, FDA is modifying the multiresidue method

for each individual compound that does not adapt to the general
procedure. These extra methods have become quite numerous.

Our laboratory is developing two approaches for multiresidue
determinations in an attempt to make the analyses encompass a
larger variety of residues.

One is quantitative TLC by reflective scanning (4), and the
other is an automated continuous flow system that utilizes
liquid-solid adsorption chromatography coupled with selective
detectors (5) (6).

Microcolumn Cleanup

This approach utilizes microcolumns of adsorbents to cleanup
samples of 10g or less. Various length glass columns of 0.5 to
1.0 cm ID can be used. They should have luer tips and can be
obtained with caps to allow low pressure to be applied or solvent
to be pumped through. Figure 1.

Hypodermic needles were used for keeping the flow rate uniform. The columns were fitted with 1 1/2" 25 guage needles when
the eluant was collected for concentration, and 1 1/2" 29 guage
needles when the eluant was spotted directly onto the thin layer
plates.

The adsorbents evaluated were: alumina [Woelm, neutral,
105°C for 1 hr.]; Florisil^R [130°C activation]; and charcoal
[Darco G-60, 12 x 20, or Nuchar +30].

When direct spotting onto a TLC plate was used, the columns
were packed with 0.5g inorganic adsorbent or 0.25g charcoal
between glass wool plugs. The adsorbent was compacted by gentle
suction, and 4 ml of eluting solvent was used. A sample size
no greater than 2g should be used.

If a larger size sample is to be cleaned-up for collection,
the sample weight is used as a factor for determining the amount
of adsorbent necessary, i.e., 10g sample, (0.5 x 10) = 5g adsorbent. The eluting volume is determined by adding 0.5 to the
weight factor, i.e., 10g sample (4 x 10.5) = 42 ml eluting solvent necessary for quantitative recovery.

When direct spotting onto a thin layer plate is used, an
automatic spotter is necessary (7). It utilizes six tubes which
transfer the eluting solvent in a dropwise manner to the origin.
The size of the spot is controlled by the flow of an inert gas
from a manifold. Three standards and three unknowns are spotted
onto a 20 x 20 cm plate.

When the sample is transferred to the cleanup column, it
should be dissolved in the same solvent used for elution or one
of lesser polarity. The volume should be no greater than 1 ml.
If necessary larger size samples can be transferred in volumes
up to 2 ml. The sample is pipetted onto the glass wood plug in
a dropwise manner and allowed to soak in. The sides are then
rinsed down with 0.5 ml of eluting solvent and allowed to sink in.
The column is then eluted with the determined volume of solvent.

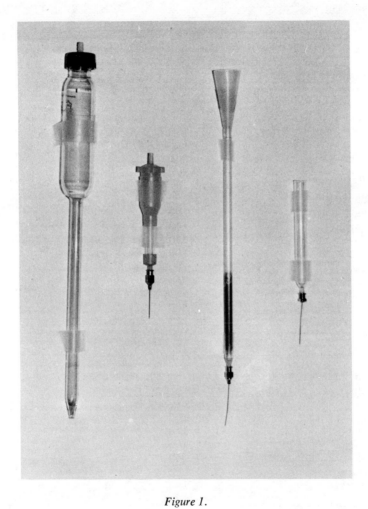

Figure 1.

When the eluants are collected, they are concentrated and can be analyzed by GLC, HPLC or TLC.

For quantitation of the TLC scan, the areas of the curves of the standards are plotted versus concentration and then the unknown concentrations are determined from this standard curve.

TLC Results

Organothiophosphates recovered from 0.5g alumina or 0.5g Florisil[R] when eluted with 4 ml 20% ethyl acetate in 2,2,4-trimethylpentane.

Pesticide	Sensitivity ng/g	Detection
Azinphosmethyl	100	TCQ
Azinphosmethyl 0-analog	100	"
Azinphosethyl	100	"
Azinphosethyl 0-analog	100	"
Imidan	100	"
Malathion	100	"
Malaoxon	200	"
Phorate	100	"
Phorate sulfone	200	"
Phorate thiol	100	"
Disulfoton	100	"
Disulfoton sulfone	200	"
Disulfoton thiol	100	"
Dicapthon	200	"
Zinophos	200	"
Coumaphos	200	"
Methyl parathion	200	"
Ethyl parathion	200	"
Zytron	100	"
Diazinon	200	"

Carbamates recovered from 0.5g alumina when eluted with 4 ml 20% ethyl acetate in 2,2,4-trimethylpentane.

Pesticide	Sensitivity ng/g	Detection
Carbaryl	100	TCQ
Aldicarb	100	"
Carbofuran	100	"
3-Hydroxy carbofuran	100	"
3-Keto carbofuran	200	"
Mesurol	100	"
Landrin	100	"

Organophosphates quantitatively eluted from 0.25g charcoal with 4 ml acetone.

Pesticide	Sensitivity ng/g	Detection
Dimethoate	100	TCQ
Malathion	100	"
Malaoxon	200	"

Phorate	100	TCQ
Phorate sulfoxide	200	"
Phorate sulfone	200	"
Phorate thiol	100	"
Phorate thiol sulfoxide	200	"
Phorate thiol sulfone	200	"
Disulfoton	100	"
Disulfoton sulfoxide	200	"
Disulfoton sulfone	200	"
Disulfoton thiol	100	"
Disulfoton thiol sulfoxide	200	"
Disulfoton thiol sulfone	200	"
Diazinon	200	"
Crufomate	100	NBP
Dicrotophos	100	"
Dichlorvos	100	"
Mevinphos	100	"
Dimethoate O-analog	200	TCQ
Trichlorfon	100	NBP
Bomyl	100	"
Crotoxyphos	100	"
Naled	100	"
Phosphamidon	100	"

The organophosphates containing sulfur were detected with the TCQ reagent under acid conditions (8).

The organophosphates without sulfur were detected with nitrobenzylpyridine (9).

The carbamates were detected with TCQ under alkaline conditions.

Recoveries ranged from 75 to 105%. Each analyst should determine the optimum elution volumes for quantitative recovery under their laboratory conditions. All new adsorbents or batches should be checked for recovery.

Continuous Flow Cleanup

A first attempt at this approach was made by adapting the gradient elution system of Bowman and Beroza (10) to a continuous flow apparatus. Hill and Jones (5) modified a Pye traveling wire detector by substituting a flame photometric detector for the conventional flame ionization one so that it would be selective for sulfur and phosphorus containing pesticides. This detector was used for monitoring the cleanup efficiency and resolving powers of different adsorbents. Figure 2 is a schematic diagram of the columns and apparatus used for this initial study.

A glass column 6.3 mm ID x 24 cm was packed with Corasil II glass beads. Standard solutions of malathion in concentration ranges of 0.1 to 1.0 μg/ml were eleuted from the column with

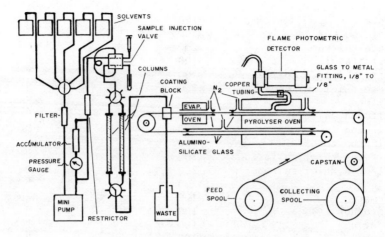

Figure 2.

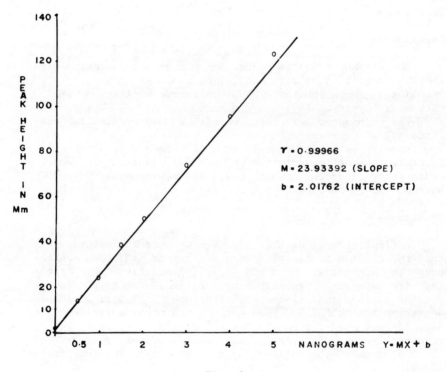

Ƴ = 0·99966

M = 23.93392 (SLOPE)

b = 2.01762 (INTERCEPT)

Figure 3.

methylene chloride + 2-propanol (80 + 20) with a flow rate of
4 ml/min. Figure 3 shows the linear response obtained.

One hundred grams of store bought string beans extracted
with methylene chloride, were concentrated to 10 ml, and 0.5 ml
equivalent to 5g of beans was placed on the glass bead column
and eluted with methylene-chloride+2-propanol solution. This
extract showed a peak corresponding to 1.5 ng of malathion.
When the same extract was fortified with a 5 ng equivalent of
malathion the resulting response was 6.5 ng malathion. Figure 4
shows the results of these comparisons.

A second column was packed with 5g of silica gel (Baker
3405). A standard solution of phorate and its five metabolites
was placed on the column and sequentially eluted with benzene,
2.5% acetone in benzene, 10% acetone, and 100% acetone. Figure
5 shows the results obtained.

Two hundred grams of Beltsville sandy loam were fortified
with the insecticide fensulfothion and its three metabolites.
The insecticides were extracted and 0.5 ml of extract equivalent
to 10g of sample was placed on the column and eluted with the
above stepwise gradient. The results are represented by Figure
6.

Two hundred grams of alfalfa from a field treated with
oxamyl were extracted and decolorized with charcoal. A 10g
aliquot was placed on the column and eluted with the acetone-
benzene gradient. The results are shown in Figure 7.

Application of Macroporous Silica Gels

The Merck Company produces a series of silica gels called
Fractosil which come in pore sizes from 200 to 25,000 nm. Conventional silica gel used for adsorption columns usually have
pore sizes of 40 to 80 nm. These Fractosil gels can be used
for gel permeation, adsorption and partition. The electron
microscope scan of Fractosil 200 compared with conventional
silica gel appears to be in the shape of uniform breakfast food
flakes, Figure 8. The adsorbent is very rigid and can be packed
dry. Although the surface area is not very great, the gel
appears to separate molecules according to polarity (12).

Hexane with increasing strengths of acetone can separate
many pesticides of varying polarities. Hundreds of food extractives were run through one column before its separation
effectiveness was impaired. Fractosil 200 is completely recyclable. It was regenerated by boiling with 10% HCl, washing
and drying at 110°C.

An adjustable glass column of 9mm ID, packed with 5g of
Fractosil 200 was connected to a Spectra Physics pump connected
to solvent reservoirs with a six way rotary switch. The sample
was injected onto the column with a rotary loop sampler. The
effluent from the column was collected with an Isco fraction
collector set so that 10 ml of eluate was collected for each

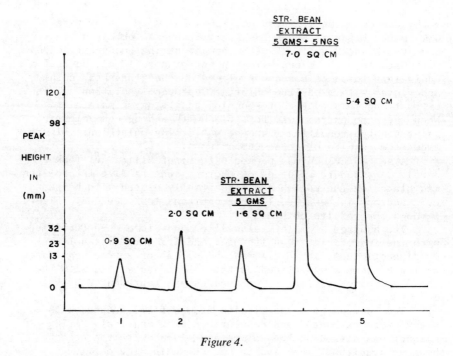

Figure 4.

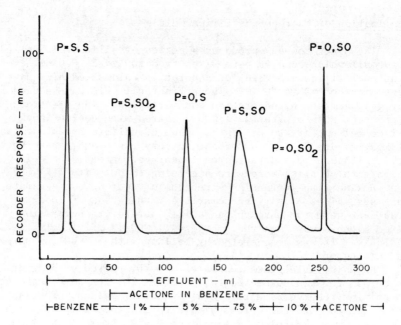

Figure 5.

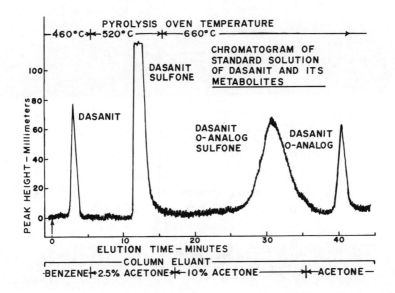

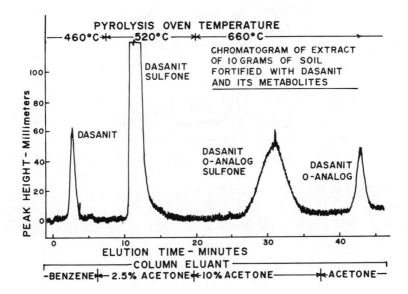

Figure 6.

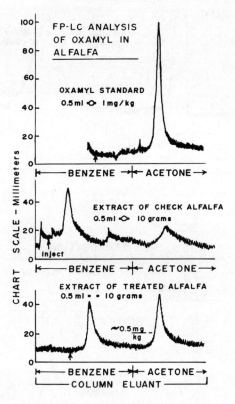

Figure 7.

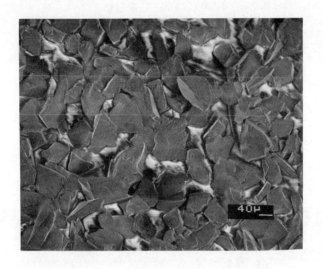

Figure 8.

2.5 minutes. Each of the six solvent reservoirs contained 100%
hexane, 10% acetone in hexane, 20% acetone in hexane, 40% acetone in hexane, 100% acetone and 100% methanol.

When the sample was injected onto the column, the pump was
started and the first 10 ml eluate collected. The pump was then
stopped and the solvent valve turned to the next gradient. This
procedure was repeated until the 100% acetone eluate was collected. After a second 10 ml eluate of acetone was taken, the
column was ready for recycling. The column was recycled by
pumping through successive volumes of 25 ml methanol, 50 ml
acetone, and 25 ml hexane.

Fortification Studies

The following matrices were fortified for the evaluation
studies of Fractosil 200: tomatoes, carrots, green apples,
animal fat, milk and greenhouse soil.

Endosulfans A, B and the sulfate were quantitatively recovered at concentration levels of 0.2 and 2.0 ppm. Aliquots
representing 2g of the original sample were injected onto columns
for clean-up. The cleaned-up eluates were analyzed by GLC and
electron capture detection to see if any interferences or
artifacts would show up under the conditions necessary to
measure 0.2 ppm of endosulfan and its metabolite. Colored
eluates were observed in the 10% fractions from tomatoes, carrots, green apples, and green house soil. Another pigment,
seen as a tight band on top of the column, was not eluted until
methanol passed through during the first stage of recycling. No
color in any fraction was observed from fat or milk samples.
However, when the eluates were evaporated down to dryness a
volume of oily material appeared in the bottom of the tubes
containing the 10% eluate. When GLC and electron capture
detection were used with the PAM GC columns (1) for organo-
chlorines, except for the 10% fraction from the animal fat,
no artifacts or base line effects were noted.

Figure 9 illustrates a typical chromatogram obtained from
a 10 µl injection of a 10 ml aliquot concentrated to 1 ml.
Figure 10 shows the rising base line and peak obtained from the
10% acetone eluate of the animal fat extract. All samples of
beef fat exhibited this peak.

The same matrices as described above were fortified with
phorate and its metabolites, and thiol demeton and its metabolites.

% pesticide in each eluate.

Pesticide	Hexane	10% Acetone	20% Acetone	40% Acetone	Acetone	Acetone
Endosulfan A	40%	60%				
Endosulfan B			96%			
Endosulfan SO4				92%		

% pesticide in each eluate. (Continued)					
		10%	20%	40%	
Pesticide	Hexane	Acetone	Acetone	Acetone	Acetone	Acetone
Phorate	80%	20%				
Phoratoxon		90%				
Phoratoxon SO					10%	90%
Phoratoxon SO$_2$					10%	90%
Demeton thiol		90%				
Demeton thiol SO					10%	90%
Demeton thiol SO2					10%	90%

A gel permeation column used in conjunction with an Autoprep 1001 was removed and replaced with a 1 cm ID adjustable glass column packed with a 15 cm length of Fractosil 200. A system for providing step gradient solvent programming was installed.

Step gradient for testing Fractosil 200 in conjunction with the Autoprep 1001.

Solvent	Volume, ml
Methanol	25
Acetone	50
Hexane	50
5% acetone/hexane	10
10% acetone/hexane	10
20% acetone/hexane	10
Acetone	20
Methanol	Rapid cycle

The use of the Fractosil 200 produced a flat almost noise free baseline with the flame photometric detector.

System testing was begun by establishing response and retention times for adjusting the solvent Program. This was accomplished by injecting samples of phorate sulfoxide and chlorpyrifos. Chromatograms with narrow well shaped peaks as shown in Figure 11 were obtained.

Two different solvent programs were used:

	Program A		Program B	
Solvent	Time (min.)	Solvent	Time (min.)
Hexane	3	Hexane	3
0.2% acetone/hexane	2	10% acetone	2
0.3% acetone/hexane	2	20% acetone	2
0.6% acetone/hexane	2	50% acetone	2
0.9% acetone/hexane	2	75% acetone	2
1.0% acetone/hexane	2	100% acetone	20
100% acetone	20	Methanol	8
Methanol	6	Acetone	16
Acetone	12	Hexane	10
Hexane	6		

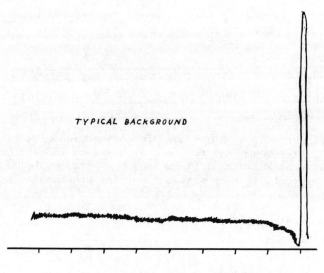

Figure 9.

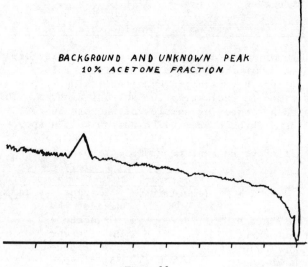

Figure 10.

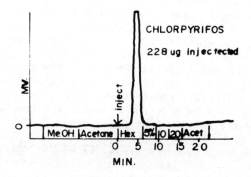

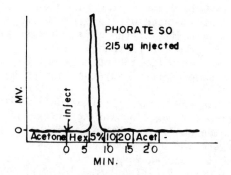

Figure 11.

Data Acquisition and Analysis System

At the initiation of this study, a computer based data system designed to plot data points versus time and determine peak areas was made operational.

The following chromatograms were obtained with the Fractosil 200 silica gel and the Pye moving wire detector as shown in Figure 2. They were plotted by the computer. The axes are the response in millivolts versus time in seconds. When the computer reconstructed the chromatogram, the Y-axis could be magnified from 0 to 520 millivolts.

Chromatogram 1 shows an extract representing 5.6g of Beltsville sandy loam soil, eluted with program B and detected by the sulfur mode. The elution flow rate was 2 ml/min.

Chromatogram 2 represents 6.0g of the same soil with similar elution and flow rate. Detection was by the phosphorus mode.

Chromatogram 3 is the response of 1.0g of soil fortified with 65.4 μg aldicarb. Small responses can be seen at 2 and 3.

Chromatogram 4 is a repeat of 3 but the computer has magnified the Y-axis to show that peaks 2 and 3 are real and not due to noise.

Chromatogram 5 is one where 0.56g of fat extracted with ethyl acetate has been injected onto the column and eluted with program B at 2.0 ml/min. with phosphorus detection.

Chromatogram 6 is 0.30g of fat that was fortified with 80.6 μg aldicarb sulfoxide eluted with program B and detected by the sulfur mode. The polarity of the aldicarb sulfoxide has allowed it to be separated from interferences.

Chromatogram 7 shows what happens when the oven tube gets dirty. This can happen after many fat samples have been analyzed.

Experience has shown that the Fractosil 200 silica gel is not very effective for removing fat interferences when lipophilic compounds are to be analyzed. However, it's very efficient when polar compounds are to be isolated.

A new synthetic carbanaceous resin (Rohm and Haas) (12) has been found to be very effective for holding up non polar interferences from fats and oils while allowing many lipophilic pesticides to be quantitatively eluted. This was adapted to automated elution and was found out to also be recyclable. Acetonitrile is used for eluting the pesticides and chloroform is used for recycling the column by displacing the adsorbed interferences.

The preliminary data has shown that the two new adsorbents are a major breakthrough for automating the cleanup step, and allowing a wider spectra of pesticides to be analyzed.

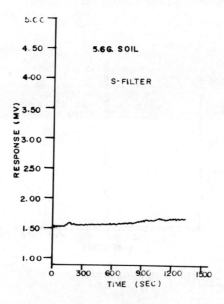

Chromatogram 1.

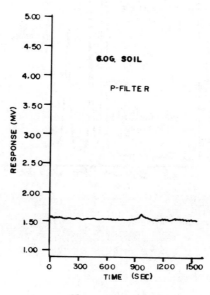

Chromatogram 2.

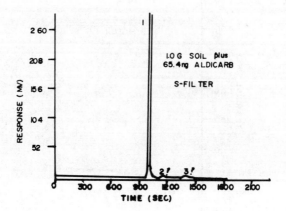

Chromatogram 3.

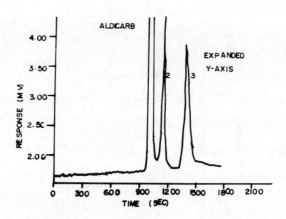

Chromatogram 4.

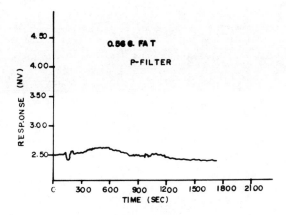

Chromatogram 5.

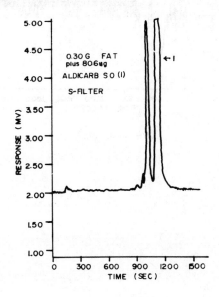

Chromatogram 6.

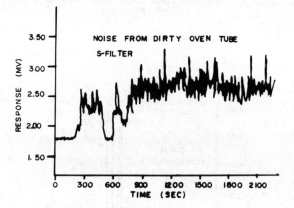

Chromatogram 7.

Acknowledgments

    We wish to thank Glenn Hanes of this laboratory for his data
collecting and development of a useful computer program.

References

(1) Pesticide Analytical Manual. Vols. I, II, and III. U.S.
Department of Health Education and Welfare. Food and Drug
Administration. Washington, D.C.

(2) Manual for Analytical Quality Control for Pesticides and
Related Compounds in Human and Environmental Samples. U.S.
Environmental Protection Agency. Health Effects Research
Laboratory. Research Triangle Park, North Carolina.

(3) McMahon, B., and Burke, J. A., J. Assoc. Offic. Anal. Chem.
61 640 (1978).

(4) Getz, M. E. Methods in Residue Analysis. Vol. IV, pg. 43.
Ed. A. S. Tahori., Gordon and Breach, New York, N. Y. (1972).

(5) Hill, K. R. and Jones, W. M. Adaptation of the Flame
Photometric Detector to Liquid Chromatography of Pesticides.
3rd International Conference of Pesticide Chemistry, Helsinki,
Finland, July 3-9, 1974.

(6) Hill, K. R., J. Chromatogr. Sci. 17, 395 (1979).

(7) Getz, M. E., J. Assoc. Offic. Anal. Chem. 54, 982 (1971).

(8) Beroza, M., Hill, K. R., and Norris, K. H., Anal. Chem. 40,
1608 (1968).

(9) Getz, M. E. and Wheeler, H. G., J. Assoc. Offic. Anal. Chem.
51, 1101 (1968).

(10) Bowman, M. C. and Beroza, M. J., Agr. Food. Chem. 16, 399
(1968).

(11) Getz, M. E., Talanta 22, 395 (1975).

(12) Getz, M. E., Hanes, G. W., and Hill, K. R. Trace Organic
Analysis, New Frontier in Analytical Chemistry, pgs. 345-53.
NBS Special Publication 519. Eds. Hertz, H. S. and Chesler, S.
N. (1979).

    Chromatograms 1-7 were obtained as part of an interagency
agreement (EPA-IAG-D6-0741) with the Chemistry Branch, Pesticide
and Toxics Substances Effects Laboratory, Health Effects Laboratory, Office of Research and Development, Environmental Protection Agency, Research Triangle Park, North Carolina. The views
expressed herein are those of the investigators and do not
necessarily reflect the official viewpoint of the Department
of Agriculture or the Environmental Protection Agency.

RECEIVED March 6, 1980.

Chemical Derivatization Techniques in Pesticide Analysis

Advances and Applications

W. P. COCHRANE

Laboratory Services Division, Agriculture Canada, Ottawa, Ontario, Canada K1A 0C5

In common with other analysts, the pesticide analyst has two
major problems facing him during the analysis of a sample, namely,
1. The IDENTITY of the pesticide present
2. The AMOUNT of pesticide or its residue in the sample.
In both these areas, chemical derivatisation has traditionally played a role and with the advent of gas chromatography an
even more important role. The reasons for preparing a derivative
suitable for GC analysis are many and varied and have been dis-
cussed thoroughly in a number of books and reviews ($\underline{1}$-$\underline{6}$). For
convenience they are summarised in Table I. As can be seen, two
different types of chemical derivatisation techniques are mentioned
under Item 4 of Criteria, There is the chemical derivatisation of
a pesticide as a pre-requisite of the method of analysis, e.g.
esterification of the chlorophenoxy acids, as well as derivatisation as a method for confirmation of identity. The former must
meet all the requirements associated with a practical, viable
analytical procedure while for the latter the emphasis is on speed,
ease of operation and reproducibility.

Apart from the above reasons derivatisation is also required
in the analyses of many of the newer pesticides which have one or
more functional groups which need protection in order to facilitate GC work. Also the increased use of high pressure liquid chromatography (HPLC) necessitates pre- or post-column derivatisation
since the choice of HPLC detectors is limited compared with GC.
The increased use of derivatisation reactions is evident from the
gradual increase in the number of publications dealing with the
subject. Figure 1 shows the yearly variation in numbers of publications dealing with derivatisation in pesticide analysis over
the period 1963 to 1978. While the interest in derivatisation
techniques in organophosphorus insecticide analysis has remained
fairly constant and a low level of activity, the OC insecticides
underwent an increased period of attention from 1968-1972 which
has stablized over the last few years. It is in the insecticidal
carbamate and herbicide areas that an overall steady increase in
the use of derivatisation reactions for quantitative and con-

Table I.   Advantages Gained by Chemical Derivatisation and Some
           Reaction Criteria

A. Advantages:

   1. Improved extractability during the clean-up procedures.
   2. Change in volatility characteristics.
   3. Increase in thermal stability of compounds.
   4. Improve chromatographic or separation behaviour.
   5. Increase in sensitivity of detection.
   6. Impart selectivity.
   7. Aid in confirmation of identity.

B. Criteria:

   1. Derivative must be formed rapidly.
   2. Preparation should require minimum of manipulation.
   3. Derivative should be relatively stable.
   4. Reaction must be quantitative and/or reproducible.
   5. Short (acceptable) GC retention time free from back-
      ground or reagent interferences.
   6. Good sensitivity.

---

firmatory analysis has occurred. This has been augmented in recent years by interest in the analysis of fungicides, notably
ethylenethiourea and the ethylenebisdithiocarbamates, as
evidenced by the trend of the overall total curve. Since a number
of reviews (4,5,7,8) cover the literature prior to 1975, this review will highlight the more recent advances in derivatisation
procedures or techniques as illustrated with applications from the
4 main groups of pesticides, namely, herbicides, OCs, OPs and fungicides.

Herbicides

     Esterification of the chlorophenoxy acids to their respective
methyl esters by reagents such as diazomethane, BF$_3$/methanol or
acid methanol mixtures have been for long the standard procedure
for the analyses of these compounds. Recently a number of workers
have been investigating the use of new and more EC sensitive
esterification reagents for general use. This has come about
since the EC detection of the methyl esters of MCPA or MCPB, which
contain only one chlorine each, are very poor. Also the methyl
ester of MCPA has a very short retention close to the solvent
front when analysed in conjunction with 2,4-D, 2,4,5-T, fenoprop,
etc. The acids and the various esters investigated are shown in
Table II. In transesterification work by Yip (9) the 2-chloro-
ethyl ester was discarded since the background pattern of the
derivative gave too many peaks. However, Woodham et al (10) used

Table II: Derivatives of the Chlorophenoxyalkyl Acids and
           Some Other Acidic Herbicides

Compound	2-chloro ethyl	2,2,2-trifluoro ethyl	2,2,2-tri- chloro ethyl	PFB[1/]
2,4-D	X	X	X	X
2,4-DB	X			X
2,4,-DP	X			X
2,4,5-T	X			X
2,4,5-TB				X
2,4,5-TP	X			X
MCPA	X	X	X	X
MCPB	X			X
MCPP				
2,3,6-TBA				X
Picloram	X			X
Dicamba				X
Fenac	X			
References	9 - 14	9 - 14	14	13, 15

1/Pentafluorobenzyl

a BCl$_3$/2-chloroethanol reagent to determine residues of 2,4-D in
soil, sediment and water samples while Gutenman and Lisk (11) used
BF$_3$/2-chloroethanol for esterification of MCPA in soil extracts.
Woodham et al (10) found that although the 2-chloroethyl ester had
the desired sensitivity, the reaction produced less interference
and gave a longer retention time, it was still subject to interferences from certain soil types (Fig. 2A). After diethyl ether
extraction a clean-up step was added employing an alkaline wash to
remove interfering substances. The lower limit of sensitivity for
this method in soils is approximately 0.01 ppm for 2,4-D (Fig. 2B).
More recently (12, 13) BCl$_3$: 2-chloroethanol was used to determine
chlorophenoxy acid residues in natural waters. It was found, as
would be expected, that this reagent produced little or no product
for 2,3,6-TBA and dicamba due to the steric hinderance of σ-Cl
atoms. The practical limits of detection ranged from a lower limit
0.01 ug/L for fenoprop to a high of 2-5 ug/L for MCPB. Yip (9)
showed that trifluoroethanol increased the ECD response of
2,4,5-T and fenoprop but not MCPA or 2,4-D. More recently,
Mierzwa and Witek (14) found that after esterification of 2,4-D
and MCPA with 20% 2,2,2-trichloroethanol in trifluoroacetic anhydride in the presence of H$_2$SO$_4$ a lower limit of 0.096 ppb
2,4-D and 0.06 ppb MCPA in 1 litre samples of water was obtained.
As in most of the derivatisation reactions, removal of excess

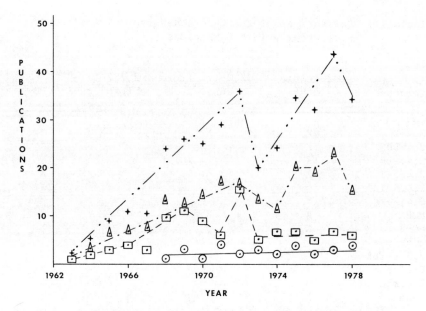

Figure 1. Variation in number of publications dealing with chemical derivatization techniques for pesticides from 1962–1978 ((+) total (including fungicides); (△) herbicides; (□) OC's (including Mirex, PCB's, etc.); (⊙) DP's)

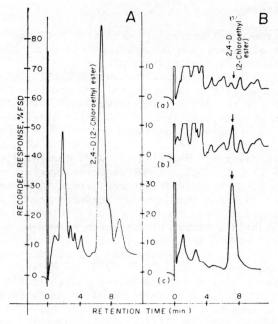

Journal of Agricultural and Food Chemistry

Figure 2. Chromatographic tracing of (A) a 2,4-D fortified soil sample (2-chloroethyl ester) without alkali treatment and (B) samples with alkali pre-wash: (a) a blank soil sample; (b) soil sample fortified with 0.013 ppm 2,4-D acid; and (c) soil sample fortified with 0.667 ppm 2,4-D acid (10)

reagents is mandatory prior to ECD analysis. It was found that
the sensitivity of the esters increased from the 2-chloroethyl to
trifluoroethyl to the trichloroethyl derivative. With MCPA the
trichloroethyl ester was 100 times more sensitive than the 2-chloro-
ethyl ester, while with 2,4-D the increase was only a factor of
10. Similarly, Chau and co-workers (15) have shown that the PFB
esters are more EC sensitive than the corresponding 2-chloroethyl
esters. For example, the PFB ester of MCPA is 2.5 times greater
than that obtained with the 2-chloroethyl ester and 1000 times
greater than MCPA methyl ester. Normally, the esterification reaction with pentafluorobenzyl bromide (PFBBr) is much simpler to
perform since it takes place in acetone which is easily evaporated
after the reaction. However, the PFBs tend to have longer retention times and this reagent reacts under basic conditions with
both phenols and carboxylic acids. It has been reported that an
OV-101/OV-210 column separated 9 PFB esters, only the MCPA and
2,4-DP over-lapped,while a 5% DC-200 or 3% DC-11 column gave complete separation of 10 PFB esters. Overall, it would appear that
the sensitivities of the PFB and trichloroethyl esters are
equivalent, with the final choice depending upon background interference from the sample and suitable column separation characteristics.

Organochlorine Insecticides and Related Compounds

Since well-established chemical derivatisation techniques already exist for the majority of the organochlorine insecticides
(5, 7) it is not surprising that recent activity in this area has
centred round compounds such as Kepone, Mirex, HCB, the PCBs, etc.
which co-interfere in both the identification and quantitation of
pesticide residues.
Kepone (chlordecone) can be converted back to Mirex by perchlorination (Fig. 3) using a 4:1 ratio of phosphorus pentachloride
to aluminium chloride in carbon tetrachloride at $145°C$ for 3 hrs.
in a closed tube (16). To eliminate any Mirex that may have been
originally present, the separation of Kepone was performed on a
micro-Florisil column prior to derivatisation. Similarly, a micro-
Florisil column was utilized after reaction to remove early
eluting peaks from the gas chromatograms, and especially when the
total amount of Kepone present was less than 25 ng. This derivatisation technique was found sensitive to ppb levels of Kepone in
environmental and biological samples. For example, Figure 4
shows an oyster extract before and after derivatisation (16). The
Kepone level averaged 0.07 ppm (upper chromatogram) in this
particular sample which was run on a 4% SE-30/6% OV-210 column.
The reaction was found to be quantitative and 8 of the more common
OC pesticides were found to disappear on reaction or give derivatives with GC retention different than Mirex.
Mirex has been associated with the control of the imported
fire ant in the south-eastern USA, where it was subsequently

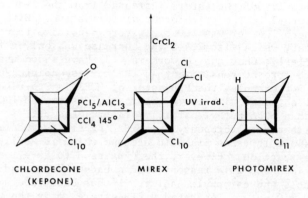

Figure 3.   Reactions of Kepone and Mirex

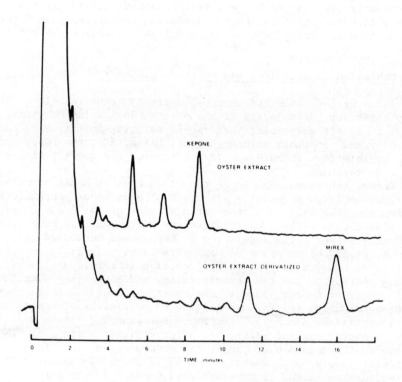

Journal of Agricultural and Food Chemistry

Figure 4.   Electron-capture gas chromatograms of an oyster extract before and
after derivatization $((PCl_5/AlCl_3),\ 0.66$ mg injected (16))

found in non-target organisms including man (17). Similarly,
Mirex and/or photomirex (8-monohydromirex) have been identified
as major contaminants in fish from Lake Ontario (18), cormorant
eggs (19) and Canadian Human milk (20). Mirex and photomirex are
particularly difficult to determine in the presence of large
quantitites of PCB without prior separation since their retention
times are similar to that of major heptachlorobiphenyls on most
GC columns. Lane et al (21) studied the photolytic and γ-ir-
radiation of duck egg tissues containing naturally occuring Mirex.
Although 7 derivatives were obtained on photolysis and 8 from
γ-irradiation the major product from both reactions was identified
as photomirex (Fig. 3). In this instance UV photolysis was carried
out at 254 nm on the actual egg homogenate and caused a 36% decrease after 48 hrs. whereas γ-irradiation resulted in a 64% loss
of Mirex. Irradiation, therefore, could be used as a confirmatory test so long as the resulting chromatograms do not become
too complex for interpretation. As postulated in a previous review (5), Mirex can also be confirmed by reduction of the gem-di-
chloro methylene group with $CrCl_2$ (22). In practise, three major
peaks are obtained, the ratios of which were found to vary with
the $CrCl_2$/acetone ratio and temperature of reaction. None of
these products appear to be photomirex. Reaction was carried out
overnight at 55-60°C or room temperature for 32 hrs. or longer.
While photomirex could be completely reacted to give a major product with a different retention time than Mirex, Kepone only reacted partially. Lewis et al (23) employed a different approach
in that Mirex-containing human tissue extracts were subjected to
a diethylamine assisted photolysis at wavelength $>$ 280 nm to
selectively eliminate PCB interferences in PCB/Mirex mixtures. A
100 min. photolysis of an Arochlor 1260/Mirex mixture in 10 ml
hexane containing diethylamine resulted in complete elimination
of the interfering heptachlorobiphenyl component and only a 0-5%
loss of Mirex. However, some interferences appear to remain in
the photomirex region of the chromatograms. Also it has been
pointed out by Mes and co-workers (20) that although Mirex can be
identified by GC, using diethylamine assisted photolysis to limit
PCB interference, quantitation at low levels (0.01-0.1 ppm) remain questionable since the workers found poor agreement between
their GC and mass spectrometry data. Other indirect approaches
to the PCB/Mirex problem have been the perchlorination of the
PCBs to decachlorobiphenyl with $SbCl_5$-containing reagents (24)
and nitration of the PCBs with a 1:1 mixture of conc. H_2SO_4/
fuming HNO_3 (25). Shown in Figure 5 is a chromatogram of an extract of a herring gull egg from Lake Ontario after clean-up on
a 1% deactivated Florosil column; note that Mirex is only
partially resolved from a heptachlorobiphenyl peak. The lower
tracing is the same extract after a 30 min. nitration at 70°C.
Mirex and photomirex are the only major constituents to survive
nitration. HCB, t-nonachlor and small amounts of mono- and di-
hydromirex compounds constitute most of the minor peaks remaining

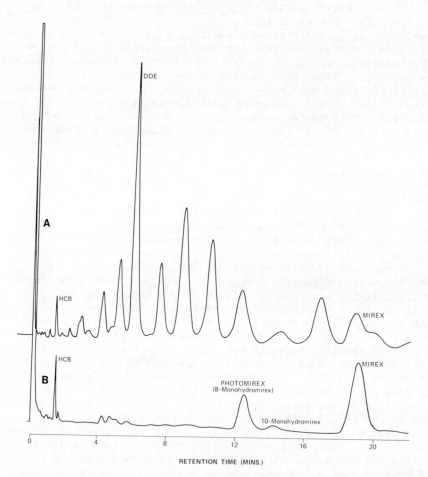

RETENTION TIME (MINS.)

Figure 5. (A) Extract of a herring gull egg from Lake Ontario after cleanup on a Florisil column; (B) same extract after nitration (represents about twice as much sample as A). Conditions: 6′ × 4 mm glass column with 1% SP-2100 on Supelcoport 100/120 mesh; flow rate 40 mL/min Argon/methane; oven 190°C; ⁶³Ni detector.

in the chromatogram. Also cis- and trans-chlordane can be
quantitated after nitration (25). HCB is recovered in low yields
(~10%) primarily due to volatilisation rather than reaction with
the acid mixture. Similarly, lindane recoveries are erratic.
After a collaborative study involving sediment, carp, eel and gull
egg samples, this nitration technique was found reliable for Mirex
and photomirex at levels ⩾10 ppb in the presence of 1,000-fold
greater levels of PCB (26).

Hexachlorobenzene (HCB) has attracted increased attention
due to its common occurence in human blood, milk and adipose
tissue, wild life samples, wheat and lake water. Various
approaches (Fig. 6) to the chemical confirmation of HCB have included the preparation of a pentachlorophenyl propyl ether when
treated with KOH in 1-propanol (27). The derivative had a longer
GC retention time than HCB. Another two step approach by Holdrinet (28) involved reaction of HCB with 0.2N KOH in ethylene-
glycol under reflux to produce pentachlorophenol which is then
esterified with diazomethane to give the PCP methyl ether derivative. It was originally demonstrated by Baker (29) that HCB is
converted to monoethoxypentachlorobenzene when refluxed with
sodium ethoxide. More recently, it has been shown (30) that extended treatment (i.e. greater than 6 hrs.) with NaOEt will further
convert the monoethoxy derivative to diethoxy compounds. Since the
purified diethoxy derivative obtained from a prep-scale reaction
gave 2 major chromatographic peaks, it is suggested a mixture of
isomers is produced. Similarly, Crist and co-workers (31) confirmed
HCB in adipose tissue by the preparation of several derivatives
from various alcohols. It was found that reaction with a KOH/2-
propanol/pyridine mixture for 10 min. at 100°C produced mono-iso-
propoxy pentachlorobenzene while after 30 min. bis-isopropoxytetra-
chlorobenzene was formed. The di-substituted derivative was less
subject to further substitution than the mono-derivative and
reproducibility of the reaction was more easily controlled such
that a lower level of 5 ppb HCB could be confirmed in fatty
samples. The flexibility of this reaction was shown by the use
of other alcohols (EtOH, 1-PrOH, 1-BuOH) and the confirmation of
0.3 ppm HCB in rat adipose tissue containing 160 ppm Arochlor
1016. Since derivatisation to the di-isopropoxy derivative was
not feasible due to interference from an Arochlor peak, confirmation based on the mono-substituted derivative was possible.

Organophosphorus and Carbamate Insecticides

Most OP insecticides may be determined directly by GC using
the phosphorus-selective flame photometric detector (FPD) which
helps to minimize clean-up. However, it must be emphasised that
the FPD is only a "selective" detector for phosphorus (at 540 nm)
or sulphur (at 394 nm) and not an "element specific detector".
This is illustrated in Fig. 7 which shows the 4 oxidation products

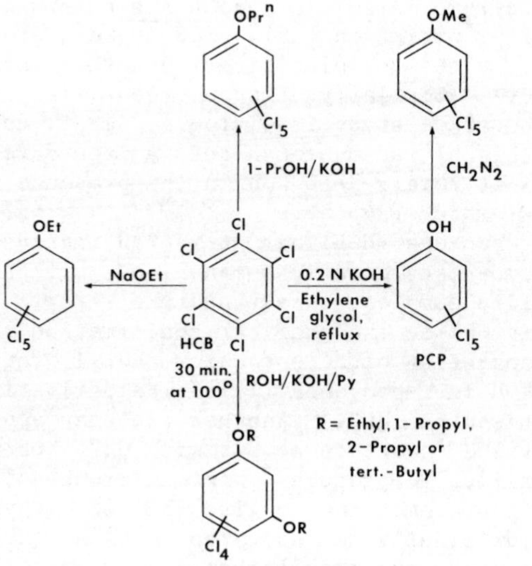

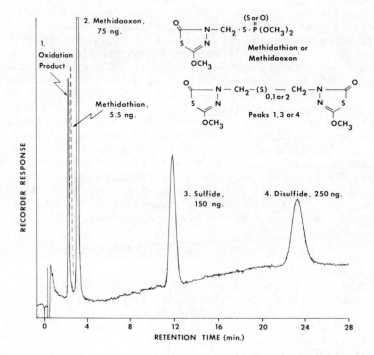

Figure 6.   Reactions used for the confirmation and/or quantitation of hexachloro-benzene (HCB)

Figure 7.   Characteristics of methidathion and its sodium hypochlorite oxidation products on 1% DEGs at 200°C with flame photometric detection (P-mode)

obtained from the macro-scale reaction of neutralised sodium hypochlorite with methidathion (Supracide^R) (32).   Generally, oxidation of the P-S groups in such insecticides as malathion, diazinon, parathion, fenitrothion, ethion, etc. yield their respective oxons. A single oxon derivative peak is the result of
hypochlorite oxidation of methidathion at the ng level but 4 products were observed at the macro-scale. Although this chromatogram was obtained using the P-mode of the FPD only one peak,
namely methidaoxon, peak 2, contains phosphorus. The other 3 products consist of 2 thiadiazolinone rings linked by methylene,
sulfide or disulfide groupings respectively. Also a 10:1 response ratio was obtained when equivalent amounts of the sulphide
and disulphide were recorded on the S- and P-channels, respectively,
of the FPD. Oxidation of phorate with NaOCl or H_2O_2 results in
phorate oxon sulfoxide and not phorate oxon. Recently the
quantitative oxidation of oxydemeton methyl with $KMnO_4$ to its
corresponding sulfone has been used for its determination in a
variety of plant and animal tissues down to a limit of 0.01 ppm
(33). An alternative approach to the analysis of oxydemeton metyl,
which contains a dialkyl substituted sulfoxide moiety, is <u>via</u>
trifluoracetylation (Fig. 8) (34). Reaction of oxydemeton methyl
with trifluoracetic anhydride at 100° for 15 min. results in a
mono-trifluoroacetoxy derivative which thermally degrades on-
column to give 2 peaks, the <u>cis</u>- and <u>trans</u>-isomers of dehydro-
oxydemeton-methyl. Other compounds with a sulfoxide moiety that
can give mono- or di-TFA derivatives suitable for confirmatory
purposes are dasanit, mesurol sulfoxide, nemocur sulfoxide,
aldicarb sulfoxide, counter sulfoxide and oxycarboxin. Due to the
poor chromatographic characteristics of sulfoxides, analysis was
previously performed following oxidation to their respective sul-
fones. Now with mesurol, all 3 oxidation products plus phenols
can be analysed simultaneously. Of course, trifluoracetylation
and methylation of NH-containing OPs and carbamates are possible
but these procedures have been extensively covered in previous
reviews (5, 7).

    However, two other areas are of interest (Fig. 9) namely, a)
derivatisation of phenols or amines to phosphorus-containing
compounds and b) on-column methylation. In the former, phosphorylation of 7 alcohols and 12 phenols with diethyl chlorophosphate
in the presence of triethylamine at 60-70°C in benzene has been
reported (35). Aliphatic alcohols, such as methanol, butanol or
isoamyl alcohol react instantaneously while phenol and cresols
require ca 1-1 1/2 hrs. and xylenols 3-4 hrs. for best results.
The resulting alkyl or aryl diethyl phosphates were selectively
detected at the 10-25 ng level with the P-mode FPD. Similarly,
Jacob <u>et</u> <u>al</u> (36) used dimethyl thiophosphinic chloride in the
presence of excess triethylamine to convert primary aliphatic,
aromatic and heterocyclic amines to the corresponding N-dimethyl-
thiophosphinic amides. The excess reagent was easily removed by
treating the reaction mixture with methanol-sodium hydrogen car-

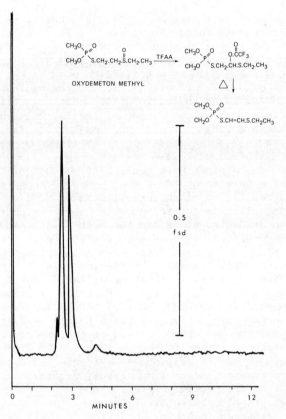

Journal of Agricultural and Food Chemistry

Figure 8. Reaction of oxydemeton methyl with trifluoroacetic anhydride (TFAA) and the GC characteristics of the two products (cis- and trans-dehydro-oxydemeton methyl) on 3% OV-17 at 190°C (34)

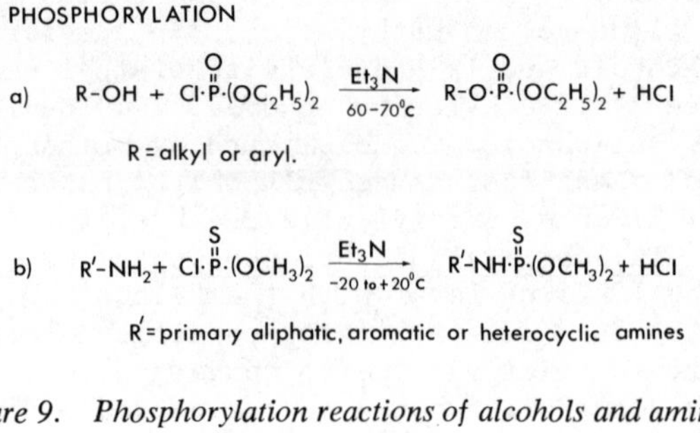

PHOSPHORYLATION

a) $\text{R-OH} + \text{Cl}\cdot\overset{\overset{\text{O}}{\|}}{\text{P}}\cdot(\text{OC}_2\text{H}_5)_2 \xrightarrow[60-70°c]{\text{Et}_3\text{N}} \text{R-O}\cdot\overset{\overset{\text{O}}{\|}}{\text{P}}\cdot(\text{OC}_2\text{H}_5)_2 + \text{HCl}$

R = alkyl or aryl.

b) $\text{R}'\text{-NH}_2 + \text{Cl}\cdot\overset{\overset{\text{S}}{\|}}{\text{P}}\cdot(\text{OCH}_3)_2 \xrightarrow[-20 \text{ to} +20°c]{\text{Et}_3\text{N}} \text{R}'\text{-NH}\cdot\overset{\overset{\text{S}}{\|}}{\text{P}}\cdot(\text{OCH}_3)_2 + \text{HCl}$

R′ = primary aliphatic, aromatic or heterocyclic amines

Figure 9.   Phosphorylation reactions of alcohols and amines

bonate and the derivatives were well separated on a SE-30 or OV-17 capillary column. Although a lower limit of detection of 0.5 pg for N-dimethylthiophosphinylaniline was obtained with a rubidium sulphate AFID, an FPD detector could also be used. This derivatisation technique was applied to twenty aromatic or heterocyclic amines as well as 6 amino acid esters. However, it should be pointed out that dimethylthiophosphinic chloride reacts and is as sensitive, if not more so, to moisture problems than TFAA. Since the resulting product contains P it too gives a FPD/AFID response.

The general subject of on-column pyrolytic methylation has been recently reviewed (37) and, in particular, this technique has been the subject of a number of papers dealing with the identification and quantitation of many organophosphorus pesticides and related dialkyl phosphorothioates (38-40) as well as carbamate pesticides (41-43). GLC on-column transesterification of OP insecticides at 260°C using trimethylphenylammonium hydroxide in methanol, results in short-chain trialkyl phosphates which can be quantitated using the P-mode FPD (Fig. 10). The exact structure of the trialkyl phosphate formed is dependent upon the structure of the original OP insecticide. Parathion is transesterified to give diethylmethylthiophosphate (DEMTP) while azinphos methyl yields trimethyl dithiophosphate (TMDTP). The minimum detectable limit for malathion was 400 pg. The technique was applied to the determination of chlorphoxim at levels of 10 ppb in fish and 0.1 ppb in water. It has been suggested that methanolic TMPAH quantitation of dialkyl phosphorothioates and phosphorodithioates could provide a convenient and safe alternative to the use of diazomethane where methylation is the final step in OP residue analysis (40).

Similarly, on-column methylation has been applied to carbamate pesticides containing an active N-H group. Wien and Tanaka (41) showed that N-aryl carbamates are methylated on-column with trimethylanilinium hydroxide-methanol to give the intact N-methyl and N-aryl derivatives. On the other hand N-methyl, O-aryl carbamates such as carbaryl or carbofuran yielded only the methyl ethers of their respective phenols. This work has now been extended to sulfur-containing carbamates such as methomyl, methiocarb, aldicarb, etc. (42-43). Here the oxime hydrolysis products of these carbamates are chromatographed as the O-methyl oximes. This reaction has been applied to the analysis of oxamyl (lower level 0.5 ug in 50 g sample) in soil using the S-mode FPD.

However, it must be pointed out that to date, the application of the phosphorylation and on-column methylation techniques have had only limited application.

Phenols and Fungicides

Pentachlorophenol (PCP) is widely used as a pesticide in agriculture and as a wood preservative in industrial and domestic products. PCP together with other chlorophenols which arise as

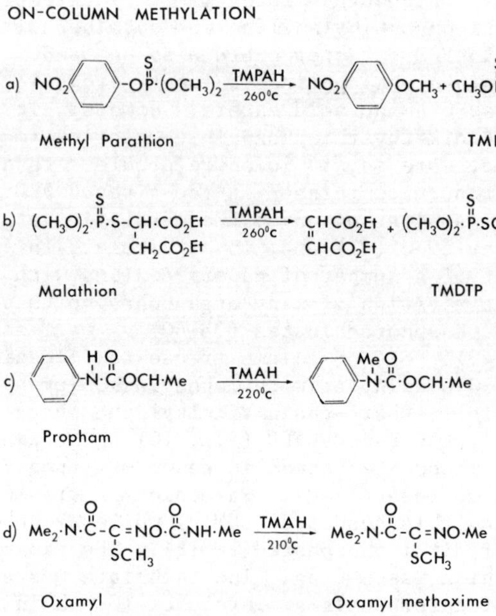

ON-COLUMN METHYLATION

Figure 10. Examples of on-column methylation reactions of organophosphorus and carbamate pesticides

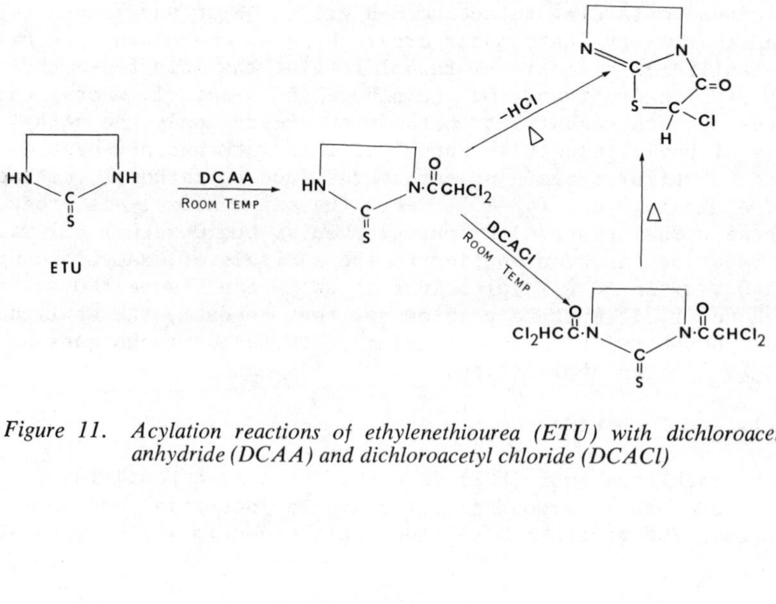

Figure 11. Acylation reactions of ethylenethiourea (ETU) with dichloroacetic anhydride (DCAA) and dichloroacetyl chloride (DCACl)

metabolites of the OP and carbamate insecticides, and phenoxy-
acids herbicides, have been variously determined in tact using FID
detection or, more commonly, as their ethers, esters or silyl derivatives (7) in environmental, agricultural and mammalian samples.
The more commonly used derivative is the methyl ether, using diazomethane, after extraction and/or an acid/base partitioning cleanup step. However, recently Edgerton and Moseman (44) compared
previous methods for the residue analysis of PCP in urine and
found an Alumina clean-up column following derivatisation was necessary for levels below 30 ppb. Also a 1 hr. closed system acid
hydrolysis step using HCl gave as much as a 17-fold higher PCP
level than did other methods, indicating PCP forms conjugates
in urine. The hydrolysis, methylation, column clean-up procedure allowed levels of 1 ppb PCP to be detected in urine. Also
a lower level of 0.01 ppm PCP in fish, shrimp and oysters has been
obtained after ethylation with diazoethane followed by a Florisil
clean-up (45). When the ethylation procedure was used on sea-
water extracts interferences were observed at concentrations lower
than 0.01 ppb using ECD-GLC. Formation of the amyl derivative
using diazopentane, increased the GC retention time sufficiently
to separate PCP from early eluting peaks. Concentrations of PCP
as low as 2 ppt could be detected on sea water. Interestingly,
the same workers also utilized HPLC with UV detection of the free
phenol with clean-up to obtain detection limits of 5 ppm in tissues
and 2 ppb in sea water. The limit of PCP acetate is about 1-2.5
pg with a ^{63}Ni-ECD (46, 47) which gave a 5 ppb limit of detectability for PCP in adipose tissue (47).

In the last few years a number of reports have appeared on
an extractive pentafluorobenzylation procedure for phenols and
carboxylic acids to enhance their electron-capturing properties.
Here the extraction and derivatization of the phenol or acid is
accomplished in one step by the partition of the phenol (or acid)
anion from the aqueous phase as an ion-pair with a quaternary
ammonium ion into an organic phase which contains the derivatization reagent (48-51). This technique has been applied to the
extractive pentafluobenzylation of 2,4-D, and MCPA in water with
a detection limit of 1-3 ug/L (52). Also, the principle has
been applied to the extractive derivatization of ethylene thiourea
(ETU) in water. Even though ETU can be GC chromatographed intact
on Versamid 900 and Carbowax 20 M packed columns (52, 53) or various
capillary columns (54), by far the most popular approach has been
the alkylation of the thiocarboxyl group and in some instances,
double derivatization involving the NH-group (7). Derivatization
has resulted in improved GC characteristics and increased sensitivity using EC detection. However, these procedures invariably
require lengthy reaction times at reflux temperatures to achieve
acceptable yields. Also, under these conditions any parent ethyl-
enebisdithiocarbamate fungicide residue that is present is converted to ETU to a certain extent. Figure 11 shows the reaction
sequence for the room temperature extractive N-acylation of ETU

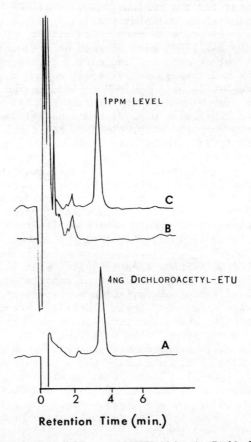

Figure 12. (A) Standard of dichloroacetyl-ETU (4.0 ng); (B) blank apple juice extract; (C) apple juice extract spiked with 1 ppm ETU after extractive derivatization with DCAA. Conditions: 6' × 2 mm i.d. glass column with 3% OV-330 on 80–100 mesh Chromosorb 750; flow rate, 30 mL/min helium; oven, 200°C; NP alkali flame ionization detector.

from water using dichloroacetic anhydride in CH_2Cl_2. To effect
efficient extraction, acetonitrile was used (as phase transfer
agent) in a 1:10 ratio with the water. Reaction is complete in
3 min. Although S-alkylation could be postulated as a possible
reaction pathway, this was discounted after the di-acylated
derivative was prepared from the mono-product using dichloroacetyl chloride. NMR and GC/MS analysis showed that elimination
of HCl together with ring-closure occurs on GC injection to give
a sharp, single peak on OV-17 or OV-330 columns. Originally,
this extractive acylation procedure was used as a rapid screening
procedure for the presence of ETU in water samples at the 0.01-
0.05 ppm level (56) and has now been extended to fruit and
vegetable juices. Figure 12 shows the chromatographic characteristics of the ETU derivative on an OV-330 column and detection
using NP heated-bead detector. Also illustrated is the before
and after chromatograms of a spiked apple juice sample at the 1
ppm level.

   In conclusion, it is obvious that chemical derivatization
continues and will continue to play an important role in both
qualitative and quantitative analysis of pesticides, their residues and metabolites. One of the major advantages being that
derivatization gives an improvement in selectivity as a result
of the formation of a characteristic derivative which responds
selectively to certain GC and HPLC detectors.

LITERATURE CITED

1. Drozd, J., J. Chromatogr., 1975, 113, 303-356.
2. Ahaya, S. J. Pharm. Sci., 1976, 65, 163-182.
3. Blau, K., King, G.S., Eds. "Handbook of Derivatives for
    Chromatography", Heyden and Sons Ltd., London, 1977.
4. Cochrane, W.P., Purkayastha. Toxicol. Environ. Chem. Reviews,
    1973, 1, 137-268.
5. Cochrane, W.P., J. Chromatogr. Sci., 1975, 13, 246-253.
6. Nickolson, J.D. Analyst 1978, 103, 1-28 (Part I); ibid 1978,
    103, 193-222 (Part 2).
7. Cochrane, W.P., Chau, A.S.Y. "Advances in Chemistry Series"
    American Chemical Society, Washington, D.C. 1971, 104, 11-27.
8. Khan, S.U. Res. Rev. 1975, 59, 21-50.
9. Yip, G., J.Assoc. Offic. Anal. Chem., 1971, 54, 343-45.
10. Woodham, D.W., Mitchell, W.G., Loftis, C.D., and Collier, C.W.
    J. Agric. Food Chem., 1971, 19, 186-188.
11. Gutenmann, W.H., and Lisk, D.J., J. Assoc. Offic. Anal. Chem.,
    1964, 47, 353-354.
12. Agemain, H., Chau, A.S.Y., J. Assoc. Offic. Anal. Chem., 1977,
    60, 1070-76.
13. Chau, A.S.Y., Terry, K., J. Assoc. Offic. Anal. Chem. 1975,
    58, 1294-1300.
14. Mierzwa, S., Witek, S., J. Chromatogr. 1977, 136, 105-111.
15. Chau, A.S.Y., Terry, K., J. Assoc. Offic. Anal. Chem., 1976,
    59, 633-36.

16. Moseman, R.F., Ward, M.K., Crest, H.L. and Zehr, R.D., J. Agric. Food Chem., 1978, 26, 965-968.
17. Anonymous, Pest. Chem. News, 1976, 4, 15
18. Kaiser, K.L.E., Science, 1974, 185, 523-525.
19. Zitko,V., Bull. Environ. Contam. Toxicol., 1976, 16, 399-405.
20. Mes, J., Davies, D.J., and Miles, W., Bull. Environ. Contam. Toxicol. 1978, 19, 564-750.
21. Lane, R.H., Grodner, R.M., and Graves, J.L., J. Agric. Food Chem., 1976, 24, 192-193.
22. Chau, A.S.Y., Carron, J.M. and Tse, H., J. Assoc. Offic. Anal. Chem., 1978, 61, 1475-1480.
23. Lewis, R.G., Hanish, R.C., MacLeod, K.E., and Sovocool, G.W., J. Agric. Food Chem., 1976, 24, 1030-1035.
24. Hallet, D.J., Norstrom, R.J., Onuska, F.I., Comba, M.E. and Sampson, R., J. Agric. Food Chem., 1976, 24, 1189-1193.
25. Holdrinet, M.V.H., Bull. Environ. Contam. Toxicol., 1979, 21, 46-52.
26. Norstrom, R.J., Won, H.T., Holdrinet, M.V.H., Calway, P.G., and Naftel, C.D., J. Assoc. Offic. Anal. Chem., 1980 (in press).
27. Collins, G.B., Holmes, D.C., Wallen, M., J. Chromatogr. 1972, 69, 198-200.
28. Holdrinet, M.V.H., J. Assoc. Offic. Anal. Chem., 1974, 57, 580-584.
29. Baker, B.E., Bull. Environ. Contam. Toxicol., 1973, 10, 279-284.
30. Rosewell,K.T. and Baker, B.E., Bull. Environ. Contam. Toxicol., 1979, 21, 470-477.
31. Crist, H.L., Moseman, R.F., and Noneman, J.W., Bull. Environ. Contam. Toxicol., 1975, 14, 273-279.
32. Singh, J., and Cochrane, W.P., J. Assoc. Offic. Anal. Chem., 1979, 62, 751-756.
33. Thirtin, J.S., Olson, T.J. and Wagner, K., J. Agric. Food Chem., 1977, 25, 573-576.
34. Greenhalgh, R., King, R.R., and Marshall, W.D., J. Agric. Food Chem., 1978, 26, 475-480.
35. Deo, P.G., and Howard, P.H., J. Assoc. Offic. Anal. Chem., 1978, 61, 210-213.
36. Jacob,K., Falkner, C., and Vogt, W., J.Chromatogr., 1978, 167, 67-75.
37. Kossa, W.C., MacGee, J., Ramachandran, S. and Webber, A.J., J. Chromatogr. Sci., 1979, 17, 177-187.
38. Dale, W.E., Miles, J.W., and Churchill II, F.C., J. Assoc. Offic. Anal. Chem., 1976, 59, 1088-1093.
39. Miles, J.W., and Dole, W.E., J. Agric. Food Chem., 1978, 26, 480-482.
40. Churchill II, F.C., Ku, D.N., and Miles, J.W., J. Agric. Food Chem., 1978, 26, 1108-1172.
41. Wein, R.G., and Tanaka, F.S., J. Chromatogr., 1977, 130, 55-63.

42. Bromilow, R.H., and Lord, K.A., J. Chromatogr., 1976, 125 495-502.
43. Bromilow, R.H., Analyst, 1976, 101, 982-85.
44. Edgerton, T.R. and Moseman, R.F., J. Agric. Food Chem., 1979, 27, 197-199.
45. Faas, L.F. and Moore, J.C., J. Agric. Food Chem., 1979, 27, 554-557.
46. Kriggsman, W., Van de Kamp, C.G., J. Chromatogr., 1977, 131, 412-416.
47. Ohe, T., Bull. Environ. Contam. Toxicol., 1979, 22, 287-292.
48. Gyllenhaal, O., Brotell, H. and Hartvig, P., J. Chromatogr., 1976, 129, 295-302.
49. Davis, B., Anal. Chem., 1977, 49, 832-834.
50. Rosenfeld, J.M., and Crocco, J.L., Anal. Chem., 1978, 50, 701-704.
51. Gyllenhaal, O., J. Chromatogr., 1978, 153, 517-520.
52. Akerbolm, M., 4th Intern. Congress of Pest. Chem. (IUPAC), Zurich, 1978, Paper No. VI-702.
53. Otto, S., Keller, W., and Dresher, J. Environ Sci. Health, 1977, B12, 179-191.
54. Farrington, D.S., and Hopkins, Analyst, 1979, 104, 111-116.
55. Hirvi, T., Pyysalo, H., and Savolainen, K., J. Agric. Food Chem., 1979, 27, 194-195.
56. Singh, J., Cochrane, W.P. and Scott, J., Bull. Environ. Contam. Toxicol., 1979, 23, 470-474.

RECEIVED February 7, 1980.

Development of Analytical Methodology for Assessment of Human Exposure to Pesticides

ROBERT F. MOSEMAN[1] and EDWARD O. OSWALD[2]

Analytical Chemistry Branch, Environmental Toxicology Division, Health Effects Research Laboratory, U.S. Environmental Protection Agency, Research Triangle Park, NC 27711

Assessment of human exposure to pesticides is important for a variety of reasons. In the occupational situation, it is necessary to know the amount of pesticide that an individual is exposed to in order to protect worker health. Formulators, loaders, pickers and pilots can experience high exposures to pesticides. Humans can be exposed to pesticides through environmental routes. The air we breath and the water we drink are but two sources of environmental exposure.

Measurement of human exposure can be done either directly or indirectly. Direct measurement involves determination of the pesticide level in the media through which the exposure occurs. Examples of this are measurement of pesticides in breathing zone air or pesticides adsorbed onto pads or clothing worn by workers (1,2). These techniques provide a direct and calculable measure of human exposure under actual conditions. Most often, however, direct measurement is not possible. In these situations indirect methods of exposure assessment must be used.

In the complex process of development of analytical methodology for the indirect assessment of exposure to pesticides, one of the first questions to be addressed, concerns what compound(s) to look for and in what sample type. In most cases the parent pesticide will be transformed to a more polar metabolite.

Exposure to organophosphate pesticides is often measured by determination of alkyl phosphate or phenol metabolites in the urine. Determination of blood cholinesterase activity can be a valuable indicator of exposure if pre-exposure cholinesterase activity is known (3, 4, 5). Since normal cholinesterase levels vary over a fairly wide range, post-exposure measurements alone do not always provide useful information. For the most part, measurement of urinary metabolites can provide positive information without pre-exposure levels. In addition, alkyl phosphate

Current address: [1]Analytical Chemistry Division; Radian Corporation; P.O. Box 9948; Austin, Texas 78766 [2]Department of Environmental Health Sciences; School of Public Health; University of South Carolina; Columbia, S.C. 29208

levels in the urine may indicate that exposure has occured even though no depression of cholinesterase can be detected. As a group, alkyl phosphate metabolites serve as an indicator of exposure to one or several organophosphate pesticides. Since several different pesticides can give rise to the same urinary alkyl phosphate, some specificity may be lost. Measurement of urinary phenols can provide a more positive indication of which pesticide an individual was exposed to. Certain halogenated organophosphates such as chlorpyrifos and ronnel may possibly be found in adipose tissue of heavily exposed individuals. In general, measurement of alkyl phosphates can serve as a better screening technique than measurement of phenolic metabolites simply because the great majority of organophosphates give rise to only about six alkyl phosphate compounds. On the other hand, almost all of the organophosphates yield a different phenol, making a truly comprehensive multi-phenol analytical procedure an almost impossible undertaking.

Assessment of exposure to carbamate pesticides can be a complex matter for a variety of reasons. Measurement of cholinesterase depression can be difficult in the case of carbamates. Not only is the methodology non-specific, but the determination itself can be questionable because of the instability of the enzyme-carbamate complex. As with cholinesterase determinations for organophosphates, the method can be highly variable even within the same laboratory.

Direct determination of intact carbamate pesticides by sensitive and specific gas chromatographic procedures can be done. Historically, these procedures employed a derivatization step to render the carbamates amenable to gas chromatography. Recent developments in column technology have allowed for the gas chromatography of some intact carbamates at nanogram and subnanogram sensitivity (6, 7, 8).

Examination of tissues and excreta from humans or animals for exposure to carbamate pesticides, will almost never result in detection of the parent compound. Exposure assessment of this nature requires determination of metabolitic products, except in extreme situations such as acute poisoning. The most widely used indicator of exposure is probably the determination of urinary phenols (9).

Tissues such as fat, blood or liver can be examined for residues of the more stable chlorinated hydrocarbon pesticides. In most cases these tissues are available as a result of elective survery, autopsy or biopsy. Exposure to DDT results in some storage of the parent compound in body fat. A large portion, however, is metabolized and stored as DDE (9). Aldrin and heptachlor are similarly transformed and stored as dieldrin and heptachlor epoxide. Levels of the urinary metabolite DDA have been used to assess exposure or body burden of DDT (10, 11, 12). Hexachlorobenzene and the various isomers of hexachlorocyclohexane are stored in fat as the parent compound but a small percentage is excreted in the urine as chlorinated phenols (13,14). Recent work in our laboratory has indicated that a significantly greater amount of urinary phenols and anilines can be recovered if the sample is hydrolyzed with acid prior to extraction. Apparently many phenols and other polar compounds are found as conjugates in liver and urine. Failure to free these compounds before extraction can lead to erroneously low results.

In the urine of experimental animals, N-dealkyl metabolites of triazine herbicides have been identified (15,16). Methodology for the determination of these kinds of compounds is currently available (15) and should be applicable for monitoring human population.

Substituted urea herbicides are a widely used group of compounds. Urinary metabolites of these compounds are halogenated anilines. Recent research has lead to the development of analytical methodology for trace levels of these compounds in the urine of experimental animals (17). The next logical step in this sequence is application to human exposure assessment.

After is has been determined which compounds are of importance, they must be available for use as analytical standards. In the absence of a reliable standard, qualitative and quantitative data cannot be obtained. In cases where standards are not available commercially, we have synthesized our own. If only the technical material is available, it may be purified to analytical quality. A very important part of the entire quality assurance portion of methods development is accurate and reliable standards.

The efficiency of an extraction procedure must be determined initially by fortifying blank (control) samples. The fortification should be done over a range to include the levels expected in real samples. If the ultimate sensitivity desired is 0.005 ppm, it does little good to establish recovery data at 0.5 ppm. Plotting recovery data versus fortification level on a log-log scale provides a good picture of the extraction efficiency. Selecting fortification levels of 0.01, 0.03, 0.10 etc. provides even spacing for data points and allows coverage of a wide range with a minimum of samples. Quadruplicate runs at each level provides sufficient data for calculating the mean and standard deviation.

A very important consideration in methods development research concerns the vast difference between recovery of compounds added to a sample in the laboratory and recovery of biologically incorporated compounds. For the most part "grown in" residues can be much more difficult to extract from sample matrices. Chemical or enzymatic hydrolysis is often required to free bound or conjugated residues. Information on recovery of biologically incorporated residues is best obtained with the use of radiolabelled compounds. In the absence of tagged compounds, quantitation of extracted residues both before and after hydrolysis can yield valuable information. Exhaustive extraction of a sample matrix using different solvents can give an indication of how vigorous the extraction must be to remove all the residue present. One must keep in mind, however, that the more polar extraction solvents can lead to more difficult cleanup and separation problems. Ideally, the solvent chosen will efficiently extract the compounds of interest with a minimum of interferences.

When dealing with human tissues, experimental dosing or feeding is not possible. Determination of pesticides in human samples taken from individuals poisoned or occupationally exposed can provide information useful in development of analytical methodology. These types of samples may contain biologically incorporated pesticides and metabolites. If human tissue samples containing the pesticides of interest are not available, the researcher must rely on animal models for establishing recovery data for pesticides and metabolites.

The choice of the determinative step to be used in an analytical procedure is dictated by the nature of the residue. The non-polar chlorinated hydrocarbon pesticides are routinely quantified using gas chromatography (GC) and electron capture(EC) detection. Alternate detectors include electrolytic conductivity and microcoulometric systems. Organophosphate pesticides which are amenable to GC are responsive to either the flame photometric detector (FPD) or the alkali flame detector (AFD). Sulfur containing compounds respond in the electrolytic conductivity or flame photometric detectors. Nitrogen containing pesticides or metabolites are generally detected with alkali flame or electrolytic conductivity detectors.

Polar compounds or compounds which decompose at elevated temperatures in gas chromatography, can often be converted to more stable derivatives prior to gas chromatography. Preparation of derivatives may also add heteroelements and render certain compounds more responsive to element selective detectors. This latter technique can sometimes ease cleanup problems by effectively reducing the background against which one must quantify. Similar techniques are useful for confirmation of results obtained by other methods.

The fastest growing instrumental technique for the determination of trace organics and pesticides is high performance liquid chromatography (HPLC). Many compounds which cannot be handled by gas chromatography are easily chromatographed and separated by HPLC (18, 19, 20, 21). Since ambient temperatures are the rule, heat-labile compounds present no particular problems. Very polar compounds are subject to chromatography without derivatization. A wide variety of column materials are available as well as reagents for paired ion chromatography. Isocratic or solvent programming may be effectively used for the desired separation.

For several years LC detectors were limited to refractive index and ultraviolet absorption systems. Recently introduced systems include the electrochemical detector and a moving belt interface allowing for chemical ionization-mass spectrometric detection. Both of these techniques provide a degree of selectivity not previously available.

Liquid chromatography has been effectively used in the residue laboratory for cleanup of sample extracts prior to the determinative step. This added dimension has allowed for detection and confirmation of trace substances at much lower levels than was possible a few years ago.

Analytical methodology used for the assessment of human exposure to pesticides should always include some form of confirmation analyses. Reliance on chromatographic peaks generated by a single detector is an open invitation to criticism. Results of analyses in a monitoring program should be confirmed at a rate of 10 to 20 percent. The confirmational procedure should provide for measurement of the residue by some independent physiochemical means. If the primary analytical technique is electron capture gas chromatography, the confirmation should be done, at the bare minimum, with gas chromatography and a selective detector such as electrolytic conductivity. Derivatization of the residue in a sample extract and subsequent chromatographic analysis provides a fairly reliable confirmational technique. Thin-layer and high performance liquid chromatography are good techniques to use in conjunction with gas chromatography. Infrared or mass spectrometric data can provide solid evidence for confirmation of results.

Some practical applications of the concepts discussed earlier have recently been encountered in our laboratory. During the initial stages of the development of a multi-residue procedure for chlorinated phenols (14), several problems were encountered. At the outset, we found that highly purified analytical standards were not easily available from commercial sources. Some of the compounds were obtained as technical grade. These were purified by recrystallization. Others did not yield good recovery even when carried through the analytical procedure without the involvement of a sample matrix. This latter problem was traced to light and oxygen sensitivity of pentachlorothiophenol, tetrachlorohydroquinone and tetrachloropyrocatechol. Decomposition of these compounds was noted during the extraction and cleanup steps and was circumvented by the addition of sodium bisulfite and wrapping the glassware in aluminum foil. Once the phenols were methylated, light sensitivity was no longer a problem and the samples could be handled in a normal manner.

Another major problem in this work concerned the gas chromatographic overlap of several compounds. The use of alternate columns was insufficient for complete resolution of all the phenols. An alumina column cleanup and separation technique was devised which overcame most of the problem.

At the present it is known that repetitive thawing and refreezing of urine samples results in disappearance of phenols (22). In order to establish the true level of urinary phenols, it is important to perform the analysis as soon as possible after the sample is collected or to freeze the sample immediately and keep it frozen until analysis.

Many of the analytical procedures for pentachlorophenol in the recent literature do not call for a hydrolysis step prior to extraction of a urine sample. During the course of our research in the development of a reliable multi-residue procedure for chlorinated phenols, we found that much more pentachlorophenol could be extracted from the urine if the sample was hydrolyzed with hydrochloric acid (23).

A similar finding was noted during the development of analytical methodology for the determination of urinary chlorinated anilines. It was found that much more 3,4-dichloroaniline could be extracted from the urine of rats fed diuron and linuron if an acid hydrolysis step was used prior to extraction (24). When using hydrolysis in an analytical procedure, the chemist must determine if the compound of interest is stable to the conditions used.

During the development of a procedure for the determination of N-dealkylated triazine herbicide, it was found that hydrolysis of urine samples was unnecessary for maximum recovery of these compounds (15).

In summary, and at the risk of repetition, it must be stressed that the development of analytical methodology for the assessment of human exposure to pesticides is a complex process. Careful attention to planning of the research is of utmost importance. As much information as possible about transformation, storage and excretion of the pesticides of interest should be gathered. Preliminary work should focus on the analytical behavior of parent compounds and metabolites. The combination of these aspects with reliable analytical standards and a sound quality assurance program should yield valid analytical methodology.

Literature Cited

1. Wolfe, H.R.; Durham, W.F.; and Armstrong, J.F. Arch. Environ. Health, 14, 622,(1967).
2. Durham, W.F.; and Wolfe, H.R. Bull. World Health Organ., 26, 75,(1962).
3. Morgan, D.P., Editor "Recognition and Management of Pesticide Poisonings", U.S.Environmental Protection Agency, Washington, D.C.,(August, 1976).
4. Ganelin, R.W.; Arizona Medicine, (October, 1964).
5. Hayes, W.J., Arch. Environ. Health, 3, 49,(1961).
6. Lorah, E.J.; Hemphill, D.D., JAOAC, 57, 570, (1974).
7. Moseman, R.F., J. Chromatogr., 166, 397, (1978).
8. Hall, R.C.; and Harris, D.E., J. Chromatogr., 169, 245,(1979).
9. Thompson, J.F., Editor "Manual of Analytical Methods for the Analysis of Pesticide Residues in Human and Environmental Samples", U.S.Environmental Protection Agency, Research Triangle Park, N.C., (June, 1977).
10. Cueto,C.; Barnes,A.G.; and Mettson, A.M., J.Agr. and Food Chem. 4, 943, (1956).

11. Durham, W.F.; Armstrong, J.S.; and Quinby, G.E., <u>Arch.</u> <u>Environ. Health</u>, <u>11</u>, 76, (1965).
12. Cranmer, M.F.; Carroll, J.J.; and Copeland, M.F., <u>Bull.</u> <u>Environ. Contamin. Toxicol.</u>, <u>4</u>, 214, (1969).
13. Chadwick, R.W.; Freal, J.J. <u>Bull. Environ. Contam. Toxicol.</u>, <u>7</u>, 137, (1972).
14. Edgerton, T.R.; Moseman, R.F.; Linder, R.E.; and Wright, L.H., <u>J. Chromatogr.</u>, <u>170</u>, 331, (1979).
15. Bradway, D.E.; Moseman, R.F., Presented at the 176th National ACS Meeting, PEST #50, Miami, Florida, Sept. 1978.
16. Erickson, M.D.; Frank, L.W.; and Morgan, D.P., <u>J. Agr. and</u> <u>Food Chem.</u>, <u>24</u>, 743, (1979).
17. Lores, E.M.: Bristol, D.W.; and Moseman, R.F., <u>J. Chrom. Sci.</u>, <u>16</u>, 358, (1978).
18. Horgan, D.F.Jr., Chapter 2 in "Analytical Methods for Pesticides and Plant Growth Regulators", Vol. VII, Sherma, J., and Zweig, G., Editors, Academic Press, New York, (1973).
19. Moye, H.A., <u>J. Chromatogr. Sci.</u>, <u>13</u>, 268, (1975).
20. Frei, R.W.; and Lawrence, J.F., <u>J. Chromatogr.</u>, <u>83</u>, 321, (1973).
21. Self, C.; McKerrell, E.H.; and Webber; T.J.N., <u>Proc. Anal.</u> <u>Div. Chem. Soc.</u>, <u>12</u>, 388, (1975).
22. Edgerton, T.R., Personal communication, August, 1979.
23. Edgerton, T.R.: and Moseman, R.F., <u>J. Agr. and Food Chem.</u>, <u>27</u>, 197, (1979).
24. Lores, E.M.; Meekins, R.W.; and Moseman, R.F., submitted to <u>J. Chromatogr.</u>, 1979.

RECEIVED February 21, 1980.

Distribution of Pesticides in Human Organs as Determined by Quantitative Thin-Layer Chromatography

S. N. TEWARI [1]

State Forensic Science Laboratory, Government of Uttar Pradesh Lucknow (India)

To control the heavy losses of crops from insects and other pests, the use of different pesticides has become extensive, and with the passing of time, more of these compounds are being developed and produced on a commercial scale. At present, apart from the inorganic compounds such as zinc phosphide and arsenious oxide, several hundred organic compounds, having widely different chemical constituents and properties, are being used throughout the world.

The frequent use of these organic pesticides has resulted in their widespread distribution in the environment. Often they are found in unwanted places, thereby causing pollution problems. Because these pesticides are toxic not only to insects and pests but also, in varying degrees, to higher animals and man, these chemicals have created a public health menace.

Today many commercial products are available as home remedies against flies, ticks, bugs, rats and other insects and pests under the trade names of 'Flit', Tick-20, Bugmar, Dalf, Bugton and many more. These pesticides are also available in concentrated solutions, powders, aerosols, emulsions, etc. for agricultural purposes. Due to their easy availability and rapid toxic action, these pesticides have been misused as homicidal agents and acted as accidental poisons. The misuse has become a problem for the toxicologist, especially in India, where poisonings by the use of pesticides are increasing annually. The investigation of these accidental deaths has become a great challenge to the forensic toxicologist.

In order to establish that the death of a victim is due to the ingestion of a specific poison, the absolute proof needed is the detection and identification of the poison in various organs of the body. Poisonous substances are rarely distributed evenly in body tissues but tend to accumulate in some particular organ due to their chemical nature and metabolic conversions. There-

[1] Director, State Forensic Science Laboratory, Govt. of Uttar Pradesh Lucknow (India).

0-8412-0581-7/80/47-136-259$05.00/0

fore to search for a poison about which no detailed information
is available, including as to how and when it was administered, a
full analysis of a variety of autopsy material is required. This
analysis provides the data on the distribution of the poison in
different organs and may be of great value in deducing the probable original dose ingested and also the time, and to some extent, the nature of administration of the poison. These data may
also help to give a clue regarding the cause of death. However,
from experience it has been found that the knowledge of (1) the
action of the pesticidal poison under consideration, (2) clinical
course of the poisoning and (3) the postmortem findings is of
great importance in the interpretation of the results obtained
from chemical analysis of the material received.

There are occasions in which death is undoubtedly due to a
pesticide, and yet the chemical analysis fails to reveal its
presence or shows only insignificant traces. This may be due to
the rapid metabolic conversion of the particular pesticide. This
is the case with parathion which is rapidly converted to para-
nitrophenol due to in vivo hydrolysis. Pesticides, especially
those that are comparatively less toxic to mammals, such as
dichlorovos, etc, may be excreted rapidly and, therefore, may
perhaps be found in the urine, though not in the tissues.

The study of the distribution of a poison, post mortem,
presents a complex problem. The estimation of the poison in the
blood presents a complex problem due to the putrefactive changes
in blood and the minute quantity available for analysis. Distribution of insecticidal poisons from the blood to the tissues and
body fluids takes place with some recirculation to the blood(1).

The body tissues are generally submitted to a toxicological
laboratory in sufficient quantities; hence these specimens are
better suited for quantitative analysis. The detection and
determination of the unchanged insecticide in the stomach, intestine, liver, kidney, spleen, lungs, brain, and in other tissues
and body fluids are useful indicators of direct insecticide
action.

In order to reach a conclusion when and how and, and to some
extent, how much of a particular insecticide was administered,
the form and quantity in which the insecticide exists in different tissues and in blood and urine may give valuable clues.
Indeed, an intelligent deduction is also based on the knowledge
of metabolic pathways and formation of other derivatives. Thus,
as parathion is excreted in the urine ultimately as p-nitrophenol, urinary p-nitrophenol levels may indicate the extent of
exposure to parathion(2). In cases of mild and moderate exposure, the excreted p-nitrophenol in urine was found to be of the
order of 0.057 to 0.322 mg. percent. In severe and fatal cases
of poisoning by parathion, the level of p-nitrophenol in urine
was from 0.16 to 1.16 mg. percent. para-Nitrophenol thus is
rapidly excreted in urine and no longer detected 48 hours after
the exposure to the pesticide.

Heyndrickx et. al.(3) have pointed out that the pesticide absorbed by mammals is transported into milk. A case in point is a one year old child who suffered from chronic parathion poisoning after drinking the milk of a cow fed on parathion-contaminated water. A review of 41 cases of organophosphorus pesticide poisoning and the quantitative data have been reported by Heyndrickx(4,5). Lewin et. al.(6) have reported the distribution data from some cases of malathion poisoning using GLC for the determination of insecticides extracted from different autopsy tissues. The data obtained by these workers are given in Table I. The distribution of sumithion, dimethoate, endrin and other organochloro- and organophosphorus pesticides have also been reported by the author and co-workers(7-12).

Once the pesticide is detected in the tissues and body fluids, careful consideration is given to the following data: (1) the amounts of the pesticide found in various body fluids and tissues examined; (2) how rapidly the poison is absorbed; (3) how rapidly and at what concentration it exerts its effects; and, (4) how the distribution through the tissues and body fluids varies with time and route of access. This information is essential, and a detailed description of studies on the distribution of pesticides in human tissues pertaining to homicide, suicide and accidental poisoning cases is described below:

Table I:
Amount of malathion estimated in different biological materials from poison cases. (from Lewin, et. al.(6)).

Case No.	Malathion (mg./100g.) present in the:		
	Stomach	Liver	Blood
1.	992	166	188
2.	172	130	29
3.	1296	170	99
4.	31	26	10
5.	357	-	-
6.	600	-	0.50
7.	5400	0.24	0.067

To study the distribution of pesticides, the amount of the pesticides present in different autopsy tissues and other biological materials have been estimated quantitatively in fatal cases of poisoning of both male and female subjects. The cases were of (1) suicides in which large excess of the poison is generally taken to ensure quick action; (2) homicides in which the poison is administered by mixing it in food or drink, and (3) accidental poisonings where large doses of pesticides are taken up while spraying.

The methods of extraction of the pesticides from tissues as reported by the author and coworkers (13-15) are described below:

Isolation of Organophosphorus Insecticides From Body Tissues.

Twenty grams of the tissue is macerated into a fine slurry by mixing it with equal amounts of anhydrous sodium sulfate and placing it into a conical flask fitted with an air condenser. Fifty milliliters of n-hexane is added, and the flask is heated on a boiling water bath for one hour. After cooling, the n-hexane layer is filtered. The extraction of the residual slurry is repeated twice with twenty-five milliliter portions of n-hexane. The filtered n-hexane fractions are combined and transfered to a separatory funnel. The hexane layer is vigorously shaken for five minutes successively with 15 ml-, 10 ml-, and 10 ml-portions of acetonitrile previously saturated with n-hexane. The acetonitrile layers are mixed and poured into another separatory funnel. In cases of dimethoate, Dasanit, phosphamidon and DDVP poisonings, the three acetonitrile fractions are combined and evaporated to dryness, and the residue is taken up in one milliliter of acetone. Further extraction of acetonitrile is avoided as this results in appreciable loss of pesticides. The acetonitrile layer is diluted with water to ten times of its original volume. Twenty-five ml. of saturated sodium sulfate solution is added to this solution and extracted three times with twenty-five milliliter-portions of n-hexane previously saturated with distilled water. The separated n-hexane layers are combined and then concentrated to two milliliters on a warm water bath. The remaining solvent is removed by passing a current of dry air over it. The residue is then dissolved in one milliliter of acetone.

The extraction from non-fatty tissues is performed as follows: The tissue sample (20 g.) is finely mixed and acidified with one milliliter of 5N sulfuric acid, then transferred to a conical flask. To this, ten grams of anhydrous sodium sulfate and fifty milliliters of acetone are added, and the mixture is refluxed on a warm water bath for thirty minutes and then filtered. The residual material is further extracted twice with fifty milliliter-portions of acetone. The acetone layers are combined, concentrated by evaporation on a warm water bath to about twenty-five milliliters and then transferred to a separatory funnel. The acetone solution is diluted with four times the volume of water. Twenty milliliters of saturated aqueous sodium sulfate is added and then extracted three times by shaking for 5 minutes with twenty-five milliliter-portions of chloroform.

The chloroform extracts are combined and shaken with a small amount of activated charcoal for decolorization and then filtered through Whatman filter paper. The charcoal on the filter paper is washed with twenty milliliters of chloroform, and the washings are collected together with the chloroform extract. The combined chloroform extract is dried by passing it through a column of anhydrous sodium sulfate and then evaporated just to dryness on a warm water bath. The remaining solvent is removed by passing a current of dry air over it. This residue is dissolved in one milliliter of acetone. Carbamate pesticides are also extracted by this method.

Isolation of Organochloro Insecticides. Twenty grams of tissue is macerated into a fine slurry and transferred to a conical flask fitted with a condenser. Thirty grams of anhydrous sodium sulfate and fifty milliliters of acetone are added. The conical flask is placed on a boiling water bath at 80° for half an hour. The flask is removed from the water bath, cooled to room temperature, and the contents are filtered. The process is repeated twice with twenty-five milliliter-portions of acetone. The filtered acetone layers are combined and transfered to a separatory funnel. One hundred milliliters of distilled water and twenty milliliters of a saturated aqueous solution of sodium sulfate is added to the acetone layer. The solution is then extracted with chloroform previously saturated with distilled water using twenty and two ten milliliter-portions. The extraction is completed by slowly shaking for five minutes (20 to 25 shakings per minute).

The separated chloroform layers are combined and placed in another 250-ml separatory funnel. Twenty milliliters of 5% aqueous potassium hydroxide solution is added and shaken for one minute. The chloroform layer is separated by discarding the aqueous layer and again shaken for one minute with twenty milliliters of 5% KOH. The separated chloroform layer is now washed with five 20 ml. portions of distilled water by slowly shaking it for one minute at each turn. The washed chloroform layer is dried by passing through a layer of anhydrous sodium sulfate previously wetted with dry chloroform. The organic solvent is then concentrated up to two milliliters on a warm water bath at 60°, and the remaining solvent is evaporated to dryness in a current of dry air. The residue which contains the pesticide is then dissolved in one milliliter of acetone.

Methods of Analysis of Pesticides. The extracted residue obtained after isolation from tissues and other biological materials is subjected to qualitative and quantitative determination of the pesticides. Sometimes, the amount of material available is so small that the colorimetric and other allied methods cannot be successfully applied as some of the residue is likely to be lost during the purification technique. Furthermore, these purification techniques required for spectrophotometry, colorimetry, and other sophisticated instrumental methods are appreciably time consuming. Therefore, other techniques were sought for the quantitative determination of pesticides. Thin layer chromatographic (TLC) techniques were found to be most suitable for toxicological analysis of pesticides. Randerath([16]) stated that

accurate results in quantitative TLC were obtained by removal of
the substance to be analyzed from the adsorbent using suitable
solvents for elution. Choulis(17,18) has also advocated the
elution technique for quantitative analyses.

The elution technique is normally used in association with
gravimetric, spectrophotometric, colorimetric, fluorometric and
other estimation methods(19). The gravimetric method being
unreliable and yielding generally high results is not suitable
for toxicological analysis. The accuracy of the fluorometric
method is generally considered to be low. Regardless, the
fluorometric method can only be applied to those substances which
fluoresce under the assay conditions or which fluoresce after
reacting with some suitable reagent. Colorimetric methods are
applicable only to those substances which form colored solutions
with a chromogenic reagent, the absorbance of which can be measured using a spectrophotometer or colorimeter. These elution
methods present several disadvantages for the following reasons:
   (i) Inaccurate marking of the spot to be eluted from TLC plate.
  (ii) Incomplete desorption of the substance in the solvent used
       for elution.
 (iii) Interference due to certain substances extracted from the
       adsorbents.

For these reasons, Purdy and Truter(20) and several other
workers have stressed methods by which measurement of the area of
the developed colored spot on TLC plates can be used directly for
quantitative analysis. Methods for measuring the spot area may
be classified as follows:
   1. Visual comparison of spot intensity and/or size to a standard amount of the same pesticide run on the same TLC plate. This
provides an estimation of the substance with approximately 20%
accuracy. A detailed description of this method of visual estimation of thin layer chromatographs has been given by Johnson(21),
who has very wisely commented that it is better to be "approximately right than precisely wrong".
   2. The area of the spot may be traced on graph paper, and the
squares on the paper counted and compared with those of known
amounts of a standard pesticide. This method gives slightly
better results than 1., but lacks accuracy.
   3. Scanning the developed colored spot on the thin layer
chromatographic plate with the help of a photoelectric densitometer.

Gaenshirt and Polderman(22) have mentioned that the errors
of the quantitative determination of a pesticide by planimetric
measurements of the area of the spots are generally high as a
result of variations in the shape of the spot and difficulties in
defining the outline of the spot. Morrison and Chatten(23) and
Oswald and Flueck(24) have pointed out the same difficulties in
the visual measurement of the area of the spot. Quantitative
analysis on TLC with a flying spot densitometer has been suggested by Schute and co-workers(25). Dallas(26) in 1965 has des-

cribed the use of a "Chromascan" for scanning colored substances
on adsorbents spread on glass plates. Genest(27) used a Photo-
volt densitometer for determination of drugs on TLC. Later,
Blunden and co-workers(28), Shellard(29) and Thomas et al.(30)
also reported on the use of a densitometer for quantitation.
Shellard and Alam(31) in 1968 have made a detailed study of the
densitometric method for TLC.

Several workers have described the thin layer chromatographic technique for the determination of organophosphorus
pesticides(32-44) as well as chlorinated pesticides(45-56).

Vylegzhanina and Kalmykova(57) have described the use of TLC
for determination of DDVP in plants, soil and water. Determination of organophosphorus insecticides in river water and effluents have also been used by Askew, et al.(58). Zycinski(59) has
determined the concentration of malathion, DDVP and other organophosphorus insecticides in water by direct measurement on TLC
plates. TLC determination of several organophosphorus insecticides such as fenitrothion, fenthion, and malathion in soil has
been suggested by Cywinska-Smoter(60) using Congo red spray
reagent for the development of the spots on TLC. Several workers
(61,62,63,64) have utilized the esterase inhibition technique for
the location of the organophosphorus insecticides on TLC and
their direct measurement on the plates. Routine determination of
organophosphorus insecticides by TLC has also been suggested by
Antoine and Mees(39). They have suggested palladium chloride
spray or 4-(4-nitrobenzyl) pyridine for location of the spots on
the TLC plate and comparison of intensity and area of the developed colored spots with those of known amounts of the insecticides on the same TLC plate. Hellomann(65) has used bovine liver
extract as spray followed by comparison of the developed yellow
colored spot for the semi-quantitative determination. Kosmatyi
et al.(66) have used bromophenol-blue/silver nitrate as chromogenic reagent for the determination of diazinon. The author
(67,68,69) has reported the densitometric determination of microgram amounts of organophosphorus insecticides in tissues.

Determination of dimethoate and malathion using palladium
chloride spray on TLC plate was described by MacNeil and co-
workers(70). Quantitative determination of organophosphorus
pesticides by TLC densitometry(71) and TLC fluorometry(72) have
also been described. Ivey and Oehler(73) have described the
determination of some organophosphorus pesticides in tissues and
urine of cattle. Getz(74) has described the quantitative estimation of chlorinated pesticides by densitometry. Dobbins(75) has
used the densitometric method for quantitation of chlorinated
pesticides using silver nitrate spray reagent. Quantitative TLC
technique for organochloro pesticides was also suggested by
MacNeil et al.(51) and Lakshinarayana and co-workers(76). Quantitation of carbamate pesticides by fluorescence densitometry on
TLC was described by some of the workers(77,78,79,80).

Zweig and Sherma(81) made an excellent review on the quali-

tative and quantitative aspects of thin layer chromatography. In
1977, a selective review on pesticide residues was written by
Thornburg(82).

From a review of the available literature it is clear that
methods based on thin layer chromatography are frequently being
used by various workers in the field for the quantitation of
pesticides. In the present paper the quantitative TLC technique
using densitometry is described in detail for the determination
of pesticides in various tissues pertaining to actual cases of
poisoning.

Preparation of Thin Layer Plates

Glass plates (20x20cm.) are well cleaned with non greasy
solvent and arranged on the aligning tray. Twenty-five grams of
the adsorbent silica gel-G (or alumina with calcium sulfate
binder) are mixed with fifty milliliters of distilled water in a
250 ml-flask and gently shaken to form a homogeneous slurry.
This slurry is transferred immediately (within sixty seconds of
mixing adsorbent with water) to the applicator. The spreader is
adjusted for a 250-μm thick layer (or other suitable thickness)
beforehand, and a thin layer of the adsorbent is coated on the
plates. In this way a homogeneous 250-μm thick layer of the
absorbent is applied to the glass plates. The plates are dried
at room temperature and then activated in an oven at 110° for 30
minutes.

Storage. The plates are stored in a desiccator over anhydrous calcium sulfate and protected from the laboratory fumes.

Application of samples. An aliquot (about one to ten μl) of
solution of the sample is spotted on the plate with a capillary
dropper or micropipette 1.5 cm. from one end of the plate. Care
must be taken to avoid diffusion of the spot; moreover, the area
of the spot should be restricted to a two- or three millimeter-
diameter. To ensure this, the solution should be applied in
stages and dried in a stream of warm air. The layer on the plate
should not be penetrated during spotting. Multiple spots should
be applied on the plate at least one centimeter apart.

Development of the plate. After ensuring that the applied
spots are dry, the spotted plate is placed in a developing
chamber containing the solvent system and the chamber is closed
with an air-tight lid. The solvent should be about 0.5 to 1 cm.
deep in the developing chamber. The chamber is saturated with
the solvent vapors for a particular separation before insertion
of the plate into the chamber. The chromatogram is allowed to
run by ascending technique at ambient temperature. The solvent
is generally allowed to run ten or fifteen centimeters up the
plate from the starting line to a pre-marked line. The plate

then removed from the developing chamber and dried to remove the
solvent.

Solvent Systems

Pure solvents of Analab grade should be used in TLC. The
various solvents possess different eluting properties. Further,
because of the differences in polarity of the organic insecticides and variations in their water solubility, several solvent
systems are tested for separation on the TLC plate. The large
number of solvent systems listed in Table II for organophosphorus-
organochlorine and carbamate insecticides provide satisfactory
alternative systems for the separation of the group of insecticides under consideration.

Location of the Spots

After separation, the plate is examined for the position of
the spots on the plate. Some of the compounds are detected on
the plate by viewing the plate under ultraviolet light (254 nm).
For location of the spots of insecticides on the TLC plate different chromogenic reagents are used.

Suitable location reagents for detecting organophosphorus-,
organochloro- and carbamate insecticides on TLC plates are given
below. Details of their composition and the treatment of the
plate necessary with each reagent are also given.

Organophosphate. For the estimation of phosphoric acid
esters, the developed plate is sprayed with 0.25% solution of
palladium chloride in 0.1 N HCl and dried in air. After exactly
ten minutes, the area of the yellow-brown colored spot developed
is measured and compared with that of the spots of standard
insecticide solutions run alongside.

Organochloride. For the estimation of chlorine containing
insecticides, the developed plate is sprayed with a 1.0% ethanolic
silver nitrate solution containing five milliliters of concentrated ammonia solution. After drying the sprayed plate in air, it
is irradiated with U.V. light (366 nm) for ten minutes. After
exactly ten minutes, the area of the developed spots is measured
and compared with that of the control insecticide run on the same
TLC plate.

Carbamate. For carbamate pesticides the plate is sprayed
with 1% ethanolic solution of Fast Blue-B, dried for thirty
minutes, and then sprayed with a 20% aqueous solution of sodium
hydroxide, and again dried at room temperature. Different
colored spots are obtained for various carbamate pesticides.
After exactly ten minutes, the spot is scanned by densitometry.

Rf Values

The most important factor in chromatography for the identi-

Table II: <u>Solvent Systems for TLC Separation of Pesticides</u>

<u>Organo-Phosphorus Pesticides</u>:

Ratio

(i)	Cyclohexane : Liquid Paraffin	(9:1)
(ii)	Cyclohexane : Chloroform	(7:3)
(iii)	Cyclohexane : Acetone : Acetonitrile . .	(17:2:1)
(iv)	Cyclohexane : Acetone : Ethanol	(95:1:4)
(v)	n-Hexane : Acetone	(17:3)
(vi)	Benzene	
(vii)	Petroleum Ether : Acetone	(7:3)
(viii)	Diethyl Ether : Methanol	(3:1)
(ix)	Chloroform : Liquid Paraffin	(9:1)
(x)	Carbon Tetrachloride : Ethylacetate . . .	(3:2)

<u>Chlorinated Organic Pesticides</u>:

Ratio

(i)	n-Hexane : Acetone	(9:1)
(ii)	n-Hexane : Liquid Paraffin	(39:1)
(iii)	n-Hexane : Butanone	(39:1)
(iv)	n-Hexane : Dioxan	(49:1)
(v)	Cyclohexane : Acetone	(99:1)
(vi)	Cyclohexane : Dimethyl Formamide	(19:1)
(vii)	Cyclohexane : Liquid Paraffin	(4:1)
(viii)	Cyclohexane : Methylene Dichloride . . .	(9:1)
(ix)	Light Petroleum : Chloroform	(97:3)
(x)	Light Petroleum : Acetic Acid	(19:1)

<u>Carbamate Pesticides</u>:

Ratio

(i)	Benzene : Ethyl-Methyl Ketone	(9:1)
(ii)	Benzene : Acetone	(19:1)
(iii)	Benzene : n-Hexane	(17:1)
(iv)	Benzene	
(v)	n-Hexane : Diethylether	(3:1)
(vi)	Cyclohexane : Ethyl-Methyl Ketone	(9:1)
(vii)	Petroleum Ether (40-60°) : Chloroform . .	(19:1)
(viii)	Chloroform	
(ix)	Chloroform : Dichloromethane	(99:1)
(x)	Chloroform : Methanol	(9:1)

fication of separated substances is the Rf value. However, the exact Rf value depends on several factors such as the adsorbent, thickness of the layer on the plate, activation of the plate, composition of the solvent system, saturation of the developing chamber with the solvent, and temperature at which the separation is carried out. Actually, thin layer chromatography is a comparative art rather than an absolute technique as commented by Clarke(83). According to Zweig and Sherma(84) an agreement of about two millimeters in the migration distances of the unknown and standard spot is considered adequate for tentative identification, since the sample spot may be affected slightly by the co-extractives despite thorough clean-up steps.

To overcome the difficulty due to non-reproducibility of Rf values, a reference standard should be run together on the same plate along with the unknown material. It makes comparison of Rf values much easier and also eliminates the chances of error due to any irregularity during the separation.

Estimation of Pesticides on TLC Plates Using Densitometry

Densitometry is a technique of direct determination of the concentration of the colored derivative on the chromatogram by absorptometry. The chromatogram is scanned with a beam of filtered light, a portion of which is absorbed by the colored spot and the remainder is transmitted by the spot on the chromatogram and measured photoelectrically.

In densitometric measurements, the colored spot is scanned along the line of solvent development. The galvanometer is at first set at zero on a portion of the sprayed and dried TLC plate near the spot to be scanned. The TLC plate is slightly shifted (say 0.5 mm), and the transmitted light through the spot is measured in terms of extinction. The spot is successively scanned, and the galvanometer records the maximum absorbance at the peak point. When the beam of light returns to the blank background of the thin layer plate, the galvanometer again reads zero. A curve is plotted of extinction versus scanning length of the spot. The area of the extinction curve is measured for each spot of different concentrations. This curve area is proportional to the concentration of the substance present in each spot. Densitometric scanning of the TLC spot is more accurate and sensitive for quantitative determination than other methods.

Data of Distribution of Pesticides in Tissues

The amount of the pesticides present in different autopsied tissues received from fatal cases of poisoning have been estimated quantitatively in our laboratory. In accidental poisoning cases where vapors of insecticides are inhaled, no insecticide is likely to be detected in the stomach or intestine but considerable amounts may be detected in the lungs, brain, and blood.

The data regarding the amount of a particular pesticide
found in various different tissues and body fluids using densitometry
on TLC plates is recorded in Tables III and IV. The age of the
victims and other particulars obtained from the records of investigating officers are also given.

Discussion

From the perusal of data recorded in Tables III and IV, it
is evident that insecticides are distributed in practically all
the visceral tissues. However, the amount of insecticide in
different body tissues depends on the time period between ingestion of the insecticide and death. At the initial onset of sym-
ptoms, the majority of the insecticides is still in the stomach.
But in cases where death is prolonged, the insecticide accumu-
lates in other tisses, e.g. liver and kidney. By conducting
control experiments on white rats which were fed lethal doses of
some of the organophosphorus insecticides (parathion and malathion) it was observed that these insecticides tended to accumulate in the body fat, and a small amount was excreted in the
urine. A small amount was also eliminated through excreta. A
higher concentration of the insecticide in the stomach is indicative that the death of the victim took place within a short period,
because the insecticide could not even be completely eliminated
from the stomach. A comparatively large concentration of pesticide
in liver tissue is suggestive of a delay in death from the time of
administration of the insecticide. However, it should be kept in
mind that the transport of insecticide to stomach from other body
tissues also depends on the media by which the insecticide is
administered. If the insecticide is mixed with milk, alcohol,
or lipid oil, the toxicity of insecticide is increased due to its
ready absorption through the gastro-intestinal tract. In such
cases, the time required is much less, and death occurs within
one or two hours of ingestion of the insecticide as in the case
shown in serial numbers 3, 11, 12, 16, 19, and 20 in Table IV
where the insecticide was administered by mixing it with liquor
or milk.

Stomach wash samples in nonfatal cases were found to contain
large quantities of insecticide, thereby indicating that the major
part of the insecticide was excreted by the body. Only trace amounts
of pesticides would have remained in the body which did not prove
fatal. The distribution pattern in different body organs is practically
the same in all cases examined for parathion poisoning, as death
generally occurred within an hour or two from administration of the
poison.

The distribution data of a case of poisoning by phosphamidon
recorded in serial number 24 in Table III indicates that comparatively large amounts of phosphamidon was detected in the lung

Table III

DISTRIBUTION OF ORGANOPHOSPHORUS INSECTICIDES IN DIFFERENT BODY TISSUES AND FLUIDS FROM CASES OF HUMAN POISONING

Insecticide	Sex & Age	Time period before death	Mode of ingestion	Concentration (milligram per cent of tissue/material)								
				Stomach	Intestine	Liver	Kidney	Spleen	Lung	Heart	Brain	Stomach Wash
1. Ethyl Parathion	M-25	1½ hrs.	in bread	8.75	6.25	4.00	3.75	2.08	2.50	3.13	-	-
2. Ethyl Parathion	F-50	2 hrs.	in bread	6.00	4.88	4.50	3.88	2.38	2.88	3.18	-	-
3. Ethyl Parathion	F-3	1 hrs.	in bread	5.13	4.88	4.13	2.63	-	4.10	4.08	-	-
4. Ethyl Parathion	F-22	2 hrs.	in bread	6.88	5.63	5.13	4.78	3.88	3.13	3.50	-	-
5. Ethyl Parathion	M-40	1½ hrs.	in liquor	6.38	5.36	4.80	4.63	2.30	5.24	5.80	-	2.60
6. Ethyl Parathion	M-28	1½ hrs.	in liquor	10.26	8.40	7.24	6.80	3.42	0.90	1.47	-	24.50
7. Methyl Parathion	F-25	4 hrs.	unknown	9.31	4.39	6.51	1.69	1.42	-	-	-	-
8. Malathion	M-28	8 hrs.	oral	46.88	-	2.31	2.86	1.67	-	-	-	-
9. Malathion	F-18	3½ hrs.	unknown	-	11.63	2.13	-	-	-	-	-	25.00
10. Malathion	M-32	5 hrs.	in liquor	75.00	-	6.00	2.86	1.34	-	-	-	8.38
11. Sumithion	M-61	3 hrs.	oral	126.40	28.40	7.40	6.82	6.20	-	-	-	65.00
12. Sumithion	F-22	2 hrs.	oral	12.5	-	12.50	6.25	4.17	-	-	-	84.20
13. Sumithion	M-43	1 hrs.	in liquor	84.6	16.38	8.32	6.28	4.20	-	-	-	-
14. Sumithion	F-25	Recovered	oral	-	-	-	-	-	-	-	-	184.8
15. Sumithion	F-18	Recovered	oral	-	-	-	-	-	-	-	-	252.0
16. Dimethoate	M-39	24 hrs.	unknown	0.84	-	2.20	1.64	0.94	-	-	-	8160.0
17. Dimethoate	F-27	16 hrs.	oral	3.25	-	4.75	2.50	4.38	2.75	1.06	0.32	28.4
18. Diazinon	M-35	2½ hrs.	unknown	6.35	2.84	3.84	2.68	3.20	-	-	-	-
19. Diazinon	M-54	3 hrs.	unknown	18.30	3.60	4.54	2.50	2.50	-	-	-	-
20. Meta-systox	F-18	1½ hrs.	with Zn3P2	16.40	4.34	3.22	2.84	1.80	-	-	-	-
21. Meta-systox	M-28	1½ hrs.	with Zn3P2	9.54	8.19	5.44	3.69	3.44	-	-	-	-
22. Baytex	M-22	Recovered	oral	-	-	-	-	-	-	-	-	-
23. Phosphamidon	F-19	Recovered	oral	-	-	-	-	-	-	-	-	61.40
24. Phosphamidon	M-50	6 hrs.	spray	0.094	0.11	0.25	0.13	0.35	0.13	-	-	32.8
25. DDVP	M-24	unknown	in liquor	14.60	-	4.32	3.84	2.96	-	-	0.31	-

In addition, subject number 22 contained 9320 mg per cent pesticide in the vomitus.

Table IV

DISTRIBUTION OF ORGANOCHLORIDE INSECTICIES IN DIFFERENT BODY TISSUES AND FLUIDS FROM CASES OF HUMAN POISONING

Insecticide	Sex & Age	Time period before death	Mode of ingestion	Concentration (milligram per cent of tissue/material)									
				Stomach	Intestine	Liver	Kidney	Spleen	Lung	Heart	Blood	Urine	Stomach Wash
1. Endrin	F-25	4 hrs.	unknown	14.50	66.00	20.00	5.17	2.17	-	-	0.85	0.35	-
2. Aldrin	F-17	3¾ hrs.	oral	-	-	4.00	1.25	-	-	-	-	-	7.88
3. Endrin	M-3	1¼ hrs.	in milk	2.63	-	-	-	-	-	1.08	0.55	0.55	-
4. Endrin	F-22	1½ hrs.	unknown	1.04	1.33	-	-	-	-	-	-	-	-
5. Endrin	M-30	5 hrs.	oral	4.46	3.19	5.69	2.31	1.20	-	-	-	-	-
6. Aldrin	M-35	4 hrs.	oral	2.90	1.31	0.84	0.75	0.54	-	-	-	-	-
7. Endrin	F-38	5 hrs.	unknown	3.75	2.56	2.13	1.34	-	-	-	-	0.25	-
8. Endrin	M-37	2½ hrs.	unknown	3.47	2.88	2.81	1.19	-	-	-	-	-	3.17
9. Endrin	F-40	3½ hrs.	unknown	1.94	1.69	4.80	1.18	-	-	-	0.43	-	-
10. BHC	F-31	2½ hrs.	unknown	5.81	3.68	3.50	2.38	0.99	-	0.56	-	-	-
11. Endrin	M-33	1½ hrs.	in liquor	4.88	1.94	1.56	1.31	1.38	-	-	-	-	-
12. Endrin	M-29	1 hrs.	in milk	6.38	1.56	1.75	1.06	0.61	-	-	-	-	-
13. Endrin	F-28	5 hrs.	unknown	5.04	-	1.94	1.44	1.19	1.06	1.79	-	-	-
14. Endrin	F-42	6 hrs.	unknown	6.29	-	2.88	1.72	1.31	0.38	0.79	0.68	-	4.69
15. Aldrin	M-35	1½ hrs.	oral	3.84	1.59	-	-	1.28	0.57	1.05	-	-	-
16. Endrin	M-32	1½ hrs.	in liquor	81.00	22.00	48.00	45.00	30.00	-	25.00	-	-	-
17. DDT	M-28	5½ hrs.	unknown	6.08	-	4.91	3.17	2.98	-	-	-	-	-
18. Endrin	F-28	1½ hrs.	unknown	60.00	25.00	22.5	12.5	12.5	-	32.15	-	-	-
19. BHC	F-31	4¼ hrs.	in liquor	7.09	3.01	4.06	3.55	3.01	-	-	-	-	-
20. Endrin	F-25	2 hrs.	in liquor	61.00	22.50	49.00	25.00	20.30	-	22.50	-	-	-
21. Endrin	F-24	3¼ hrs.	unknown	13.88	8.06	10.91	6.59	4.91	1.72	-	-	-	-

In addition, 8.12, 3.25, 20.00 mg per cent was found in the vomitus of subject numbers 2, 9, and 20 respectively. Also, subject number 2 showed 6.38 mg per cent pesticide in the stomach contents.

and brain tissues and only very small amounts of phosphamidon could be found in the stomach. These findings suggest that phosphamidon was most probably administered through inhalation rather than by mouth. The distribution data was supported by the report of the investigation officer that the victim was accidentally exposed to the insecticide vapor during spraying of his agricultural field.

Similarly, large excess of sumithion present in stomach wash samples and in the stomach tissues is recorded in serial numbers 11, 13, 14, and 15 in Table III. This is suggestive of suicidal poisoning by oral comsumption of Tik-20, a household insecticidal preparation containing sumithion.

Data recorded in Tables III and IV shows the amount of the insecticide in the unchanged form. However, most of the tissues also contain the metabolite of the insecticide. From the above discussion it is evident that the study of the distribution of pesticides in different tissues and body fluids can help in resolving the questions of how and when the pesticide was administered. Furthermore, the clue regarding the nature of poisoning whether suicidal, homicidal or accidental can also be obtained by careful study of the distribution pattern of the insecticides in body tissues and fluids derived by quantitative TLC.

Abstract

The absolute proof of fatal cases of poisoning is the detection of the poison in different organs of the body. Due to the chemical nature of the poison and its metabolism, the suspected chemical is rarely evenly distributed in body tissues but tends to accumulate in some particular organ. Hence, to conduct a thorough forensic investigation, it is necessary to analyze different visceral tissues qualitatively and quantitatively for poison residues. In this paper, the quantitative TLC technique using densitometry for the determination of traces of pesticides has been described in detail. The concentration of pesticides in various human organs from poisoning cases has been recorded as well as data on tissue distribution. The distribution pattern of several important pesticides, like parathion, malathion, sumithion, and endrin has been studied. The results of these studies have aided in solving the questions in the probable causes of death or poisoning-- how and when the pesticide was probably administered, and if death occurred by suicide, homicide or accident. Several actual cases of poisoning and subsequent tissue analyses by quantitative TLC have been described.

Acknowledgement

Editing and final typing of this chapter was done by Douglas Smith, Linda A. Keola Pereira and Gunter Zweig, to whom the editors are greatly indebted.

Literature Cited

1. Horncastle, D.C.J. Med. Sci. Law, 1977, 17(1), 37.
2. Arterberry, J.D.; Durham, W.I.; Elliott, J.W.; Wolfe, H.R. Archiv. Envir. Health, 1961, 3, 476.
3. Heyndrickx, A.; Vercruysse, A. J. Pharm. Belg., 1964, 12, 161.
4. Heyndrickx, A.; Vercruysse, A.; Nee, M. J. Pharm. Belg., 1967, 22, 127.
5. Heyndrickx, A.; Vanhoff, F.; Dewolf, L.; Van Petghom, C. J. Forensic Sci. Soc., 1974, 14, 131.
6. Lewin, J.I.; Norsis, R.J.; Hughes, J.T. J. For. Sci., 1973, 2, 101.
7. Tewari, S.N.; Harpalani, S.P. J. Ind. Acad. Forens. Sci., 1977, 16, 61.
8. Tewari, S.N.; Harpalani, S.P. Chemical Era, 1976, 12, 397.
9. Tewari, S.N.; Sharma, I.C. Proc. Nat. Acad. Sci. (India), 1976, 41(A I), 59.
10. Tewari, S.N.; Harpalani, S.P. "Toxicological study of the distribution of organophosphorus insecticides in different tissues in cases of human poisonings", Presented at the Scientific Mtg. at Hyderabad, Dec. 9, 1978.
11. Tewari, S.N.; Sharma, I.C. Chemical Era, 1978, XIV(6), 215.
12. Tewari, S.N.; Harpalani, S.P. J. Ind. Acad. Forensic Sci., 1977, 16, 4.
13. Tewari, S.N.; Harpalani, S.P. J. Chromatog., 1977, 130, 229.
14. Tewari, S.N.; Sharma, I.C. J. Chromatog. 1977, 131, 275.
15. Tewari, S.N.; Singh, R. "TLC technique for the separation and identification of carbamate pesticides in post-mortem material", J. Chromatog., 1979, 172, 528.
16. Randerath, K. "Thin Layer Chromatography"; Academic Press: New York and London, 1963.
17. Choulis, N.H. Chimik. Chronika, 1965, 30A, 52.
18. Choulis, N.H. J. Pharm. Sci., 1967, 56, 196.
19. Court, W.E.; Shellard, E.J., Ed. "Quantitative Paper and Thin Layer Chromatography"; Academic Press: London and New York, 1968.
20. Purdy, S.J.; Truter, E.V. Analyst, 1962, 87, 802.
21. Johnson, C.A. "The Visual Assessment of TLC"; Academic Press: London and New York, 1968.
22. Gaenshirt, H.; Polderman, J. J. Chromatog., 1964, 16, 510.
23. Morrison, J.C.; Chatten, L.G. J. Pharm. Sci., 1964, 53, 1205.
24. Oswald, N.; Flueck, H. Pharm. Acta. Helv., 1964, 39, 293.
25. Schute, J.B.; DeJong, H.J.; Dingjam, H.A. Pharm. Weekly, 1970, 105, 1025.
26. Dallas, M.S.J. J. Chromatog., 1965, 17, 267.
27. Genest, K. J. Chromatog., 1965, 18, 531.
28. Blunden, G.; Hardman, R.; Morrison, J.C. J. Pharm. Sci., 1967, 56, 948.

29. Shellard, E.J. Lab. Pract., 1964, 13, 290.
30. Thomas, A.E., III; Scharoun, J.E.; Ralston, H. J. Am. Oil.
     Chem. Soc., 1965, 42, 790.
31. Shellard, E.J.; Alam, M.Z. J. Chromatog., 1968, 33, 347.
32. Salo, T.; Salminen, K.; Fiskari, K. Z. Lebensmittelfunters-
     U-Forsch, 1962, 117, 369.
33. Luckens, M.M. J. Forensic Sci., 1966, 11(1), 64.
34. Jadwiga, Z. Roczniki. Paust. Zakl. Hig., 1965, 16(4), 397.
35. Beckman, H.; Winterlin, W. Bull. Environ. Contain. Toxi-
     col., 1966, 1(3), 78.
36. Hill, K. R. Chem. Spac. Mfr. Ass. Proc. Ann. Meet., 1968,
     55, 172.
37. Ramasamy, M. Analyst, 1969, 94, 1075.
38. Villeneuve, D.C.; Butterfield, A.C.; Grant, D.L.; McCully,
     K.A. J. Chromatog., 1970, 48(3), 567.
39. Antoine, O.; Mees, G. J. Chromatog., 1971, 58(2), 257.
40. Midio, A.F. Revta. Farm. Bioquim. Univ. S. Paulo., 1973,
     11(1), 105.
41. Tewari, S.N.; Harpalani, S.P. Z. Anal. Chem., 1977, 285,
     48.
42. Tewari, S.N.; Harpalani, S.P. Proc. Nat. Acad. Sci., India,
     1972, 42A, 287.
43. Tewari, S.N.; Harpalani, S.P. Microchimica Acta., 1973, 2,
     321.
44. Tewari, S.N.; Harpalani, S.P. J. Ind. Acad. Forens. Sci.,
     1977, 16(2), 14.
45. Fischer, R.; Otterbeck, N. Sci. Pharm., 1959, 27, 1.
46   Curry, A.S. "Poison detn. in human organs."; Charles C.
     Thomas: Springfield, 1963.
47. Mills, P.A. J. Ass. Offic. Agr. Chem., 1959, 42, 734.
48. Kovacs, M.F., Jr. J. Ass. Offic. Anal. Chem., 1966, 42(2),
     365.
49. Klisendo, M.A.; Yurdova, Z.F. Khim. Sel. Khoz., 1968, 6(8),
     593.
50. Bublik, L.I.; Kosmatyi, E.S. Zev. Lab., 1970, 36(10), 1194.
51. MacNeil, J.D.; Frei, R.W. J. Chromatog., 1975, 13(6), 279.
52. Ballschmiter, K.; Toelg, C. Z. Anal. Chem., 1966, 215(5),
     305.
53. Dobbins, M.F.; Touchstone, J.C. "Quantitative TLC."; Ed. by
     J.C. Touchstone, Wiley: New York. 1973.
54. Tewari, S.N.; Sharma, I.C. Z. Anal. Chem. 1976, 278, 127.
55. Tewari, S.N.; Sharma, I.C. Z. Anal. Chem. 1976, 282, 143.
56. Tewari, S.N.; Sharma, I.C. Microchimica Acta., 1976, 2(3-
     4), 323.
57. Vylegzhanina, G.F.; Kalmykova, R.G. Gig. Sanit., 1969,
     34(4), 75.
58. Askew, J.; Ruzicka, J.H.A.; Wheals, B.B. Analyst, 1969, 94,
     275.
59. Zycinski, D. Roc. Panst. Zakl. Hig., 1971, 22(2), 189.
60. Cywinska-Smoter, K. Roc. Panst. Zakl. Hig., 1972, 23(5), 505.

61. Stijve, T.; Cardinale, E. Mitt. Geb. Lebensmittelunters V.
    Hyg. 1971, 62(1), 25.
62. Schutzmann, R.L. J. Assoc. Off. Anal. Chem., 1970, 53(5),
    1056.
63. Sheifert, J.; Devidek, J. J. Chromatog., 1971, 59(2), 446.
64. Zadiozinska, J. Roczn. Panst. Zakl. Hig., 1972, 23(1), 23.
65. Hellomann, B. Roczn. Panst. Zakl. Hig. 1973, 24(2), 175.
66. Kosmatyi, E.S.; Kavetskii, V.M.; Lebedinska, L. Ukr. Khim.
    Z., 1973, 39(10), 1053.
67. Tewari, S.N.; Harpalani, S.P. Agra University Journal of
    Research (Science), 1977, XXVI(II), 47.
68. Tewari, S.N.; Harpalani, S.P. Pro. Nat. Acad. Sci. (India),
    1977, 47(B), 157.
69. Tewari, S.N.; Harpalani, S.P. Agra University Journal of
    Research (Science), 1977, XXVI(I), 13.
70. MacNeil, J.D.; MacLellan, B.L.; Frei, R.W. J. Assoc. Off.
    Anal. Chem., 1974, 57(1), 165.
71. Stefanac, Z.; Stengl, B.; Vasilic, Z. J. Chromatog., 1976,
    124, 127.
72. Zakrevsky, J.G.; Mallet, V.N. Bull. Environ. Contain.
    Toxicol., 1975, 13, 633.
73. Ivey, M.C.; Oehler, D.D. J. Agric. Food Chem., 1976, 24,
    1049.
74. Getz, M.E. Pestic. Chem. Proc. Int. Congr. Pestic. Chem.
    2nd, Ed. by Tahori; Gordon; and Breach, 1971, 4, 43.
75. Dobbins, M.F.; Touchstone, J.C. Quant. Thin Layer Chrom-
    atog., 1974, 90, 396.
76. Lakshinarayana, V.; Menon, P.K. Res. Ind., 1974, 19(2),
    65.
77. Frei, R.W.; Lawrence, J. Pestic. Chem. Proc. Int. Congr.
    Pestic. Chem. 2nd, Ed. by Lahori; Alexander; Gordon and
    Breach, 1971, 4, 193.
78. Lawrence, J.F.; Laver, G.W. J. Assoc. Off. Anal. Chem.,
    1974, 57(5), 1022.
79. Francoeur, Y.; Mallet, V. J. Assoc. Off. Anal. Chem.,
    1976, 59, 172.
80. Ernest, G.F.; Roder, S.J.; Tjan, G.H.; Jansen, J.T. J.
    Assoc. Off. Anal. Chem., 1976, 58, 1015.
81. Zweig, G.; Sherma, J. Anal. Chem., 1976, 48(5), 66.
82. Thornburg, W. Anal. Chem. 1977, 49(5), 98.
83. Clarke, E.G.C. Isolation and Identification of Drugs"; The
    Pharmaceutical Press: London, 1969.
84. Zweig, G.; Sherma, J. "Thin Layer and Liquid Chromatography
    of Pesticides of International Importance", Vol. VII,
    Academic Press: New York, 1973.

RECEIVED May 27, 1980.

The Determination of 2,3,7,8-Tetrachlorodibenzo-*p*-dioxin in Human Milk

L. A. SHADOFF

Analytical Laboratories, Dow Chemical, Midland, MI 48640

Recent studies have indicated that humans may be exposed to 2,3,7,8-tetrachlorodibenzo-p- dioxin(2,3,7,8-TCDD), a toxic and teratogenic substance in laboratory animals [1]. A fraction of beef fat samples from cattle known to have grazed on pasture treated with herbicide 2,4,5-T (which contains trace quantities of 2,3,7,8-TCDD) have been reported to contain low part per trillion (ppt) levels of 2,3,7,8-TCDD [2,3]; Two studies of bovine milk reported no detectable chlorodioxins however [4,5]. Combustion processes have been reported to produce chlorinated dioxins which enter the air as fly ash and soot [6,7,8]. Three different studies of human milk have been carried out by various workers to determine if humans contain detectable concentrations of 2,3,7,8-TCDD [9,20,22].

In the first reported study [9], six samples of milk were obtained in an area of Texas where 2,4,5-T is used for range-land management. The sample preparation scheme was caustic saponification, extraction, chromatography on silica followed by chromatography on alumina. The extract was evaporated to 10-20 μl and TCDD determined by gas chromatography/mass spectrometry (GC/MS). This technology is similar to that reported for beef fat sample preparation [2,3]. No 2,3,7,8-TCDD was detected with detection limits of 1-6 ppt.

Preliminary results of the determination of TCDD in twenty-one human milk samples [10] indicated the possible presence of TCDD in three of the samples at the 0.6-1.6 ppt level which was equal to the detection limit. The sample preparation procedure [12] did not utilize a saponfication step. Saponification had been used in prior procedures due to the possibility that TCDD was strongly bound to the matrix as has been the case for some pesticides [14]. In order to demonstrate that all the TCDD may be recovered from milk by extraction, a comparison of results between saponification and extraction without saponifiction was performed. Such a test must be

0-8412-0581-7/80/47-136-277$05.00/0

performed for every matrix, however; recovery from milk does
not insure that it may be recovered from liver, for example.
The advantage to eliminating saponification from the sample
preparation procedure is that fewer materials will be extracted
from the matrix along with the TCDD, since many substances are
indeed bound.

The determination of TCDD in the final extract was by high
resolution mass spectrometry multiple scan averaging. No gas
chromatography was employed; the sample was vaporized via the
direct insertion probe (12,13). This method is inherently more
prone to interferences than GC-MS (17,18); the response
observed may not be due to TCDD. The results do serve as an
indication, however, that if TCDD is present in human milk it
is at 1 ppt or less. Thus any methods developed must be
capable of this detection limit.

A recently concluded study of thirteen samples from the U.S.
was a collaborative effort sponsored by NIEHS. The samples
were prepared and TCDD determined by three different
laboratories. There was no confirmed detection of TCDD in any
of the samples (11). A summary of the results of completed
studies of the determination of TCDD in human milk is given in
Table I.

EXPERIMENTAL

CAUTION, 2,3,7,8-TCDD is an extremely toxic substance which
should be handled with care (13).

The methodology used for the sample preparation in our
laboratory of the NIEHS samples employed new technology in many
of the stages and is an adaption of that reported for fish (8).
The initial extraction involves completely absorbing the sample
onto cellulose gauze (X-tube, Analytichem International). The
milk is not treated in any way except for the addition of
^{13}C-2,3,7,8-TCDD as an internal standard. The capacity of the
gauze is 15g of milk and at that loading, recovery experiments
indicate that approximately 10% of the TCDD is extracted for
each 10 ml of hexane passed through the gauze at ambient temperature. The volume of extract would be in excess of 100 ml.
Therefore, the gauze is placed in an extraction apparatus
and hexane refluxed through the tube for a sufficient time to
pass at least 1 ℓ through the gauze. The plastic container
supplied with the X-tube is not suitable for hot extraction
since it deforms and contaminates the extract with Ionol and
low molecular weight polypropylene. A glass thimble was
constructed with an outlet at the bottom of 1 cm of 0.5
m.m.i.d. tubing to control the flow through the gauze.

The approximately 75 ml of hexane extract is transferred
directly onto a column containing 6 g of activated silica
(Bio-Sil, Biorad, Inc.) coated with 25% w/w conc. H_2SO_4 (8).
This column has a high capacity for the fats from milk, leaving

TABLE I. Summary of the results of completed studies on the Determination of 2,3,7,8-TCDD in human milk as of March, 1979.

2,3,7,8-TCDD IN HUMAN MILK

DONOR LOCATION	STUDY BY	RESULTS (detection limit ppt)
Texas	Dow	6 spls not detected (1-6)
Missouri	Harvard	6 spls not detected (0.7-3.3)
Texas	Harvard	2 spls 0.6-1.5 (0.6-1.6)
		4 spls not detected (1.2-1.9)
Oregon	Harvard	1 spl 0.8-1.4 (0.7-1.5)
		6 spls not detected (0.4-2.9)
Massachusetts	Harvard	6 spls not detected (0.3-1.6)
Texas	NIEHS[*]	9 spls not detected (0.4-4)
Kansas	NIEHS[*]	4 spls not detected (1-10)

[*]Collaboration Study - Dow, Harvard, NIEHS(RTP).

a residue of less than 10 mg when 1g of fat is passed through
the column. The progress through the column may be easily
monitored since it turns deep orange as the material is
absorbed. This column replaces the sulfuric acid extraction
employed in previous procedures (12,15,16,17) (see Figure 1).
    The effluent from this column plus a 30 ml rinse
containing the TCDD is passed directly onto a column containing
5g of activated basic alumina (Biorad, Inc.) which traps the
TCDD. The hexane passing through the column is discarded. The
TCDD is eluted from the column with 10 ml of methylene
chloride. This alumina column transfers the TCDD from over 100
ml of hexane to 10 ml of methylene chloride allowing a much
shorter time for solvent evaporation and the use of smaller
glassware. Recovery of TCDD at this stage is in excess of
85%.
    The eluate is evaporated just to dryness, taken up in
hexane and placed on a column of 10% $AgNO_3$ on activated silica
(8)(which removes DDE, many aliphatic halogenated substances
and sulfur containing substances). The eluate containing the
TCDD drops directly onto a high aspect ratio basic alumina
column (8). PCB's are eluted and TCDD retained on this column
by addition of 50 ml of 50% carbon tetrachloride in hexane.
The TCDD is then collected by the addition of 15 ml of 50%
methylene chloride in hexane.
    To this stage the sample preparation is similar to that
previously used to determine TCDD in environmental samples.
The detection limits reported were generally in excess of 5 ppt
(3,17) due to interferences from substances not removed in the
sample preparation. An additional, highly efficient separation
step was employed to allow the lower detection limits required
for meaningful determinations of TCDD in human milk.
    Reverse phase high pressure liquid chromatography (HPLC)
was used to further fractionate the sample and add another
dimension of specificity (8,21). The extract was evaported to
dryness and taken up in $CHCl_3$. The entire extract was injected
onto a DuPont Zorbax ODS column at 40°C using 2 cc/min. CH_3OH
mobile phase. Typical chromatograms are shown in Figure 2.
Note that a considerable amount of U.V. absorbing material
present in the extract is separated here. The TCDD fraction is
extracted into hexane after the addition of water. After
evaporation to dryness the sample is ready for the
determination of TCDD.
    Gas chromatography-low resolution mass spectrometry was
used to determine if TCDD is in the samples (12). The GC column
used was a 6 ft. x 2 mm i.d. glass 2.5% BMBT liquid crystal on
100/120 mesh Chromosorb WHP (Altech Assoc.) at 225°C. In
Figure 3 is shown the output from a standard and typical human
milk extract run on an LKB-9000 with a SI-150 (System
Industries) data system equipped with a computer controlled
multiple ion detector (Ledland, Inc.). The entire extract was

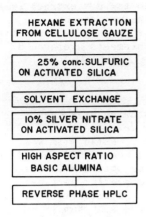

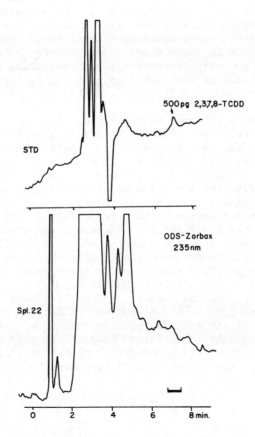

Figure 1.   Sample preparation scheme for the determination of 2,3,7,8-TCDD in human milk as performed for NIEHS study

Figure 2.   High Pressure reverse-phase liquid chromatography trace of 2,3,7,8-TCDD standard and a typical human milk extract

injected by adding 7 μl of iso-octane to the dry extract in a
0.3 ml cone shaped vial, washing down the walls, and
withdrawing as much as possible (the syringe needle was first
filled with iso-octane). Recovery measurements averaged 88 %
efficiency of removal from the vial (see Figure 3).

RESULTS AND DISCUSSION

As indicated in Table I, several studies have been
performed to determine whether there is detectable
concentrations of 2,3,7,8-TCDD in humans. Human milk was
chosen because it has a high fat content where TCDD should
reside since it is lypophyllic, it may be obtained in large
enough quantities to allow sufficient sample for part per
trillion determinations, and it does not require violation of
the body and the legal problems associated with sample
collection.

The geographic areas were chosen in all cases to determine
if the presence of TCDD correlates to the use of 2,4,5-T
herbicides which contain trace levels of 2,3,7,8-TCDD. In no
case was there a confirmed detection of TCDD. The detection
limits achieved in these studies were in the low part per
trillion range with the Harvard study indicating the possible
presence of TCDD at or below 1 ppt setting an upper limit to
the concentration of TCDD which a method must be capable of
detecting.

The sample preparation methodology for the determination
of TCDD at these low levels is an active area of development as
indicated by the improved procedure reported here and used to
analyze the milk for the NIEHS study. It uses reagent modified
adsorbants, a higher efficiency basic alumina column than
previously (19) and a new degree of separation, reverse phase
HPLC as an integral part of the procedure. All of these are
directed towards improving the specificity of the sample
preparation for 2,3,7,8-TCDD. Increased specificity is needed
since the lowest detection limit achievable in all previous
studies of TCDD in environmental samples has been dictated by
the presence of other substances in the samples which have not
been removed by the sample preparation.

Recently, all the isomers of TCDD have been synthesized
and separated (21). This allows the estimation of TCDD isomer
specificity of the methods. The separation of components by
gas chromatography is discussed by Jennings (22) who relates
relative retention times, theoretal plates and degree of
separation in the formula (rearranged here to separate out the
term α):

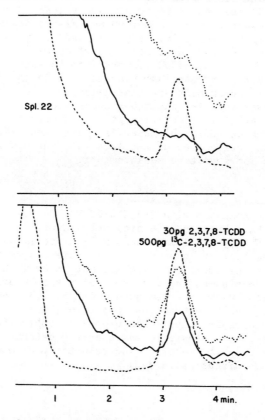

Figure 3. *Gas chromatography–low resolution mass spectrometry mass chromatograms for 2,3,7,8-TCDD standard and human milk extract ((————) 320; (· · ·) 322; (— — —) 334)*

$$\alpha = \frac{N}{N-4R}$$

α = relative retention times
      of the two components
N = theoretical plates
R = degree of separation
      (1.0=completely separated)

With a good packed column of 4000 theoretical plates and a degree of separation of 0.5 (not separated) substituted into the formula, all peaks with relative retention indices between 0.968 and 1.033 are indicated as not separated. Based on this the GC peak measured as 2,3,7,8-TCDD may also contain 1,2,6,9-, 1,2,3,6-, 1,2,3,9-, 1,2,3,7-, 1,2,3,8-, and/or 1,2,3,4-TCDD (21). The reverse phase HPLC step reported here would separate the 1,2,6,9- and 1,2,3,4-TCDD isomers from the others.

The analytical accomplishments achieved in the studies of 2,3,7,8-TCDD in human milk are considerable. The detection limits are lower than for any other non-radioactive compound with a high degree of selectivity gained through multi-step sample preparation and highly specific detection. None the less, it is an active area of research into the development of more specific and sensitive measurements.

Abstract

Are humans contaminated with 2,3,7,8-TCDD? Several laboratories have been and are presently involved in determining this by examining human milk. Preliminary results have indicated that it is necessary to achieve detection limits below one part per trillion. The attempt to achieve this requires the development of very selective sample preparations capable of preserving picogram quantities of 2,3,7,8-TCDD. In general, they involve extraction, partition, and flow-through liquid chromatography steps. The prepared samples have been analyzed by the most sensitive and specific detectors available, mass spectrometers. The results to date and the methodology are discussed in detail.

LITERATURE CITED

1. Schwetz, B. A., Norris, J. M., Sparschu, G. L., Rowe, V. K. Gehring, P. J., Emerson, J. L., Gerbig, C. G., Adv. Chem. Ser., 120, 55 (1973).
2. Kocher, C. W., Mahle, N. H., Hummel, R. A., Shadoff, L. A., Getzendaner, M. F., Bull. Environ. Contam. Toxicol., 19, 229 (1978).
3. Personal Communication, Ross, R. T., Dioxin Project Manager Office of Pesticide Programs, U. S. Environmental Protection Agency, December 16, 1975 and June 25, 1976, to Analytical Collaborators: Pest. Chem. News. 17 (June 23, 1976).
4. Mahle, N. H., Higgins, H. S., Getzendaner, M. E., Bull. Environ. Contam. Toxicol., 18, 123 (1977).
5. Lamparski, L. L., Mahle, N. H., Shadoff, L. A., J. Ag. Food Chem., 26, 1113 (1978).
6. Olie, K., Vermeulen, P. L., Hutzinger, O., Chemosphere, 1977, 455.
7. Buser, H. R., Bosshardt, H. P., Rappe, C., ibid., 1978, 165.
8. "The Trace Chemistries of Fire", Chlorinated Dioxin Task Force, Dow Chemical Co., 1978.
9. Shadoff, L. A., Hummel, R. A., Lamparski, L. L., Davidson, J. H., Bull. Environ. Contam. Toxicol., 18, 478 (1977).
10. Meselson, M., O'Keefe, P. W., Personal Communication to U. S. Congressman J. Weaver, January 26, 1977.
11. Personal Communication, McKinney, J. D., 1979.
12. O'Keefe, P. W. and Meselson, M., J.A.O.A.C., 61 621 (1978).
13. "Summary of Safe Handling of 2,3,7,8-Tetrachlorodibenzo-p-dioxins in the Laboratory", Biochemical Research Laboratory, The Dow Chemical Company, Midland, MI 48640.
14. Clark, D.E., Palmer, J.S., Radeleff, Rd., Crookshank, H.R., Farr, F. M., J. Ag. Food Chem., 23, 573 (1975).
15. Baughmann, R. W., Meselson, M., Environ. Health Persp., 5, 27 (1973).
16. Baughmann, R. W., Meselson, M., 166th National Meeting, American Chemical Society, Chicago, Illinois (1973), Abstract Pest. 55.
17. Shadoff, L. A., Hummel, R. A., Biomed. Mass Spectrom., 5, 7 (1978).
18. McKinney, J. D., "Chlorinated Phenoxy Acids and Their Dioxins", Ramel, C. (ed), Ecol. Bull. (Stockholm), 27, 53 (1978).
19. Hummel, R. A., J. Agric. Food Chem., 25, 1049 (1977).
20. Pfeiffer, C. D., Nestrick, T. J., Kocher, C. W. Anal. Chem., 50, 800 (1978).
21. Nestrick, T. J., Lamparski, L. L., Stehl, R. H., Anal. Chem., 51, 2273 (1979).
22. Jennings, W., "Gas Chromatography with Glass Capillary Columns", Academic Press, New York, 1978, pp 13 ff.

RECEIVED March 30, 1980.

Reduction of Radioactive Metabolic Data Using A Desk-Top Computer Network

W. L. SECREST, W. C. FISCHER, D. P. RYSKIEWICH, J. E. CASSIDY, and
G. J. MARCO

Ciba–Geigy Corporation, Agricultural Division, P.O. Box 11422, Greensboro, NC 27409

In order to register a chemical with the Environmental Protection Agency (EPA) for use as a pesticide, sound scientific data must be presented and strict requirements be met. The average time required to generate data which satisfy these criteria is presently estimated to be nine years from the original synthesis to the appearance of an agricultural chemical in the marketplace. Extensive metabolism information is necessary for complete understanding of toxicological, environmental, and biochemical effects of the chemical and to support the residue data. A single herbicide, Dual®, for use on a single crop, soybeans, required 46 volumes of data submitted to the EPA for review. These volumes represented a stack of paper 10 feet tall. In order for the metabolism group to meet their tight deadlines for our extensive contribution to this volume of information, we extensively use computers for collection, reduction, summarization, and interpretation of data.

Three general questions about pesticides in agricultural products and the environment must be answered: where is it and its metabolites, how much of these are there and what are the metabolites. The Metabolism Section of the Biochemistry Department uses a variety of analytical techniques to answer these questions. Metabolism studies are generally done with radioactive carbon (^{14}C), which provides a convenient method for detection and quantitation of a radioactive pesticide. A preferred way to measure radioactivity is with a liquid scintillation counter (LSC). From 1/2 to 22 minutes can be required per sample vial. Between 300 and 600 vials are counted on an average day.

Outputs from the LSC must be corrected for background radiation and counting efficiency. The parts-per-million (ppm) equivalent to the radioactive chemical in a biological, soil or water sample can then be determined by accounting for aliquot size, specific activity, and other correction factors specific for the sample. Partitioning characteristics and chromatographic data can be treated in a like manner. Although many

0-8412-0581-7/80/47-136-287$05.00/0

modern LSC(s) offer options which calculate dpm from raw counting
data, these options offer little or no advantage in this application because of the many additional calculations and appropriate formatting of results necessary.

Based on past experience, the following times would be
required to reduce the data from 100 samples; more than four
hours for hand calculations, 3 hours for a simple programmable
hand calculator, 30 minutes for a time-share computer service,
and less than 2 minutes for directly interfaced minicomputers.

Hardware

The desk top computer network is shown in Figure 1. The
liquid scintillation counters which we routinely use to generate
metabolism data include two Beckman LS-200 series liquid scintillation counters and two Tracor Mark III liquid scintillation
counters, however, other LSC makes and models may be interfaced.
The Tracor counters are equipped with two standard communications
output connectors. One is used to drive the printer, the other
is available to the user. The Beckman counters use a unique
teletype driver but an interface produced by William Palmer
Industries can split the signal and provide a standard communication output. Therefore, all data output from the liquid scintillation counters are available in an electrical standard format,
RS-232C. Hardware interfacing was therefore simplified and data
collection was a matter of software development.

The data reduction hardware is based on a Hewlett-Packard
9825A desk top computer. It is supported by one megabyte of
flexible disk storage, a printer/plotter and the necessary interface equipment for on-line LSC data collection. Reliability was
a prime factor when the hardware was chosen. The LSC(s) and data
system run virtually unattended, 24 hours a day, 365 days a year.
Samples are typically counted for two minutes each plus one
minute for the external standard. Therefore, data from the four
counters are received by the HP9825A at an average interval of
45 seconds. With such a demand on the system, computers require
good service support, more so than other instruments. Since
installation in May, 1977, there have been less than two work
days of cumulative downtime.

The system has since been upgraded to a desk top computer
network because of the increased load on the initial HP9825A
system. An HP9835A has been added so that data reduction and
analysis will not interfere with data collection. This second
desk top computer is supported by one megabyte of flexible disk
memory, a 180 CPS printer, and a four color X-Y plotter. An
HP2647 graphics terminal was added to provide additional remote
access to the data bases being generated. The system lends itself to expansion and continued modernization.

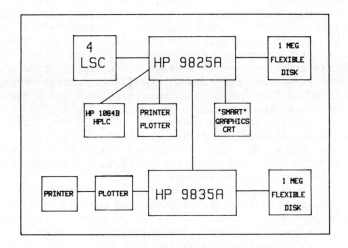

Figure 1. Desk-top computer network

Software

LSC Data Collection. The LSC data are collected by the
HP9825A using what is known as buffered I/O (input/output). A
portion of the computer's memory is set aside for the data from
each LSC. As the LSC(s) output data, it is stored directly in
the buffer assigned to the particular counter without otherwise
affecting the program which is running. After a predetermined
number of characters have been received or a specific character
has been received, the computer jumps to a subroutine which stores
the counter data. Each LSC is automatically assigned a priority,
and data output by two or more counters are handled simultaneously according to the priority.

Data storage actually involves reformatting the information
from each LSC, and storing it in a particular sequence as com-
pactly as possible. Each LSC manufacturer outputs data in a
different format and frequently several different output formats
are available on an individual counter. The computer is programmed to convert these different outputs to a standard format
for use in later computations. Nonessential data in each output
line are discarded to improve storage efficiency. After the data
have been reformatted, the computer reads the corresponding line
of data from the flexible disk. The oldest data are discarded
and replaced with the new data. The entire line is then restored
which completes the sequence. This procedure always retains the
last two passes of data. With the completion of storage, the
computer returns to the main program and continues from where
the interruption occurred. The time lost during the interruption
is typically less than two seconds.

Because this equipment runs continuously, provisions must be
made for the inevitable power failures and occasional deviations
from the expected data formats. The computer has a built-in
provision for power failures. The software is written so that
the program which was running at the time the power went off is
reloaded and restarted. The only data which is lost is that
which had been stored in the buffer but not stored on the
flexible disk at the time of the power failure.

Errors may arise from a variety of sources, i.e., when the
data collection program detects that something has gone wrong.
The software is written to assume the worst possible circumstance. The error, the line number in which the error occurred,
and the data are printed on the internal strip printer. It is
also assumed that the program might have been loaded incorrectly.
Therefore, the computer is directed to reload and restart the
program. This feature allows errors in the software or hardware
to be rapidly traced and corrected.

At the present time, the additional sample data required
for the calculations are entered through the computer keyboard.
These data include aliquot sizes, moisture factors, combustion
efficiencies, partitioning volumes, etc. Prompts appear on the

computer display asking for each entry. This makes the system
easy to use and new personnel are able to learn the system
quickly. Figure 2 is an example of these data entries. The
data may be checked and corrected if necessary. The example
shown is tissue combustion data from an animal balance study.

The computer not only puts numerical data into formulas
and produces an answer, it can make judgments on the data and
produce more meaningful information. For example, a tolerance
of $\pm$ 10% is set for the two passes of LSC data. When this
tolerance is exceeded, the computer assumes the most recent data
are the most accurate. The output is noted so that the vials
can be left in the LSC for another pass.

The computer not only corrects for background radiation,
sample quenching and instrument efficiency, it calculates the
statistics for LSC counting as presented by Currie (1). The
data outputs are, therefore, more meaningful, because detection
and quantitation limits are known.

Output. The computer produces a hard copy output for each
data set entered. This output includes the "raw" counting data,
laboratory personnel entries, intermediate calculations, and the
final results. The format is easy to use and allows the results
to be hand calculated to check the computations. The computer
has a real time clock which dates each output and provides a
space for the individual to sign. This satisfies the proposed
EPA guidelines for "raw" data. An example of a hard copy output
which may be obtained is shown in Figure 3. In this example,
the entered data (Figure 2) is used with the LSC data to calculate the ppm and percent of ^{14}C dose in various animal tissues
after combustion. Another hard copy output is shown in Figure 4.
In this example, ion exchange column chromatographic data are
expressed in histogram form. The upper histogram is the ^{14}C
elution pattern of metabolites in a crop extract and the lower
histogram is a ^{3}H labeled standard which was added to the extract
for cochromatographic purposes.

These formatted outputs are stored on a flexible disk
exactly the same as printed. Later the data on the disk are
stored by sample number and project. All raw and reduced data
for a project are stored on a single flexible disk. After the
data have been stored on a flexible disk, it is a simple matter
for the computer to make summaries or do trend analyses. An
example of one of these summaries is shown in Figure 5. In this
example, the percent of ^{14}C remaining in the peel and fruit at
various time periods after spraying apples with a ^{14}C labeled
compound, was analyzed by a curve fit program.

Expansion. As data processing equipment becomes more
"bench" oriented, data from other instruments can be input
directly to the network. A Hewlett-Packard 1084B HPLC is
presently interfaced allowing reduced data from its processor

```
File Name: wsl
Line Number                      Description              Entry
-----------                      -----------              -----
     7                           Background            -30.0000
     8                          First Sample            25.0000
     9                                                  18.0000
    10                                                  35.0000
    11                    Combustion Efficiency           0.9600
    12                       Specific Activity           43.6000
                              Sample Name             Back Fat
    13                       1st Aliquot Weight          0.2281
    14                       2nd Aliquot Weight          0.1898
    15                       Total Sample Size          82.0000
                              Sample Name             Omental Fat
    16                       1st Aliquot Weight          0.1788
    17                       2nd Aliquot Weight          0.1564
    18                       Total Sample Size         215.0000
                              Sample Name             Leg Muscle
    19                       1st Aliquot Weight          0.2165
    20                       2nd Aliquot Weight          0.1730
    21                       Total Sample Size         270.0000
                              Sample Name             Tenderloin
    22                       1st Aliquot Weight          0.1546
    23                       2nd Aliquot Weight          0.1709
    24                       Total Sample Size         100.0000
                              Sample Name             Brain
    25                       1st Aliquot Weight          0.1967
    26                       2nd Aliquot Weight          0.1518
    27                       Total Sample Size          47.0000
                              Sample Name             Heart
    28                       1st Aliquot Weight          0.1839
    29                       2nd Aliquot Weight          0.1624
    30                       Total Sample Size         116.0000
    31                          Sample Name              0.2432
    32                       2nd Aliquot Weight          0.1780
    33                       Total Sample Size         183.0000
                              Sample Name             Kidney
    34                       1st Aliquot Weight          0.1924
    35                       2nd Aliquot Weight          0.1406
    36                       Total Sample Size         426.0000
                              Sample Name             Liver
    37                       1st Aliquot Weight          0.2536
    38                       2nd Aliquot Weight          0.2192
    39                       Total Sample Size        2800.0000
    40                       Daily Dose (dpm)     338800000.0000
    41                     Number of Days Dosed         10.0000
```

Figure 2.   Tissue combustion data entered for an animal balance study

Specific Activity = 43.60
Daily Dose = 338800000
Total Dose = 3388000000
Harvey Efficiency = 0.97

EKCD	CPM1	CPM2	EFF	DPM	ALIQ	TOT WCT	DPM/GM	TOT DPM	%DOSE	PPM	PPM EFF
Back Fat											
73	206	226	76.38	187	0.2281	82.0	797	65362	0.00	0.008	0.004
73	86	103	78.59	27	0.1398	82.0	190	15583	<* 0.001	<*0.0040	
Omental Fat											
73	148	149	79.24	95	0.1788	215.0	518	111266	0.00	0.005	0.005
73	102	1055	79.10	639	0.1564	215.0	3968	853157	0.03	0.041	0.011
Leg Muscle											
73	97	95	77.87	30	0.2165	270.0	132	35770	<* 0.003	<*0.0040	
73	83	80	77.35	11	0.1730	270.0	62	16656	< 0.001	<0.0013	
Tenderloin											
73	95	92	77.87	26	0.1546	100.0	165	16536	<* 0.002	<*0.0040	
73	109	111	78.08	47	0.1709	100.0	269	26929	<* 0.001	<*0.0040	
Brain											
73	293	309	77.73	293	0.1967	47.0	1448	68066	0.00	0.015	0.006
73	248	247	78.14	223	0.1518	47.0	1429	67151	0.00	0.015	0.007
Heart											
73	104	102	77.82	39	0.1839	116.0	204	23616	<* 0.002	<*0.0040	
73	107	102	77.85	40	0.1624	116.0	242	28069	<* 0.002	<*0.0040	
Kidney											
73	395	400	77.55	418	0.2432	183.0	1671	305765	0.01	0.017	0.006
73	346	341	77.46	349	0.1780	183.0	1905	348642	0.01	0.020	0.007
Liver											
73	927	900	77.85	1080	0.1924	426.0	5450	2321610	0.07	0.056	0.011
73	695	700	77.67	804	0.1406	426.0	5554	2365834	0.07	0.057	0.013
Int Contents											
73	1980	2026	77.52	2490	0.2536	2800.0	9534	26694203	0.79	0.098	0.012
73	1693	1625	77.76	2040	0.2192	2800.0	9036	25302119	0.75	0.093	0.013

Figure 3. Output using entered data (Figure 2) with LSC collected data to calculate the ppm and percent of ^{14}C dose in various animal tissues after combustion

to be summarized and reported off-line. An example of a
specific activity determination using this off-line reporting
feature is illustrated in Figure 6. The microprocessor in the
HPLC calculates the amount of compound present and reports this
to the computer. The computer calculates the dpm(s) in the
collected peak. The specific activity is then calculated and
reported along with all counting data to the HPLC printer.
Analytical equipment with compatible outputs, i.e., balances,
are easily interfaced into the network. Direct inputs from this
equipment save technician time and minimize transcription
errors.

The desk top computer network is planned to be interfaced
to larger data reduction equipment, either directly or by modem
(Figure 7). Present plans call for directly interfacing the
desk top computers, an FT-IR, and a GC-MS into an in-house
Hewlett-Packard 1000E which is currently servicing analog
instruments. Data from these sources can be compiled into a
single report which can be used to support product registration.

A minicomputer, such as a HP-1000E, can also be used for
chemical inventory, ^{14}C license requirements and structure
library.

The HP-1000E is presently interfaced, by modem, to the
central corporate computer, an IBM 3033. The capacity of the
corporate processor is, therefore, available for operations on
large data bases and archival storage. This hierarchy maximizes
the use of the strengths of each processor by assigning tasks to
the equipment which can handle them best.

Conclusions

The desk top computer network is a cost effective laboratory
data reduction system. In addition, laboratory personnel have
the computing equipment necessary for scientific data collection,
reduction, retrieval and reporting. They also control priorities
as needed with any scientific data generating instrument. If
rapid turnaround time is needed to make a scientific decision
for the next experimental step, they can obtain the necessary
computer results within minutes. Experimental design is not
dictated by computer availability; instead, the computer is a
tool which allows for better summarization and interpretation
of results. This improves both experimental design and completion necessary to support the scientists' overall strategy for
obtaining the data to complete the objectives of the experiment.
As data processing becomes more available at the "bench" level,
its use and usefulness increase. The scientist has control of
the experiments but when greater computer capacity is required,
he can interact with computer professionals in an effective
manner.

Figure 4. Output of ion-exchange column chromatographic data expressed in histogram form. The upper histogram is the ¹⁴C elution pattern of metabolites in a crop extract and the lower histogram is a ³H-labeled standard added to the extract for cochromatographic purposes.

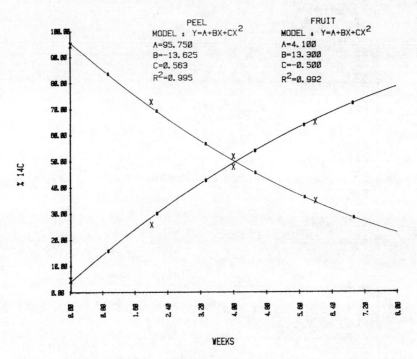

Figure 5.   Data summary showing the percent ^{14}C remains in the peel and fruit at various time periods after spraying apples with a ^{14}C-labeled compound

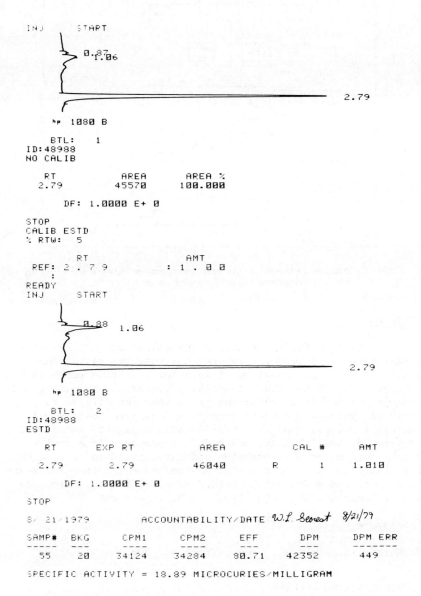

Figure 6. Specific activity determination using a HPLC with an off-line reporting option

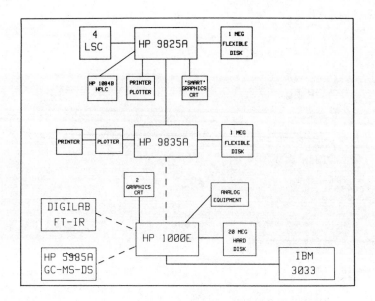

Figure 7. Planned expansion of the desk-top computer network ((———) existing conditions; (– – –) planned connections)

Abstract

Direct interfacing of liquid scintillation counters with a Hewlett-Packard 9800 series desk top computer network has greatly simplified reduction of data for pesticide metabolism studies. The system frees laboratory personnel from using hand calculators and from collecting paper tape for time-share computer systems. Standard formats are easy to review, audit and use for final reports. Experimental design is not dictated by computer availability, instead, the computer is a tool which allows for better summarization and interpretation of results. This improves both experimental design and completion necessary to support the scientists' overall strategy for obtaining the data to complete the objectives of the experiment.

Literature Cited

1. Currie, L. A., Anal. Chem., 1968, 40, 586.

RECEIVED February 21, 1980.

Some Applications of Fourier Transform Infrared Spectroscopy to Pesticide Analysis

STEPHEN R. LOWRY[1] and CHARLES L. GRAY

Diamond Sharmock Corporation, P.O. Box 348, Painesville, OH 44077

Fourier transform infrared spectroscopy (FTIR) has provided support to a number of areas in Diamond Shamrock's pesticide program. Commercially available FTIR spectrometers offer a number of advantages over dispersive instruments. Although some of the advantages are related to the ability to perform computerized data manipulations, the basic design of the FTIR system does provide superior capabilities in infrared spectroscopy (1).

A schematic diagram of a Michelson interferometer, the heart of the FTIR system, is shown in Figure 1 (2). The Michelson interferometer modulates each wavelength in the infrared region at a different frequency in the audio range. The modulated radiation passes through a sample chamber and is measured by the detector. This signal, called an interferogram, is converted by means of the Fourier transform into a single beam spectrum. An interferogram and corresponding spectrum are shown in Figure 2. A normal infrared spectrum is obtained by ratioing the single beam spectrum to a reference spectrum. The transmittance and absorbance spectra are shown in Figure 3.

Although this seems to be an elaborate method of obtaining a spectrum, several theoretical advantages result from rapid scanning FTIR. The first, Fellgett's advantage, results from the fact that data from all frequencies are being measured by the detector simultaneously. This is in contrast to dispersive instruments where only the small range of frequencies emerging from the exit slit at any one time are being measured. The second, Jacquinot's advantage, arises due to the increased energy throughput because no slits are used in the interferometer. The third, Conne's advantage, results from the high degree of wavelength accuracy found in most FTIR spectrometers due to the laser reference system. This accuracy allows a large

[1]Current address: Nicolet Instrument Corporation
5225 Verona Road, Madison, Wisconsin 53711

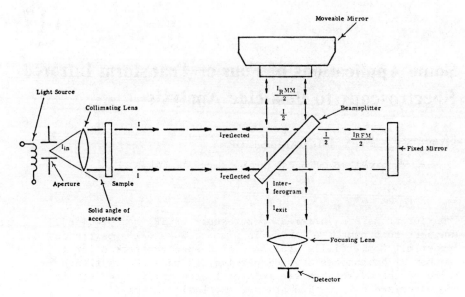

Figure 1.   Diagram of a Michelson interferometer

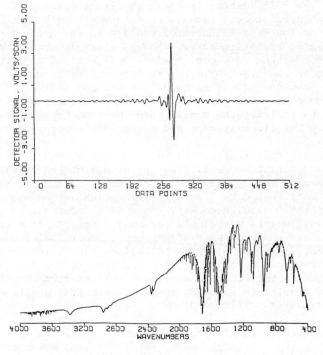

Figure 2.   Interferogram (top) and resulting single beam spectrum after Fourier transformation (bottom)

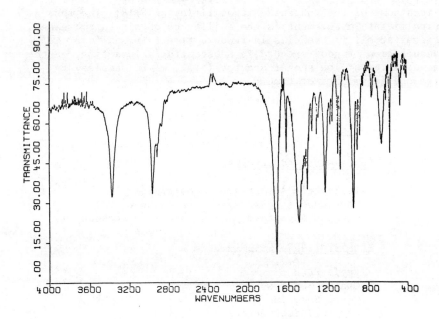

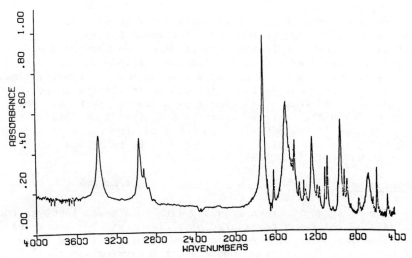

Figure 3. Normal IR spectrum plotted in transmittance (top) and absorbance (bottom)

number of scans to be co-added, which reduces the spectral noise.
    These three advantages are the basis of many of the
improvements of FTIR over dispersive spectroscopy. Table I
lists some of the increased capabilities of FTIR. The most
significant improvement is the ability to obtain an increased
signal at all frequencies in reduced time. This allows the
acquisition of good spectra in substantially less time, or the
acquisition of low signal spectra (i.e., micro samples) in
reasonable amounts of time.

Table I

Advantages of FTIR

Fundamental Advantages

Fellgett's Advantage	Multiplex
Jacquinot's Advantage	Throughput
Conne's Advantage	Frequency Accuracy

Practical Applications

Rapid Scan
Opaque Samples
Micro Sampling
Spectral Subtraction

    The remainder of this paper will describe specific appli-
cations of FTIR in support of one of Diamond Shamrock's
pesticide programs. All the results described here were obtained from studies of a single compound, thiofanox (P), 3,3-
dimethyl-1-methylthio-2-butanone O-[(methylamino)carbonyl]oxime.
Thiofanox is a potent systemic and contact carbamate insecticide.
The infrared spectrum of thiofanox is shown in Figure 4. The
major peaks in the spectrum are: the N-H stretch at 3380 cm^{-1},
C-H stretches in the 3000-2800 cm^{-1} region, the carbonyl band at
1720 cm^{-1}, the C=N at 1620 cm^{-1}, CNH at 1500 cm^{-1}, and two
other bands at 1235 cm^{-1} and 950 cm^{-1}.

Formulations

    The first application we would like to discuss involves the
analysis of thiofanox formulated on a clay carrier. Two specific
questions for which FTIR provided answers were: (1) are there
interactions between the compound and the carrier material?;
and (2) is the active ingredient completely removed by solvent
extraction? The answer to these questions required a comparison
of spectra from the blank carrier and the formulated material
before and after extraction. These spectra were obtained by
thoroughly grinding each sample, mixing with KBr, and pressing

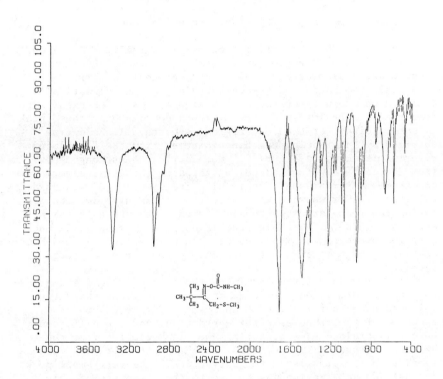

Figure 4.   IR spectrum of thiofanox obtained as a KBr pellet

into a disc. One hundred 8192-point interferograms were co-
added before a 16,384-point transform was performed. The
spectrum of the formulation is dominated by strong silicate
peaks from the carrier material. Figure 5 shows the spectrum
of the formulation (top), and the blank carrier (bottom). The
only thiofanox peak which stands out clearly is the carbonyl
peak at 1720 cm^{-1}. Since all the spectral data is stored in a
digital format, a point-by-point weighed subtraction can be
performed between two spectra. The subtraction of one spectrum
from another is based on the following equation:

Sample spectrum × FCS − Reference spectrum × FCR = 0   (1)

where FCS and FCR are scaling factors. Generally, FCS = 1 and
FCR is continuously varied until peaks disappear. On our
instrument, a Nicolet Model 7199, this process is carried out
by visual examination on a scope display, which automatically
displays the results of the subtraction as FCR is changed. The
resulting difference spectrum will then contain only those peaks
unique to either of the two original spectra. The spectrum
obtained by subtracting the blank carrier from the formulated
material is shown in Figure 6. Considering the strong bands
from the carrier which had to be subtracted, this is a very good
spectrum. A comparison of this with the spectrum of standard
thiofanox in KBr shows that most of the peaks have not changed
during the formulation process. The only exception is the N-H
stretch at 3380 cm^{-1}. This peak is completely missing in the
difference spectrum. The loss of this peak suggests the N-H may
be binding to some of the groups on the carrier or to residual
water.

The procedure for determining the amount of active ingredient in the formulated pesticide involves a solvent extraction
followed by either IR or HPLC analysis. Figure 7 shows a
spectrum of the granules after extraction, and the difference
spectrum obtained by subtracting the blank carrier from the
extracted granules. It can be seen that little or no thiofanox
remains on the granules after extraction. The very small peak
in the 1700 cm^{-1} region may be due to trace amounts of thiofanox,
but the level is below the accuracy limit of the assay. Also,
the negative peaks in the difference spectrum indicate that a
component of the carrier is removed by the extraction process.
These peaks appear to correspond to a carbonate salt, but an
exact assignment has not been made.

Although the FTIR is not used routinely for the assay of
thiofanox, in cases where problems from interferences arise or
when discrepancies between results occur, the subtraction capabilities of FTIR can be used to obtain quantitative information.
Figure 8 contains the spectrum of the extract from the granules,
a spectrum of the extraction solvent, methylene chloride, and
the difference spectrum. The difference spectrum contains a

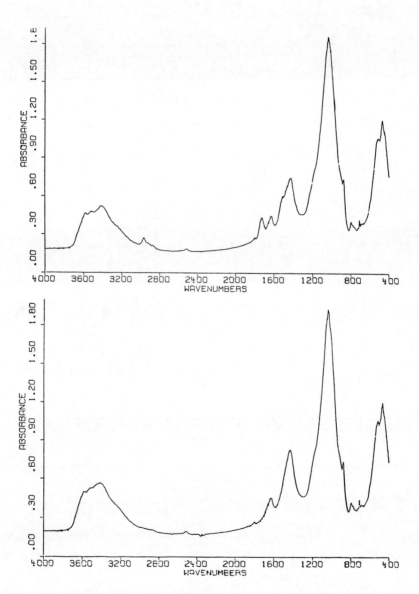

Figure 5. Spectra of a thiofanox formulation (top) and the blank carrier (bottom)

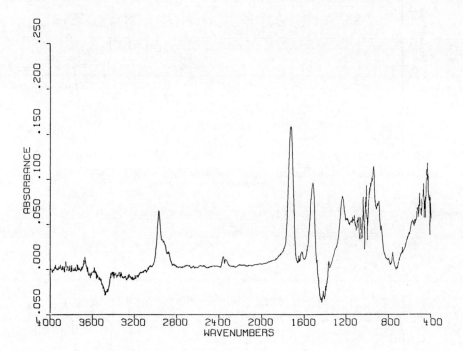

*Figure 6.   Difference spectrum showing the thiofanox formulation with the carrier
subtracted*

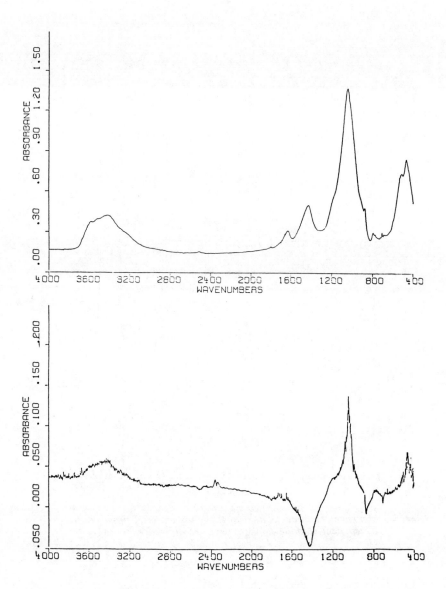

Figure 7. Spectra of a thiofanox formulation after solvent extraction (top) and the extracted formulation with the blank carrier subtracted (bottom)

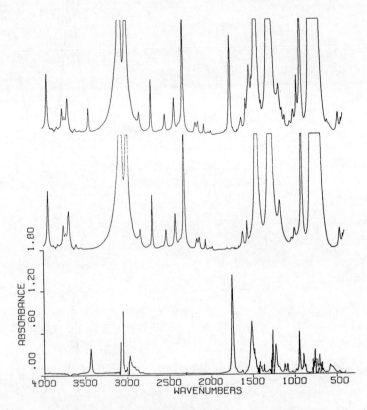

Figure 8. Methylene chloride solution of thiofanox extracted from carrier (top),
a methylene chloride blank (center), *and difference* (bottom)

number of peaks which are far enough from the opaque regions of the solvent to easily permit a quantitative analysis. Figure 9 shows the spectrum of the extract from the granules with the solvent already subtracted, a similar spectrum of a standard thiofanox solution, and the resulting difference spectrum.

Normally, the infrared assay of thiofanox is performed by comparing the intensity of the peak at 945 cm^{-1} of the standard and sample solutions. The amount of thiofanox in the sample is calculated using the following equation:

$$\text{Wt. Sample} = \text{Wt. Standard} \times \frac{A_{sample}}{A_{standard}} \qquad (2)$$

where A is the baseline corrected absorbance for the sample and standard, respectively, Using the spectral subtraction technique would require that the intensity of the 945 cm^{-1} band in the difference spectrum between the sample and standard go to zero. The equation would be:

$$A_{sample} \times \text{FCS} - A_{standard} \times \text{FCS} = 0 \qquad (3)$$

since the scaling factor FCS = 1, then

$$\frac{A_{sample}}{A_{standard}} = \text{FCR} \qquad (4)$$

or, based on the spectral subtraction, the weight of thiofanox in the sample would be calculated:

$$\text{Wt.}_{sample} = \text{Wt.}_{standard} \times \text{FCR} \qquad (5)$$

The advantage of the subtraction method occurs when FCR is determined using the entire spectrum rather than a single peak. This means that an interference at a single peak will not affect the analysis to a significant degree. Based on the spectra shown in Figure 9, the weight of extracted material was 0.54 g. This represents an assay of 8.9% active ingredient, compared to 9.0% thiofanox applied during the formulation. This is well within the accuracy for this method.

Gas Evolution

Because of the known thermal instability of carbamates at high temperatures, a method of identifying any volatile species would be desirable. The rapid data acquisition possible with FTIR would enable changes in both concentration and composition to be monitored as a function of time. A 10 cm gas cell was modified to enable the measurement of volatile components from

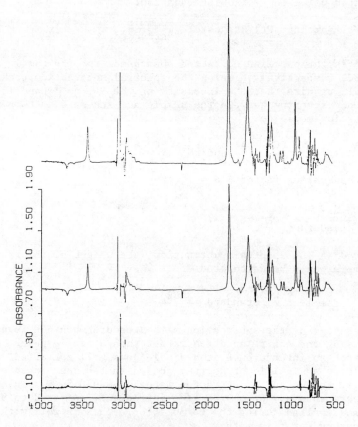

Figure 9. Extract from formulation with solvent subtracted (top), thiofanox standard with solvent subtracted (center), and difference between sample and standard with FCR = 1.376

either technical thiofanox or formulated material at various
temperatures. A small opening was made in the bottom of the
cell, and a short neck attached. A sample is placed in a small
sample cup which is clamped to the neck of the cell. The entire
cell is then evacuated. A heating block is placed around the
cup, and the cell is mounted in the instrument. After a background spectrum is acquired, the cell is heated. The temperature
is controlled and monitored by means of a thermocouple attached
to a thermal control unit. Figure 10 shows the initial spectrum
of a sample of high quality technical material at 100°C. The
positive peaks in the spectrum correspond to methyl isocyanate,
which indicates the carbamate linkage is being cleaved. The
negative peaks in the spectrum correspond to a decrease in the
level of water vapor and CO_2 outside the cell as the instrument
is purged with nitrogen. A spectrum obtained at 150°C is
shown in Figure 11. While substantial amounts of methyl isocyanate remain, numerous other peaks also appear. These are due
to the presence of isobutane, isobutene, and CO_2.

Experiments with lower quality technical material and a
formulation showed a very different behavior. When these
samples were examined at 70°C, the major gaseous product was CO_2.
Although small amounts of methyl isocyanate were detected, it
appears that impurities in these samples are reacting with
thiofanox before the carbamate is cleaved. Because the CO_2 is
the dominant volatile material in these situations, a method
was developed to quantitate the evolution of CO_2 as a function
of time. This data should give the kinetic rate for the thermal
decomposition and provide a rapid stability indicating assay.
Figure 12 shows the CO_2 spectra from three commercially prepared
standard gas mixtures. These spectra indicate CO_2 can be easily
detected down to 5 ppm by weight in nitrogen. The calibration
curve based on these standards is shown in Figure 13. The
absorbance of the CO_2 band clearly follows Beer's law in this
region. Figure 14 shows a plot of CO_2 concentration as a
function of time for a technical thiofanox sample at 70°C. This
graph indicates the evolution rate is constant over the time
period of the experiment. The rate of decomposition can be
calculated from this plot and is 1.15 µℓ/g/min.

Micro Sampling With Applications To Metabolite Studies

The final application of FTIR to be discussed is the
analysis of microsamples. A major problem encountered in the
infrared examination of microsamples is the reduction in energy
reaching the detector (3). The sensitivity of FTIR proved
beneficial in work performed in support of metabolite studies of
thiofanox. Samples were prepared by dissolving the compound in
a small amount of solvent (10-20 µℓ), followed by withdrawing
the solution into a syringe. The tip of the syringe was coated
with a small amount of KBr, and the solution was slowly released

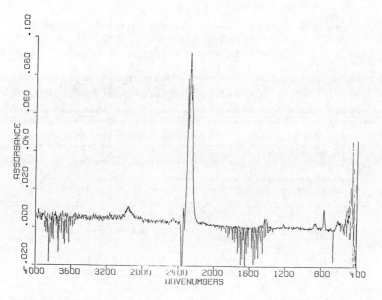

Figure 10. Volatile product obtained from heating high quality thiofanox to 100°C

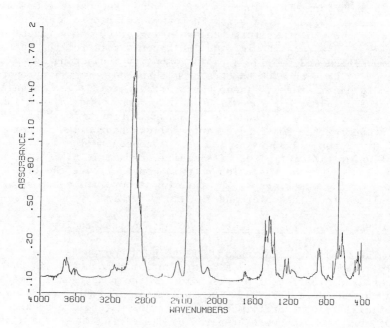

Figure 11. Volatile products obtained from heating high quality thiofanox to 150°C

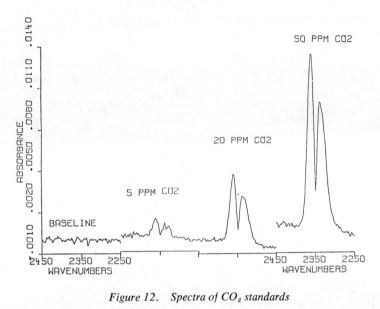

Figure 12.   Spectra of CO₂ standards

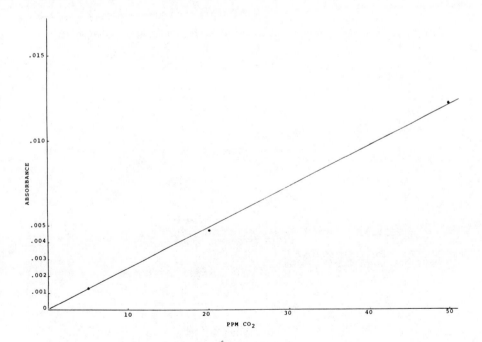

Figure 13.   Calibration curve from CO₂ standards

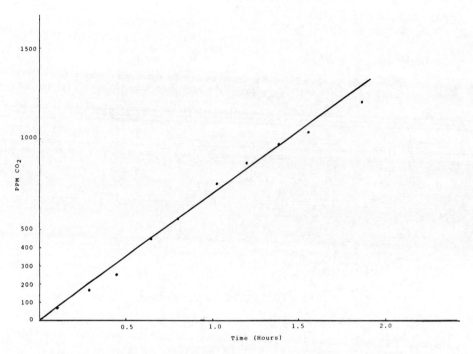

Figure 14.   CO₂ concentration vs. time for a technical thiofanox sample at 70°C

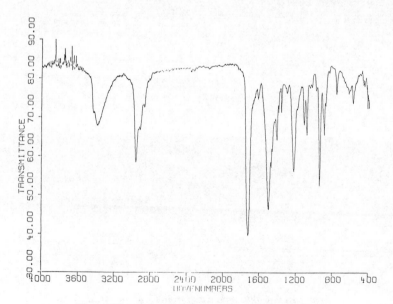

Figure 15.   Spectrum obtained from 10 μg thiofanox

into the KBr as the solvent evaporated. The resulting powder
was placed in a micro die (1.5 mm diameter), and a pellet prepared. Figure 15 shows the spectrum of 10 µg of thiofanox
prepared in this fashion.

   With detection limits this low, FTIR appears to be capable
of detecting quantities frequently encountered in TLC experiments. To evaluate this possibility, TLC plates were spotted
with 10 µg and 20 µg quantities of thiofanox. The plates were
developed in an acetone-hexane (1:1) solvent system, and the
spots were detected under UV light or with iodine vapors.
Initially, the spot was scraped from the plate and analyzed
directly in KBr. However, the silica gel produced too much
interference, and the thiofanox could not be detected. A
second set of spots were scraped from the plate, extracted with
chloroform, and the solution was dried on KBr using the procedure described above. Figure 16 shows the extract from a 20
µg spot. This spectrum is still dominated by the silica gel
peaks. By utilizing the subtraction capabilities of the FTIR
system, the peaks from the silica gel can be mathematically
removed from the spectrum of the sample. Figure 17 shows a
spectrum of a blank spot from the TLC plate, and the difference
spectrum from these is shown in Figure 18. This spectrum
closely matches the spectrum obtained from 10 µg of thiofanox
shown earlier.

   Previous studies at Diamond Shamrock have indicated that
one degradation pathway for thiofanox is shown in Figure 19 (4).
TLC experiments were performed on several of these metabolites,
and the spots were analyzed by FTIR. Several of these spectra
are shown in Figure 20. These results have demonstrated that
FTIR can provide useful information for the identification of
metabolites.

Impurity Study

   With the ever-increasing government regulations concerning
pesticides, a substantial effort is generally required to
identify impurities present in a technical product. Because
thiofanox cannot be analyzed directly by the GC/MS method
frequently used in this type of study, preparative HPLC was used
to locate new impurities. As new impurities were found, the
fractions were further purified using analytical columns. These
fractions were then examined by FTIR, NMR, and mass spectrometry.
Since the amount of sample isolated using these columns is
extremely small, the sensitivity of FTIR proved extremely valuable in providing an overall characterization of several
impurities.

Conclusions

   Although this paper has described several applications of

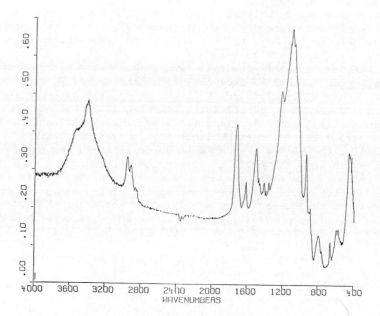

Figure 16.   Spectrum from a 20-µg TLC spot of thiofanox

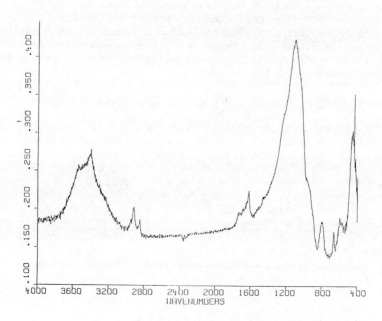

Figure 17.   Spectrum of a blank TLC spot

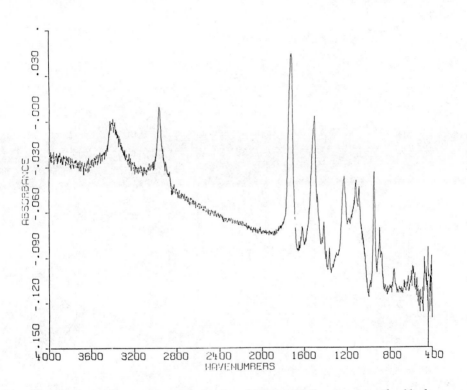

Figure 18.   Difference spectrum showing thiofanox TLC spot minus the blank

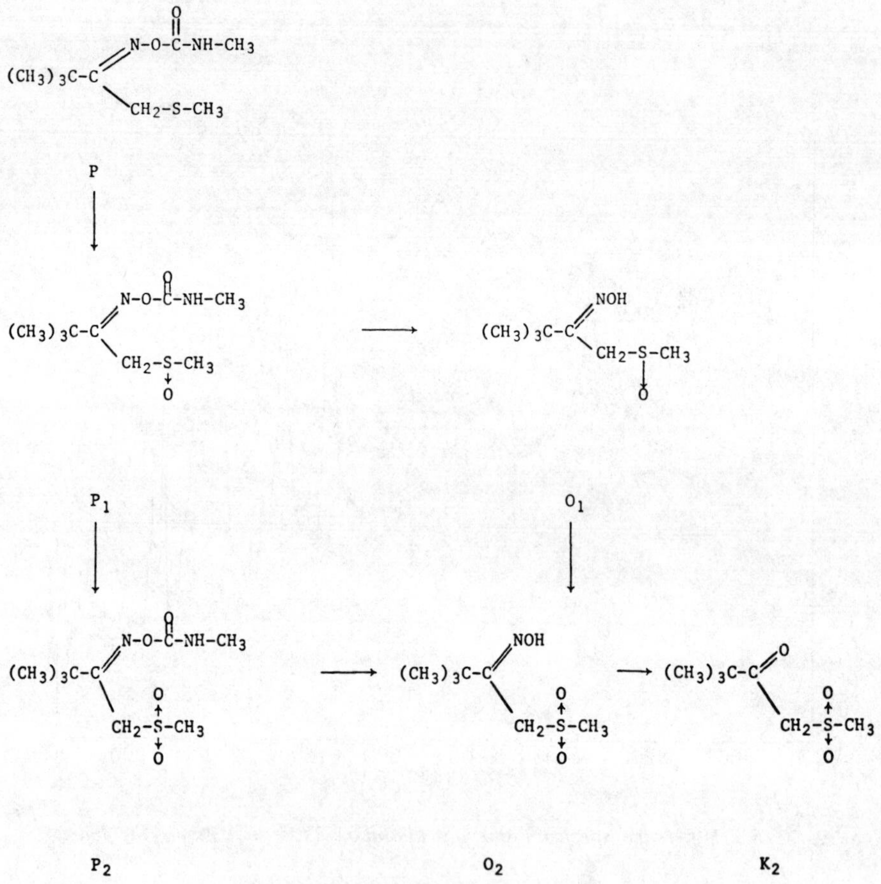

Figure 19.   Degradation pathway for thiofanox

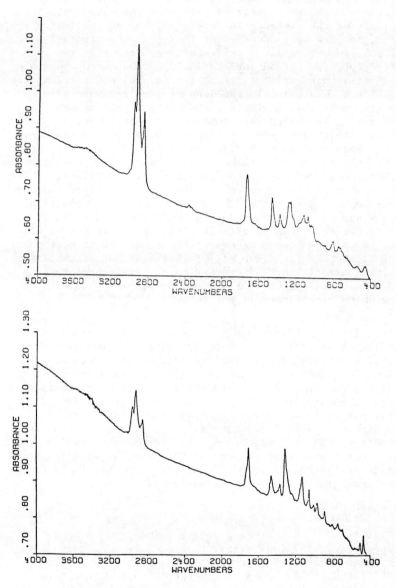

Figure 20. Spectra of 10 μg P₁ (top) and 20 μg K₂ (bottom) metabolites from TLC spots

FTIR spectroscopy to a particular pesticide, the methods have general applications to numerous compounds. Most of these utilize the high sensitivity of FTIR, and the data manipulation capability of the system. In several of the gas evolution studies, spectra were acquired at less than one-minute intervals. While this is not really "rapid scanning," the high resolution required for vapor phase spectra would not have been possible with a normal dispersive instrument. Several other techniques using FTIR show promise in the area of pesticide analysis. These are GC-IR (5), LC-IR (6), and diffuse reflectance (7). On-the-fly GC-IR systems are commercially available, and lower detection limits are being continually reported. While GC-IR may not replace GC/MS in residue and metabolism work, it can provide valuable data in these areas. On-the-fly LC-IR systems have been developed and are also commercially available. The major problem in these systems is the strong infrared absorbence of many common LC solvents. However, with proper selection of solvents and the development of LC conditions specifically designed for the LC-IR experiment, these problems may be over-come. Recent reports on diffuse reflectance measurements by FTIR indicate the technique may provide a method of examining formulated material or TLC spots with no sample preparation. While this technique is still in the development stage, it may become quite significant in the future.

Abstract

The high sensitivity and computerized data manipulation capabilities of FTIR spectroscopy provide the chemist with a powerful analytical tool. FTIR techniques have been applied to problems in a number of areas in our pesticide program. Specific areas where FTIR has provided valuable information include: quantitative analysis of active material; impurity identification in technical material; analysis of volatile components from formulated material; and the identification of metabolites. In this paper, we will discuss the results from these studies and describe some of the problems we encountered. We will also discuss some of the new developments in FTIR that might prove useful in pesticide analysis.

Literature Cited

1. Griffiths, P. R.; Sloane, H. J.; Hannah, R. W. Appl. Spectrosc., 1977, 31, 485.
2. Nicolet Model 7199 Instrument Manual, pp. 2-3.
3. Griffiths, P. R. "Chemical Infrared Fourier Transform Spectroscopy"; Wiley-Interscience: New York, NY, 1975; p. 257.
4. Chin, W. T.; Duane, W. C.; Ballee, D. L.; Stallard, D. E. J. Agric. Food Chem., 1976, 24, 1071.
5. Coffey, P.; Mattson, D. R.; Wright, J. C. Am. Lab., 1978, 10(5), 126.
6. Vidrine, D. W.; Mattson, D. R. Appl. Spectrosc., 1978, 32, 502.
7. Fuller, M. P.; Griffiths, P. R. Anal. Chem., 1978, 50, 1906.

RECEIVED March 3, 1980.

Potential of Immunochemical Technology for Pesticide Analysis

BRUCE D. HAMMOCK

Division of Toxicology and Physiology, Department of Entomology,
University of California, Riverside, CA 92521

RALPH O. MUMMA

Pesticide Research Laboratory, Department of Entomology, Pennsylvania State
University, University Park, PA 16802

In the fields of clinical chemistry and endocrinology,
immunochemistry is often the analytical method of choice.
Immunochemical methods of analysis offer many advantages including sensitivity, specificity, speed of analysis, ease of automation, cost effectiveness, and general applicability. The
importance of immunochemical assays was recognized by Rosalyn
Yalow sharing the Nobel Prize in Physiology and Medicine based,
in part, on her pioneering work in immunoassay development ([1],
[2]). Surprisingly, immunochemistry has found little or no practical application for the analysis of pesticides or other environmental contaminants ([3]). This fact is surprising because the
chemical classes currently assayed by immunochemical techniques
([2],[4]) are not fundamentally distinct from many classes of fungicides, herbicides, insecticides, nematocides, or plant growth
regulators. Possibly the tremendous success of gas liquid chromatography (GLC) and ion selective detectors in the analysis of the
chlorinated hydrocarbon insecticides fostered a generation of
pesticide analytical chemists who were experts in and disciples
of GLC. The phenomenal success of immunochemistry, specifically
radioimmunoassay (RIA), was possibly analogous in fostering a
generation of clinical chemists who look first to immunochemistry
for the analysis of hormones and pharmaceuticals, even in cases
when RIA is not necessarily the technique of choice. Biological
techniques, in contrast to physical or chemical techniques for
residue analysis, have been criticized by analytical pesticide
chemists. It is a common misconception that immunochemical
methods can be classed as biological techniques of residue
analysis. Although a living organism or, at least, a cell line
is required for antibody production, immunoassays using these
antibodies are based on physical and chemical properties, and
immunoassays can be explained in terms of the law of mass action.
A tremendous immunochemical technology has developed especially
in clinical chemistry, and it is time that this technology was
exploited to solve new and pressing problems in environmental
chemistry.

0-8412-0581-7/80/47-136-321$08.00/0

In this chapter the potential of several immunochemical techniques for residue analysis will be explored. Sufficient background methodology will be presented to allow the reader to evaluate the advantages and disadvantages of immunochemical techniques and their potential application to residue problems. A chapter of this length provides, at best, superficial treatment of immunochemical methodology and theory, but this overview, in conjunction with the included references, should offer the reader ready access to the specific immunochemical literature. Hopefully, this article will assist pesticide residue laboratories in applying existing immunochemical technology to specific problems in pesticide analytical chemistry.

The most common, but by no means the only or even the most promising, immunochemical assay for small molecules is radioimmunoassay (RIA). As an overview, an immunoassay involves chemically attaching the small molecule of interest (or a derivative of it) to a carrier protein and raising specific antibody titers to it in the serum of an animal. Very dilute antibody solutions are then used to bind the small molecule which has been radiolabeled. The competition of varying known concentrations of unlabeled material is measured and the resulting standard curve used to determine unknown concentrations (Table I). The steps leading to the development of an RIA are outlined below followed by a description of other immunochemical procedures and an analysis of the attributes and limitations of immunoassay.

Table I

Steps in the Development of an RIA

Synthesize hapten	Prepare radioligand
Couple hapten	Choose method for bound/free separation
Purify antigen	Optimize assay conditions
Characterize antigen	Develop standard curve
Immunize animal	Characterize assay
Titer antibody	Determine assay reliability
Characterize antibody	

Methodology of Antibody Formation

Hapten Synthesis. Antibody titers are raised in an experimental animal in response to an antigen or immunogen. In general, an effective antigen must be rather large and foreign to the animal to be immunized; (proteins of greater than 10,000 mw are common antigens). By comparison, most pesticides are rather small molecules, and therefore they must first be conjugated to a protein or other large antigenic molecule before they can be used as antigens. Such small molecules which become immunogenic after attachment to a large carrier molecule are called haptens. If the pesticide has a reactive functionality suitable for conjugation it may itself be the hapten. Otherwise, a derivative of the pesticide must be synthesized suitable for attachment to the carrier. Coordination between a hapten and carrier protein may be sufficient for raising antibody titers, but covalent linkages are more reliable and definitive (4). For many pesticides, potentially useful haptens have already been described as metabolite standards or environmental degradation products.

The choice of the hapten and the conjugation procedure used may profoundly affect the ultimate sensitivity and specificity of the immunochemical assay. Generally, antibody specificity is highest for the part of the molecule distal or furthest from the carrier protein. This knowledge has frequently been utilized to develop immunoassays which will, on one hand, detect general classes of compounds which have common functionalities and to develop other assays which are highly specific. In an hypothetical system (Fig. 1), three similar molecules are represented. If molecule I is used as a hapten and it is conjugated to the protein through functionality b, the resulting antibody population is likely to cross-react with the closely related molecules II and III. Such an antibody population might find utility in developing an assay to the class of compounds represented by molecules I, II and III. Alternatively, if molecule I is conjugated through functionality a, the resulting antibody population is likely to distinguish among the three molecules and be useful for a specific assay of molecule I with minimal interference from related compounds II and III.

The importance of the site of conjugation of a hapten to a protein has been demonstrated many times with steroids and pharmaceuticals such as the barbiturates (5), and it was recently demonstrated with the insecticide S-bioallethrin (1R,3R,4'S allethrin) (Fig. 2) (6,7). The S-bioallethrin was conjugated to a carrier protein via an hydroxyl functionality of the propene side chain of the rethrelone moiety (Fig. 2A). RIA based on the resulting antibody population indicated a high degree of specificity for the absolute configuration of the chrysanthemate moiety distal from the point of conjugation and much lower specificity for the more proximal chiral center in the allethrelone moiety. In addition, the antibody could not distinguish S-bioallethrin from pyrethrin I probably because pyrethrin I has an identical configuration and differs from allethrin only in the propene side chain (6,7). If it were important to raise an antibody titer capable of distinguishing between allethrin and the pyrethrins I, the hapten could have been conjugated through its carbo-

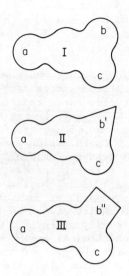

Figure 1.   Illustration of importance of hapten selection on immunoassay specificity.

A hapten molecule (I) coupled to a protein through functionality "a" would be expected to raise an antibody titer useful for an assay of molecule I but not II or III. A hapten molecule (I) coupled to a protein through functionality "b" would be likely to raise an antibody titer useful for the assay of the class of molecules represented by I, II, or III.

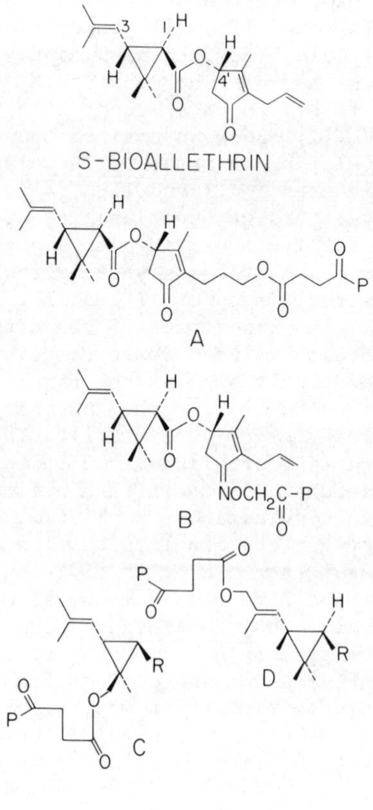

Figure 2.   The structure of S-bioallethrin (1R, 3R, 4'S allethrin) and possible haptens for the formation of antigens for allethrin.

The hemisuccinate of an alcohol derivative of allethrin's propene side chain (A) illustrates the use of a spacer arm between the carrier protein and the molecule of interest. Antibodies to this antigen demonstrated the greatest specificity for the chrysanthemate end of the molecule. The allethrin CMO derivative (B) was prepared at the 1' ketone. Haptens attached through a gem dimethyl group (C) or the isobutenyl group (D) would be expected to lead to antibodies with a greater specificity for the allethrelone end of the molecule. P indicates protein.

methoxyoxime (CMO) derivative at C-1 of the allethrelone moiety
(Fig. 2B), or better through a hydroxyl substituent on a gem
dimethyl group or a functionality on the isobutenyl side chain of
the chrysanthemic acid moiety (Fig. 2C,2D). Several studies have
emphasized that it is often important for maximal specificity to
have the hapten separated from the carrier protein by a spacer
arm. A hemisuccinate moiety was used for this purpose in the
case of S-bioallethrin. Several of the conjugation procedures
discussed in the following paragraphs insert a spacer arm between
the hapten and the carrier protein due to the nature of the
conjugation reagent involved, while in some cases a more deliberate attempt to insert a spacer may be made (8,9).

   Hapten Coupling. The functionalities on a protein usually
used for coupling haptens include NH, SH, OH, and COOH. Numerous
coupling techniques have been utilized and are described in
detail in the pharmacology and endocrinology literature. Coupling techniques for affinity chromatography are also often applicable to hapten-carrier coupling (10). An overview of the most
widely used coupling techniques is presented below. This overview is not exhaustive; rather, it is intended to illustrate some
of the many synthetic routes open to the pesticide analytical
chemist. When possible, examples have been drawn from the areas
of entomology or pesticide or environmental chemistry. Langone
and Van Vunakis (11) used an N-hydroxysuccinimide (NHS) (12)
active ester of a carboxyl substituted analog similar to aldrin
and dieldrin formed by dehydration with N,N'-dicyclohexylcarbo-
diimide (DCC) to conjugate with human serum albumin (Fig. 3, Rn
1). Similar active ester methods have been used to conjugate
to proteins carboxylic acid derivatives of allethrin (6), diflubenzuron (13), juvenile hormone (14,15), ecdysone (15),
polypodine β-oxime (16), and numerous compounds of medicinal
interest. The active ester can be purified (12,17,18), and it is
fairly stable under acidic conditions.
   Alternatively, water-soluble carbodiimides such as 1-cyclo-
hexyl-3-(2-morpholinyl-4-ethyl) carbodiimide methyl p-toluene sul-
fonate (CMC) or 1-ethyl-3-(3-dimethyl-aminopropyl) carbodiimide·
HCl (EDC) are available which allow a direct coupling of an amine
and carboxylic acid without the isolation of an active ester
(Fig. 3, Rn 2) (19,20,21). Such procedures may be very useful
with rather water-soluble or unstable haptens, and the resulting
cross-linking of the protein may actually increase its anti-
genicity (4) although solubility is commonly reduced. In designing subsequent assays one should remember that water-soluble
carbodiimides and some other coupling agents may react directly
with a protein and subsequent antibodies may be directed, in
part, against the resulting guanidino or acyl urea derivatives
(Fig. 3, Rn 3). When using immunodiffusion (discussed later)
for estimation of antibody titers, this laboratory has used
haptens coupled to different proteins by chemically distinct

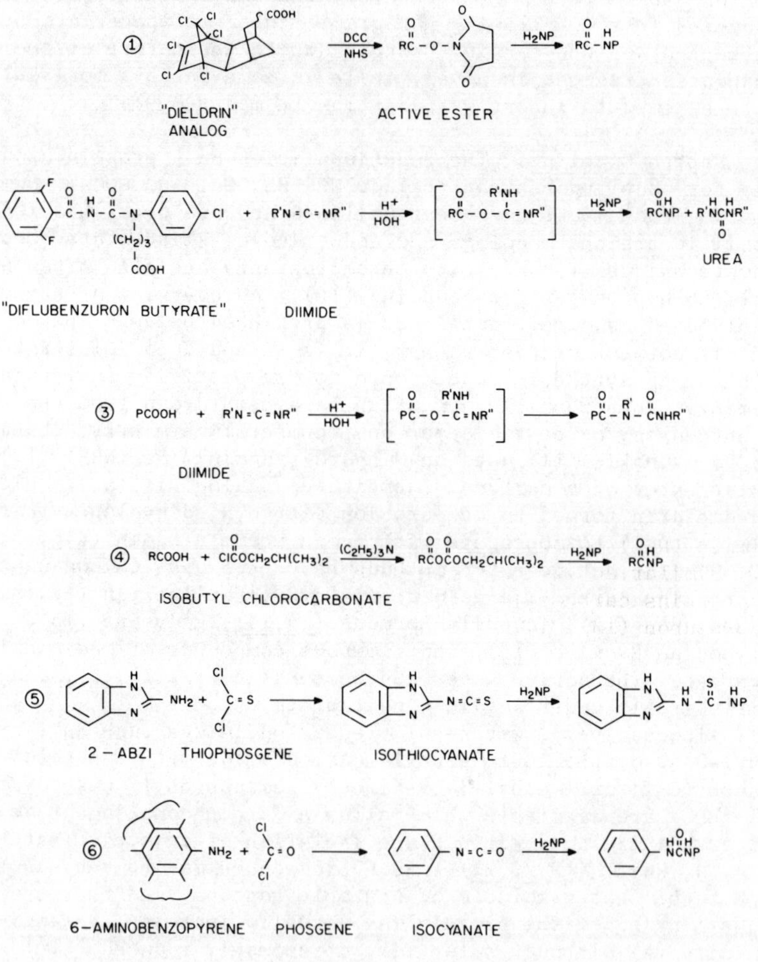

Figure 3.   Some methods of hapten–protein coupling. Except for phosphorous in parathion, P indicates protein. See text for a description of reagents.

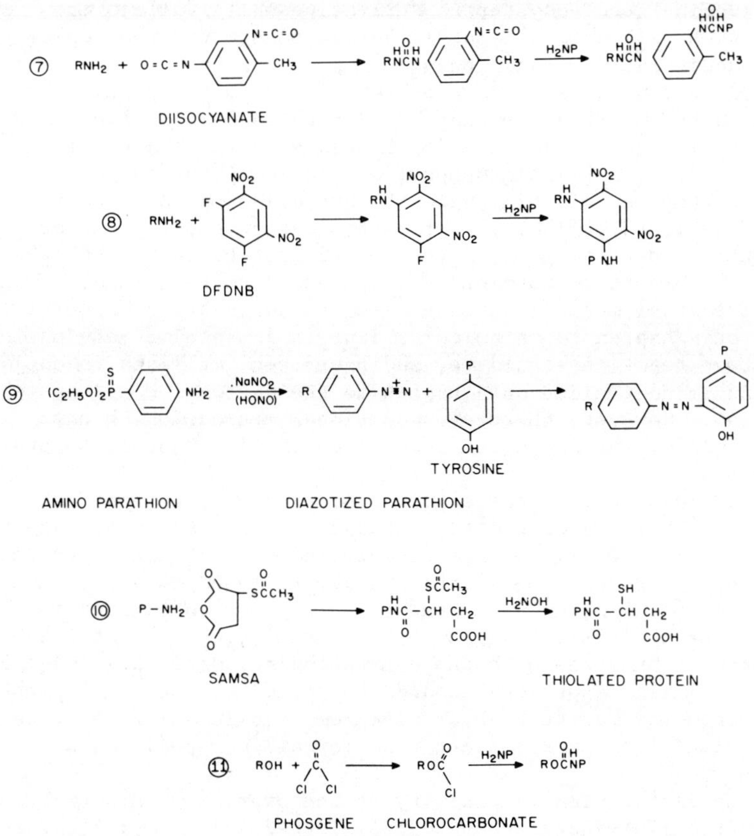

Figure 3. Continued

procedures. For instance, diflubenzuron derivatives were coupled
using both the purified NHS active ester and via a water-soluble
carbodiimide (Fig. 3, Rn 1,2) (13) to several different proteins.

Several other routes resulting in conjugation of a carboxylic acid group to an amine include reaction of the acid with an
alkylchlorocarbonate (chloroformate) (Fig. 3, Rn 4). The ethyl
and isobutylchlorocarbonates are commonly used; for instance,
deReggi et al. (22) used ethylchlorocarbonate to make a conjugate
of a succinylated ecdysterone derivative while Vallejo and
Ercegovich (23) used sec-butylchlorocarbonate for the conjugation
of a succinylated solanidine.

Pioneering work on immunochemical assays for pesticides
involved the synthesis of haptens for DDT and malathion. Haas
and Guardia (24) used the acid chlorides of malathion half ester
and DDA (2,2-bis-[p-chlorophenyl]acetic acid) for conjugation
while Centeno et al. (25) used the anhydrides of DDA and malathion diacid (0,0-dimethyl S-[1,2-bis-carboxyethyl]phosphorodi-
thioate). In retrospect, more specific antibodies of a higher
titer may have been obtained had a spacer arm been used.

The above methods were utilized to conjugate a carboxyl
group on a hapten to an amino residue on a protein. Obviously,
the above reactions could be, and have been, utilized to conjugate an amino residue on a hapten to the carboxyl residues on
proteins. However, there are additional methods which have
proven useful for conjugating amine containing haptens to proteins.

For instance, Lukens et al. (26) attached 2-aminobenzimi-
dazole (2-ABZI - a degradation product of the carbamate fungicide
Benomyl) to ovalbumin by reacting the amine of 2-ABZI with thio-
phosgene to produce the isothiocyanate followed by addition of
ovalbumin (Fig. 3, Rn 5). Benzo[a]pyrene was conjugated to
bovine serum albumin (BSA) by forming the isocyanate at C-6 by
reaction of phosgene with the corresponding amine (27) (Fig. 3,
Rn 6). Similar approaches could be applied to the development of
an immunoassay for tetrachlorodibenzo-p-dioxin using the recently
synthesized 1-amino-2,3,7,8-tetrachlorodibenzo-p-dioxin as a
hapten (28).

The differential reactivity of the sterically hindered and
unhindered isocyanate groups of tolylene-2,4-diisocyanate facili-
tates the stepwise conjugation of hapten (R) and protein (P)
amino groups (Fig. 3, Rn 7). p,p'-Difluoro-m,m'-dinitrobenzene
(DFDNB) reacts with numerous functionalities including primary
and secondary amines, imidazoles, and phenols to yield mixtures
of conjugated materials (Fig. 3, Rn 8). This reaction is apparently harder to control than the diisocyanate reactions, but it
is much more versatile.

Aromatic amines may be converted to their diazonium salts
with nitrous acid. The hapten may then be bound via azo linkages
to the tyrosine (shown), histidine, lysine, and possibly arginine
and tryptophane residues of the carrier protein by mixing the

protein and diazonium salt under basic conditions. This method
was used in the classic immunochemical studies by Pauling et al.
(29) and Landsteiner (30) and more recently to coupled aminoparathion to bovine serum albumin (BSA) (Fig. 3, Rn 9) (31). The
sulfhydryl residue is commonly encountered in pesticides, and it
can be utilized to conjugate a hapten to a protein via a disulfide bridge. Most proteins do not have numerous free sulfhydryl
groups, so the free SH groups can be "enriched" by reacting the
protein with N-acetyl-homocysteine thiolactone or more recently
S-acetylmercaptosuccinic anhydride (SAMSA) (Fig. 3, Rn 10)
followed by addition of the hapten (32). Thiolated proteins can
be used for reaction with any compound capable of forming covalent bonds with sulfur. Glutathione or other conjugates of
pesticide metabolites could also possibly be used for coupling to
proteins.

Other functionalities on a hapten can be directly linked to
a protein by a variety of methods or they can be converted to
compounds containing a free amine or carboxyl group and then conjugated by the above methods. By reacting aldehydes or ketones
with carboxymethoxylamine hemihydrochloride (CMA) the resulting
oxime with a free carboxyl group can be formed as shown for the
allethrin CMO derivative (Fig. 2B) (6). This procedure has also
been used in coupling reactions leading to antibodies for insect
molting hormones (33,34,35). The allethrin CMO derivative was
found to be quite unstable, and this fact emphasizes the need for
rigorous structural proof of hapten structure.

Hydroxylated pesticides are common metabolites and thus, a
choice of hydroxylated materials are often available for conjugation. Exposure of metabolites with primary or secondary alcohols
available to succinic anhydride in pyridine leads to a hemisuccinate as shown for allethrin derivatives in Fig. 2A. This
method has been used to derivatize many compounds of biological
interest including ecdysone and solanidine (15,22,23). Alternatively, hydroxyl groups can be exposed to equimolar phosgene
resulting in a chlorocarbonate which will react with amino groups
of proteins (Fig. 3, Rn 11) or reacted with ethyldiazoacetate
followed by hydrolysis to give a carboxymethyl ether (36).

Phenols and diazotized p-aminobenzoic acid react to introduce a free carboxyl group (32). Ethylbromoacetate was used to
derivatize phenolic metabolites of the insecticide diflubenzuron
and model pyrethroids under anhydrous conditions. The resulting
ethyl ester could be hydrolyzed in dilute methanolic base without
hydrolyzing diflubenzuron. Longer spacers can be introduced by
using bromopropionates and buterates, but harsher conditions are
required for these less reactive bromides. The bromoacids can be
used for more water-soluble haptens and chloroacetic acid has
been used for estrogen (13,37). Phenols can also be coupled
using other divalent reagents such as tolylene-2,4-diisocyanate
or cyanuric chloride. A conjugated olefin can be reacted with 3-
mercaptopropionate to also yield a free carboxylic acid at the

end of a convenient spacer arm (38).

Numerous other methods of conjugation are available and will likely be obvious to the chemist familiar with the properties of the pesticide of interest. In addition to many methods in the literature, numerous reviews give either detailed conjugation procedures (32) or references to these procedures (4,10,39). As will be discussed below, organic chemists may find the difficulty of establishing the structure(s) of the final conjugate disconcerting. In contrast to numerous papers in the literature where "recipes" for conjugation are simply followed, it is important to verify the structure of the hapten at each step of the synthesis. It may also be important to adapt the conditions of conjugation to the specific reaction in question. The stability of the hapten, the resistance of the carrier protein to denaturation, and the relative solubilities of the hapten and the carrier protein in the reaction medium should be considered.

Antigen Purification and Characterization. The antigen (hapten-carrier protein conjugate) is usually separated from low molecular weight by-products based on its large size. Dialysis is an obvious method of separation, but some lipophilic molecules pass through a dialysis membrane with great difficulty. Gel filtration provides another convenient method of separation. If the protein is not too badly denatured, repeated precipitation with an organic solvent such as ethanol is a rather certain way of removing lipophilic impurities.

The most quantitative methods of determining the moles of hapten bound per mole of carrier protein include the use of radiolabeled haptens or the monitoring of a change in the absorbance of the hapten-carrier conjugate in a spectral region where the protein itself does not strongly absorb (4). For the parathion conjugate a phosphorus determination proved to be a useful method (31). These methods are often not appropriate, so alternate methods such as the monitoring of the proteins' reactive groups (such as free amine) before and after conjugation must be employed (4,6,32,40,41). Careful controls are necessary with these procedures because self conjugation or denaturation of the protein may decrease or even increase the apparent functionalities available for binding.

There is no consensus on the optimum number of hapten molecules per carrier, but at least two molecules are required for subsequent immunoprecipitin tests. Many early studies used very high loading and useful antibody titers continue to be raised using heavily loaded antigens. Some workers feel that antibody titers with a higher average specificity for the hapten can be raised using low loading (4,42). Numerous proteins have been successfully used as carriers. Bovine and human serum albumin are very commonly used because they have numerous free amine groups and are remarkably soluble when cross-linked or even when heavily loaded with haptens. Many workers have found that mollusk

hemocyanin is phenomenally immunogenic. Although hemocyanin has
been successfully used as a carrier for pesticide haptens, one
often encounters solubility problems. Many other commercially
available and exotic proteins have also been used as carriers.
One can be relatively certain that a protein will be immunogenic
if it has a molecular weight >10,000 and if it is immunochemi-
cally foreign to the animal receiving the antigen. One should
also consider the ultimate use of the antibody when choosing the
carrier protein. For instance, human serum albumin would be a
poor carrier for a hapten if the resulting antibody were to be
used to monitor human blood samples by immunodiffusion.

   Choice of Animal for Antibody Production. Numerous verte-
brates have been used as the source of antibodies. As techniques
become more sensitive, less antibody is needed. Guinea pigs and
rabbits are thus commonly used. Even mice are used, especially
since the major cell lines now available for cloning antibodies
are derived from mice (43). If larger quantities of antibodies
are needed, one can move to either larger numbers of small mammals or to goats, sheep and larger mammals. The use of avian
species is not common when haptens are used, but they may yield
high, broad spectrum antibody titers against mammalian proteins.
The nature of the antibodies obtained will vary somewhat with the
species used. For instance, goats are known to often produce
antibodies with very high affinity for haptens while guinea pigs
often yield a high titer of complement.

   Immunization Procedures. The antigen is usually injected
into the recipient animal in Freund's complete adjuvant. This
water-in-oil emulsion provides a slow release formulation for the
antigen, protects the antigen, and with dead Mycobacteria, it
stimulates the immune system. Subsequent booster injections are
usually given in adjuvant without the Mycobacteria in order to
avoid severe allergic response. The resulting antibody titer in
the serum is monitored and when it has reached an acceptably high
level, blood is withdrawn and the serum isolated for use in assay
development. Many of the numerous immunization protocols are
referenced in Parker (4) while Williams and Chase (32) give
detailed instructions on the handling of animals.
   Although the assays using antibodies have reached a high
state of sophistication, a definitive work on immunization procedures is still lacking. It is not generally possible to
reproduce the exact titer and specificity of antibodies even in
apparently identical animals. This lack of reproducibility in
the raising of antibody titers may have led to the reluctance on
the part of pesticide analytical chemists to embrace immunochemical techniques. However, the animal is only the tool used
to obtain the antibody, and once the antibody is in hand, the
assays are physical in contrast to biological assays. Most
radioimmunoassays use serum dilutions of 1:5,000 to 1:100,000 so

that a single rabbit will yield enough antibody for a staggering
number of assays. If properly handled and frozen, antibodies may
be stored for long periods. Ultimately, the serum from a single
animal will be exhausted. Although it may be difficult to obtain
another batch of serum of phenomenally high titer, affinity and
specificity, numerous studies have shown that for most molecules
one has a very high probability of obtaining useful sera following the injection of a limited number of animals (44) by standard
procedures. For instance, out of 8 rabbits injected with several
diflubenzuron antigens, antibody titers were detected in all
rabbits against the carrier protein and in 7 against the hapten
(13). Monoclonal antibody technology (see below) promises to
improve the consistent availability antibodies as reagents (45).

The antibody titer is monitored in the serum by any of the
numerous analytical procedures discussed below. For instance,
antibodies were detected to an allethrin-hemocyanin conjugate by
immunodiffusion studies using among other molecules, an alleth-
rin-BSA conjugate (6) and passive hemagglutination was similarly
used following immunization with a parathion-BSA conjugate (31).
In addition to immunodiffusion, a radioimmunoassay was developed
using a low specific activity ^{14}C diflubenzuron label for the
monitoring of diflubenzuron antibody titers (13). In order to
determine antibody titer, the serum is generally diluted until a
serum concentration is reached which will bind 50% of a constant
amount of hapten (44).

Antibody Characterization. In attempting to characterize
the antibodies in a serum sample it should be kept in mind that
one is dealing with a heterogeneous population of antibody molecules of varying specificity and affinity. There are undoubtedly
antibodies present which recognize the carrier protein, but
contribution of these antibodies to assay binding can be eliminated by using a different carrier or a tagged hapten. Even when
only the hapten is recognized in the assay, one is dealing with a
heterogeneous antibody population in a serum sample. An estimation of the average affinity constant (K_a) is often useful in the
optimization of competitive binding assays (46), and one estimation of sensitivity is taken as one tenth of the reciprocal of
the average binding affinity (44). Estimates of the average K_a
are obtained by plotting a function of the hapten which is antibody bound vs a function of the concentration of the hapten.
Such plots include Michaelis Menten curves, Scatchard plots, and
Sips plots. The later two plots will also give an estimate of
the heterogeneity of the antibody population (4,44,46).

The specificity of an antibody titer refers to the degree of
cross-reactivity one sees with the antiserum used. By using
different tagged haptens one can vary the specificity of the
resulting assay; however, there is an intrinsic specificity of
the antiserum which is difficult (although possible) to improve.
The specificity of an antiserum is usually established by compe-

titive binding studies. Specificity is often expressed as the concentration of a substance needed to displace 50% of an antibody bound hapten. The specificity of an antiserum may be very high, requiring many-fold higher concentrations of very closely related molecules to displace the radioligand. Such specificity is the basis of the major advantages of immunochemical assays over many classical procedures; however, it may be misleading. Although an antibody may effectively discriminate among several very closely related molecules, it may bind quite tightly to an unknown molecule in an extract. Also, even a 1000X selectivity may be overcome if very high levels of even poorly reactive contaminants are present. Such problems are most common when lipophilic haptens are used. In classical GLC assays one is usually looking at a weak electrical response indicating the presence of, for instance, a mass fragment or electron capturing material. Such observations are only indicative of the presence of a pesticide if careful control runs have been performed, and similar controls are also necessary in assays in which antibodies are used.

Competitive Binding. Competitive binding provides the principle upon which most immunochemical assays are based. Enough antibody is added to a small, constant amount of radiolabeled antigen to bind 35-50% of it (the same principles apply regardless of the tag used to identify the antigen). As increasing amounts of unlabeled antigen are added, one decreases the amount of bound radiolabeled antigen which is then separated by one of a variety of techniques from the free radiolabeled antigen (Fig. 4). By monitoring the percentage bound and/or free radiolabeled antigen as the concentration of unlabeled antigen is increased, one can establish a standard curve. This standard curve can then be used to determine the concentration of an unknown (Fig. 4).

Bound/Free Separations. Usually the most time-consuming part of a radioimmunoassay involves the separation of the bound and free radiolabeled antigen. There are several promising new techniques which avoid this separation step, but the most sensitive assays still employ separation. Equilibrium dialysis is an esthetically pleasing method of separation, but it does not lend itself to the processing of a large number of samples. Gel permeation chromatography works on the same principle because the large antibody bound antigen elutes ahead of the smaller free antigen. Gel permeation is usually too slow for routine immunoassay procedure, although it forms the basis for several very rapid automated procedures. Nitrocellulose membranes will allow small antigens to pass through while retaining antibodies and they form the basis of several rapid analytical procedures. The use of dextran-coated charcoal to precipitate unbound antigen is commonly used, and it should be generally applicable to pesticides (6,7). The dextran coating on the charcoal and/or the

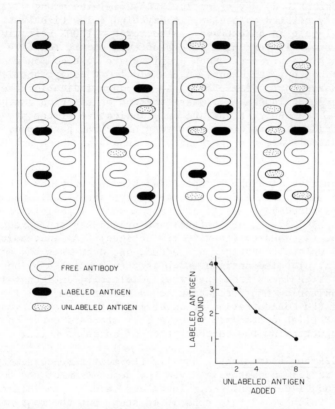

Figure 4.   Illustration of the principle of competitive binding.

An increasing amount of unlabeled antigen displaces a constant amount of labeled antigen from a constant amount of antibody. Separation of antibody bound and free material results in a standard curve that can be used to determine the amount of unlabeled antigen in unknown samples.

presence of nonspecific sera greatly reduce the precipitation of antibodies and may increase the ease with which a charcoal suspension is handled. Florisil may sometimes be substituted for charcoal. Although techniques which bind the antigen are generally applicable to lipophilic molecules, they will shift the bound/free equilibrium with time. Thus, assays are often more time-dependent with antigen binding rather than antibody binding techniques.

The binding of a charged antibody to ion exchange resins or hydroxyapatite is also commonly used for separation. The antibody can be precipitated using polyethylene glycol or ammonium sulfate and/or a second antibody such as goat anti-rabbit leaving unbound antigen in solution. More recently a surface protein, protein A, on the surface of some Staphylococcus aureus cells has been found to specifically bind and precipitate many antibodies. Solid phase systems in which the antibody is coated on tubes or attached to polyacrylamide or dextran particles lend themselves to very rapid analysis. Larger amounts of antibody are generally required for solid phase assays and some additional effort in assay optimization is often needed.

There are several lines of research which may lead to very sensitive, rapid immunochemical assays which do not require separation of free and bound antigen. The ELISA procedure discussed below does require separation, but direct inhibition of a hapten-substituted enzyme by antigen binding (EMIT procedure) may alleviate a separation requirement. A sensitive method for the analysis of 2-ABZI has been demonstrated based on fluorescence polarization (26). Binding can also be measured without separation by attaching electron spin resonance (ESR) probes to antigens and monitoring the ESR band width of the nitroxide signal. Metal tagged haptens which are then analyzed by atomic absorption spectrometry show some promise (47,48). Lasers are increasing the sensitivity of turbidity methods but, at best, these methods are of moderate sensitivity.

Optimization of the Assay. As discussed earlier, competitive binding assays are based on the law of mass action where the affinity of the antibody for an antigen is $K_a = Ab \cdot Ag/(Ab)(Ag)$ and when 50% of the total antigen is bound $K_a = 1/(Ab)$. If an antibody population were homogeneous, the mathematics used to describe the binding would be rather straightforward. However, the heterogenicity of the antibody population complicates the situation so that there is no mathematical treatment that will completely describe all antibody-antigen interactions. The mathematical bases of immunoassay have been discussed by a number of workers (49-54), and work in this direction is continuing with a trend towards the development of computer programs universally adaptable to data from a variety of immunoassays (53-56). One can optimize an assay based on a logical progression of experiments using the physical constants intrinsic to the assay (46,

56), or one can approach the optimum assay conditions empirically
by determining the amount of antibody needed to give 35-50%
binding of ~10,000 CPM of the radiolabeled antigen at experi-
mentally determined incubation times (4,46). Most assays are
run under conditions approaching or at equilibrium. However, the
theoretical assay sensitivity and assay speed can be increased by
using nonequilibrium conditions. Nonequilibrium conditions are
often used in automated procedures.

Numerous methods exist for plotting competitive binding
data. In choosing a method, one should keep in mind that the
selection of an optimum plotting technique may simplify data
handling or facilitate quality control evaluations, but plotting
methods cannot enhance the intrinsic sensitivity or accuracy of
an assay. The amount of antibody bound ligand is usually
measured because small changes in antigen concentration will
yield larger relative changes in the antibody bound radioactivity
than in the unbound radioactivity. In theory, assay precision
should be enhanced by monitoring both bound and free antigen, but
this course is seldom followed.

Parker (4) pragmatically suggests plotting counts precipitated on the ordinate against the logarithm of the total unlabeled antigen concentration on the abcissa rather than spending
an inordinate amount of time in selecting the "optimum" plotting
procedure. The standard curve can usually be made more linear by
using logit, probit, or arc sine functions (52). Unless automated data reduction is used, such plots provide adequate
standard curves for most assays. There are numerous commercial
products for RIA data reduction as well as a variety of published
programs.

With constant random error, the precision of an RIA
increases as the slope of the dose-response curve increases and
decreases as the error increases with constant slope. Future
availability of monoclonal antibodies may greatly increase the
steepness and improve the shape of the resulting dose-response
curve. As with any analytical techniques, it is crucial to
appreciate the confidence intervals which one has at various
points of the dose-response curve, in addition to the many
measurement and collection errors which may be made before the
immunoassay is employed.

Choice of Radioligand. A ^{14}C radiolabel will probably exist
for most pesticides which will be considered for radioimmunoassay
development. Such an intrinsic radiolabel will prove very
valuable in titering antisera and possibly in numerous other
steps from antigen synthesis through assay development. Unfortunately, for the actual assay, the commonly available ^{14}C
radiolabels may not be of high enough specific activity. The
theoretical limit on the specific activity of a single carbon
atom is ~63 mCi/mmole, and few pesticides have a specific
activity of over 50 mCi/mmole even when they are labeled in

numerous positions. Although many factors influence radioimmunoassay sensitivity, assay sensitivity generally increases
with the square root of the specific activity of the radioligand.
Thus, ^{3}H and ^{125}I are commonly used. ^{3}H may be incorporated into
the structure of a pesticide directly (an intrinsic radioligand).
High levels of incorporation are possible by a wide variety of
procedures, and carrier free tritium will yield about 29 Ci/
mmol/atom incorporated. Isotope effects are much more common
with ^{3}H than with ^{14}C, and seemingly trivial radiosyntheses may
become very difficult when high specific activities are desired.

It is not necessary that the tracer or radioligand is struc-
turally identical to the pesticide of interest. The same considerations used in deciding where to attach a hapten to a
protein should be applied to attaching a hapten to either a
commercially available labeled compound or to a compound which is
easily labeled in a subsequent step. For instance, the conjugation of the hemisuccinate of S-bioallethrin to commercially
available ^{3}H tyramine (p-[2-aminoethyl]phenol) led to a useful
radioligand (6,7).

The most common isotope used in radioimmunoassays is ^{125}I.
Incorporation of a single atom of ^{125}I will result in a specific
activity of ∿2400 Ci/mmol. Since introduction of an iodine will
usually cause a tremendous change in a hapten, the iodine is
usually introduced on a separate moiety such as histamine, tyrosine, or tyramine which is attached to the hapten. ^{125}I offers
many advantages over ^{3}H as a tracer. It is relatively easy and
inexpensive to introduce, and its high specific activity leads to
greater theoretical assay sensitivity. As a gamma emitter it is
seldom subject to quench, and it can be efficiently detected with
a solid scintillation counter. Solid scintillation counters are
usually less expensive than liquid scintillation counters of
similar sophistication. ^{125}I does not require the use of scintillation solution which makes assays easier, cheaper, and
faster. However, as a gamma emitter which can undergo bio-
accumulation, ^{125}I must be handled very carefully, and its 2-
month half-life (vs 12 yrs for ^{3}H) necessitates repeated radiosyntheses. An additional problem is that many laboratories new
to radioimmunoassay do not have solid scintillation counters even
though liquid scintillation counters are commonly available.

There are numerous commercial adapters which increase the
efficiency of liquid scintillation counters for gamma emitters.
A modification of a suggestion by Beckman Instruments has proven
quite useful in this laboratory. Thin-walled glass tubes were
permanently attached through a hole in the cap of a standard
scintillation vial with epoxy-cement. The vial was filled with a
standard scintillation solution such as Omniflour® or 1.5% butyl
PBD in toluene containing, in addition, tetraethyl or tetrabutyl
lead. For the conditions in our laboratory, 3% v/v of tetraethyl
lead in the scintillation solution resulted in >55% counting
efficiency for ^{125}I on the ^{3}H window of a Beckman LS230 for

samples added in 6 x 50 mm glass test tubes. The scintillation
vials were permanently sealed under N_2 in order to avoid the
decomposition of the solution and the release of toxic vapors.
There are several other suggestions for the counting of ^{125}I in
liquid scintillation systems (57,58). Iodine is often introduced
ortho to the phenol of tyrosine or a tyrosine-like material or
into histamine attached to a hapten under mild oxidizing conditions. Chloramine T (N-chloro-p-toluenesulfonamide sodium salt)
or lactoperoxidase-H_2O_2 are often used as oxidizing agents. A
solid phase chloroamide has been recently reported (59). These
procedures are only suitable if the molecule is stable to oxidizing conditions. Alternatively, a separate molecule may be
labeled with iodine and then attached to the hapten (18). Some
such compounds are commercially available and detailed procedures
are available from most suppliers of radioactive iodine.

The same philosophies which apply to the choice of a radiolabel for radioimmunoassay generally apply to the attachment of
any tracer or indicator molecule. These indicators may include
such things as a fluorescent, electron spin resonance, metallic,
or enzymatic markers. If an intrinsic radiolabel is not used,
the method by which the label is introduced may effect the assay
specificity and sensitivity, just as the choice of hapten does
(Fig. 1). If the same hapten derivative is used for preparing
the antigen and the radioligand, the resulting antibody may have
a higher affinity for the radioligand than the molecule to be
assayed. This situation will reduce the theoretical sensitivity
of the assay (see Figure 5).

Enzyme-linked Immunosorbent Assay. A promising alternative
to the RIA procedure is an enzyme-linked immunosorbent assay
(ELISA) which depends upon the conjugation of a functional enzyme
to either an antigen or antibody. The amount of enzyme present
in a competitive binding assay is quantitated instead of the
amount of radiolabeled compound. The concentration of the enzyme
can be determined through its subsequent reaction with a substrate which results in a measurable spectroscopic change.

Conjugation of enzymes to antigens or antibodies were first
developed for histochemical techniques and were used for locali-
zation of antigens and antibodies in tissue sections (60).
Enzymes were quickly adapted for immunoassays, and Engvall and
Perlmann (61) developed a procedure for the quantitation of an
antigen. Alkaline phosphatase was conjugated via glutaraldehyde
to rabbit IgG (antigen). Sheep antibody against rabbit IgG was
coupled to microcrystalline cellulose by cyanogen bromide and the
amount of antigen binding to the antibody was a direct relationship with the amount of phosphate ester cleaved by the coupled
enzyme in a given period of time. This technique was widely
adapted for the quantitation of various proteins and infectious
agents (62-69).

A number of enzymes have been used with immunoassays. These

include lactic dehydrogenase, mushroom tyrosinase, glucose oxidase, acid phosphatase, alkaline phosphatase, and horse-radish peroxidase. The latter two enzymes have received most of the attention and peroxidase is usually preferred because of its low cost (70) although other reactions may be more sensitive and reproducible (68,69).

Bi- or multi-functional reagents have been used to link enzymes to other proteins. These include various carbodiimides, bisdiazotized amines, cyanuric chloride and glutaraldehyde. Enzymes linked to rabbit IgG antibodies from sheep, goats, and horses are commercially available and greatly facilitate the ELISA procedure. The preparation of the antigen and the development of the corresponding rabbit antibody (IgG fraction) have been described previously.

The ELISA procedure recently has been used for the analysis of parathion (31). Since this procedure has considerable potential a more detailed description of the analysis of parathion is in order. The conjugation procedure using amino parathion (AP) was described earlier (Fig. 3, Rn 9), and this conjugate was then administered to rabbits for development of a population of specific antibodies (Ab_1) against BSA or AP. Ab_1 demonstrated immunological activity only for the hapten when AP was conjugated to rabbit serum albumin (RSA). This antigen (RSA-AP) was rendered insoluble via attachment to the polystyrene surface of microtiter plates under basic conditions (Fig. 6.1).

Following the removal of excess antigen the specific antiserum (Ab_1) was allowed to react with the surface bound antigen (Fig. 6.2). After washing away excess antiserum, an enzyme (E, horse-radish peroxidase) conjugated to goat γ-globulin (Ab_2), produced against rabbit γ-globulin of the antiserum (Ab_1), was added (Fig. 6.3). The binding of the enzyme complex to the solid phase was a measure of the amount of bound RSA-AP and Ab_1. The enzyme concentration was measured spectrophotometrically by means of its catalyzing the oxidation of hydrogen peroxide in the presence of 5-aminosalicyclic acid in a given period of time (Fig. 6.4).

Analysis of parathion by this technique is based on the competition between the free form of parathion (P) and its conjugated form (AP) for the binding sites on the first antibody (Ab_1) (Fig. 6.5). Due to this competition, there is a decrease in the binding of the conjugated form as the concentration of the free form (P) increases. Complete inhibition of the binding of Ab_1 for BSA-AP may result when P is present in greater quantity than Ab_1. The concentration of the parathion in an unknown sample can then be determined by comparing the degree of antibody inhibition caused by the addition of the sample extract with that resulting from the addition of known amounts of the same substance. For parathion analysis no cleanup of the extracts of fruits or vegetables were necessary.

Various parameters of the ELISA procedure need to be

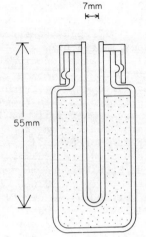

Figure 5.   Cross section of a scintillation vial illustrating a system for counting ^{125}I in a liquid scintillation system.

The sample to be counted is inserted into the 7 × 55 mm glass well immersed in heavy metal charged scintillation cocktail in a permanently sealed vial.

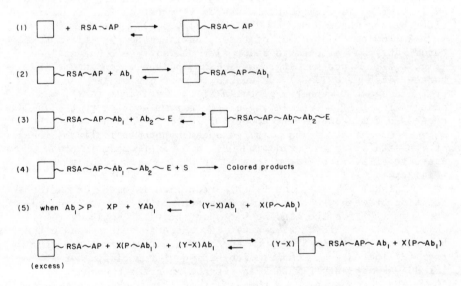

Figure 6.   Schematic representation of ELISA.

(□) polystyrene surface; (RSA) rabbit serum albumin; (AP) conjugated aminoparathion; (Ab_1) first antibody (rabbit anti-parathion); (Ab_2) second antibody (goat anti-rabbit); (E) enzyme (horse-radish peroxidase); (S) substrate; and (P) free hapten (parathion).

optimized for each analysis. These parameters include the concentration of antibodies, time of incubations, temperature, and the composition of the washing fluids. Optimum time of incubations for analysis of parathion ranged from 2-3 hours. A solution of 0.9% NaCl and Tween 20 was determined to be most suitable for washing the microtiter plate in all steps of the procedure.

The antiserum exhibited high specificity for the functionalities of parathion, e.g., 58 ng/ml of parathion produced 50% inhibition of the oxidation of 5-aminosalicyclic acid. Changing of the ethyl groups to methyl groups as in methyl parathion (2000 ng per 50% inhibition) or replacement of the sulfur atom with oxygen as in paraoxon (1850 ng per 50% inhibition, 23), greatly reduced the competitive binding. Amino parathion did exhibit a significant cross reaction (275 ng per 50% inhibition, 23) but p-nitrophenol bound poorly to the antibody (5000 ng per 50% inhibition). The lower limit of detection of parathion by the ELISA procedure was found to be 5.0-10.0 ng/ml which corresponded to 0.025-0.05 ppm in crude extracts of fruit, vegetables, and human serum.

The procedure gave good reproducibility as expressed in the coefficient of variation (CV%) of results of between-run (6.2-8.6) and within-run (4.8-6.5) variations. Accuracy of the ELISA procedure was tested by comparing results of parathion analysis in extracts of fortified and field samples with results obtained by a GLC method. Correlation coefficients ranged in almost all cases between 0.93-0.99.

The ELISA procedure for the analysis of parathion as described above requires nearly eight hours, although many samples can be simultaneously assayed. However, incubation times can be shortened to one-half hour, in most cases, resulting in only a 10% reduction in sensitivity. Also the polystyrene microtiter plates containing bound RSA-AP can be mass produced and stored in a freezer. Since the enzyme-linked antibody can be purchased, the limiting factor of the applicability of the ELISA procedure, as well as the RIA procedures, for other pesticides is the development of the antiserum to the pesticide.

The ELISA procedure shares many of the advantages of RIA, and it has additional advantages of requiring only inexpensive equipment and of being well adapted to automated or partially automated methods. For instance, Ruitenberg et al., (71) has mechanized the ELISA procedure for screening of 4000 sera samples daily. A number of disadvantages of the ELISA procedure also can be sited. These include the nonstability of the developed color requiring daily analysis (not necessary in the RIA method), nonlinearity of color development, and less sensitivity than some other immunochemical methods.

Other Immunoassay Methods. Other immunoassay methods can be used to quantitate the hapten; these include homogeneous enzyme immunoassay (EMIT), radial immunodiffusion, immunoelectrophoresis and passive hemagglutination tests. These techniques are often used to characterize the antibody, but they also can be used to quantitate the hapten through inhibition experiments. The EMIT procedure involves direct inhibition of enzyme activity when an antibody binds to a hapten conjugated near the enzyme's active site. Thus, it is particularly useful with small molecules, and it is very rapid because no separation steps are required (69). This technique is promising, but it has not been widely used and it is often of lower sensitivity than ELISA.

Radial immunodiffusion procedures are varied but all depend upon the diffusion of the antigen or antibody in a gel producing a precipitate which is proportional to the quantity of reactants (often sensitive to 25 ng protein with visual methods (32,72, 73)). Modifications using radiolabeled antigens or antibodies may increase the sensitivity fifty-fold (74,75). If the antibody is first mixed with the hapten, the concentration of free unbound antibody will decrease proportionately and result in a decrease in the precipitate formed with the antigen which can be observed visually or with radiolabeled methods.

The hemagglutination test is often used to express antibody titer, but it also can be used to quantitate the hapten. This test is based on the fact that erythrocytes, when treated with a dilute solution of tannic acid, acquire the property of being able to adsorb protein (the conjugated antigen). Such protein-coated red blood cells are agglutinated by specific antiserum directed against the hapten. The agglutination titer is expressed as the reciprocal of the highest dilution of the serum that causes agglutination of the red blood cells. The antiserum can be incubated with the sample extract containing the specific hapten prior to conducting the hemagglutination test. The amount of unbound antiserum is reduced by the amount equivalent to the free hapten and consequently the dilution of antiserum necessary to produce agglutination is inversely proportional to the amount of hapten in the sample (32).

Monoclonal Antibody Technology. Based on pioneering work by Köhler and Milstein (76) a new technology is evolving which may greatly improve the availability, specificity, and sensitivity of antisera. Monoclonal antibody technology has been the subject of a technical compendium (43) and a nontechnical review (45). Simplistically, spleen lymphocytes immunized in vivo or in vitro are fused with a myeloma cell line, and the resulting hybrids are selected on the basis of nutritional requirements and then cloned. Those clones producing monoclonal antibodies of the desired specificity are injected into mice where the resulting ascites tumor fluid may contain gram quantities of a monoclonal antibody. Alternatively, antibodies can be collected directly from a cell culture medium. Since clones are used, it is not necessary for the antigen to be highly pure and antibodies

selective for optical isomers conceivably could be obtained from
a racemic hapten. Although in its infancy, monoclonal antibody
technology may offer many advantages. As earlier discussed, the
serum of an animal contains a large population of antibodies with
varying specificities and affinities. With monoclonal antibodies
one or more antibody types of high affinity and optimum specificity can be selected and propagated for use in immunoassays.
Immunoassays using monoclonal antibodies are characterized by a
very steep dose-response curve which translates as greatly
enhanced assay precision. Although the clonal hybrids may have
an unstable chromosome complement, with proper technical care the
lines could be considered "immortal". Thus, a single, uniform
antibody reagent of defined specificity could be provided to many
laboratories from frozen cell lines which are occasionally thawed
to produce antibodies. On the negative side, monoclonal technology is new and has not been widely applied to the analysis
of haptens. Relatively simple cell culture facilities are needed
and since culturing must currently be done in antibiotic-free
medium, a high level of technical skill is necessary. Several
companies have recently entered the field, and it may soon be
possible to obtain antibodies on a contract basis. Some of the
potential advantages and disadvantages of monoclonal technology
as applied to residue analysis are listed in Table II.

Table II

Potential Advantages and Disadvantages of
Monoclonal Antibodies as Applied to Pesticide Residue Analysis

Advantages	Disadvantages
Cell lines "immortal"	Chromasome complement often
Very high specificity	unstable
possible	Specificity too great
Steep dose-response curve	Dose-response curve too steep
(very precise assays)	(small linear region)
Technology advancing rapidly	New technology
Large amount of antibodies	Not widely used with haptens
Extraordinary high titer	Simple cell culture facilities
Uniform antibodies	needed
Activate in vivo or in vitro	Antibiotic-free medium used
Pure antigen not necessary	Precipitation assays difficult

Attributes and Limitations of Immunoassay

Immunoassay Sensitivity. Yalow (2) points out that as
little as 0.1 picogram (0.05 picomolar) gastrin can easily be
detected by immunoassay in a milliliter of incubation medium.
Immunoassays to small lipophilic molecules are generally less

sensitive than those to proteins and peptides, and molecules
having several polar functionalities separated by nonpolar areas
often lend themselves most readily to highly sensitive radioimmunoassays (4). With steroids, sensitivities on the order of 1
picogram are not uncommon (44) and useful assays for steroids and
drugs have been devised at much lower sensitivities. As discussed earlier, the actual assay sensitivity depends upon the
affinity of the serum, the incubation volume, and the amount of
tracer and antibody used (which translate, in part, to the
specific activity of the tracer).

The sensitivity of the overall analytical procedure depends
upon many factors obviously including the type of sample to be
analyzed and the skill of the analytical chemist. If an immunoassay is used to measure the amount of pesticide in a water
sample by adding the water sample directly to the immunoassay,
very high sensitivity may not be obtained although the assay will
require very little time to perform. Alternatively, if the water
sample is extracted and the immunoassay is employed only after
several highly efficient cleanup steps, phenomenal sensitivity
may be obtained at the expense of a large investment in time. In
some situations, immunochemical methods may decrease the limit of
detectability of a pesticide residue (77), but more importantly
they may, in some cases, decrease the time and cost needed to
reach a level of detectability as has been demonstrated with
parathion (31).

Specificity. The specificity of an immunoassay is related
in some respects to the sensitivity. The high specificity of
immunoassays often allows samples to be analyzed with a minimum
of cleanup. The remarkable specificity of antigen-antibody
interactions has been reviewed in the classic text by Landsteiner
(30). More recently, Al-Rubae (31) demonstrated a high level of
specificity in an ELISA procedure for parathion. As previously
discussed, parathion could be easily distinguished from methyl
parathion and p-nitrophenol and the assay demonstrated little or
no cross reactivity with a number of other pesticides. A radioimmunoassay for S-bioallethrin showed no cross reaction for any
of several other pyrethroids tested (except pyrethrin I), and it
was capable of distinguishing S-bioallethrin (1R,3R,4'S) from the
other 7 optical and geometrical isomers of allethrin. Since
biological activity and biodegradation may depend upon the configuration of an insecticide (6,7) the ability of immunoassays to
distinguish among chiral isomers (78,79) may become of great
importance to future pesticide metabolism and residue investigations. The high potential specificity of immunochemical methods
may prove very useful, in conjunction with other methods, in the
confirmation of the presence of residues. Monoclonal technology
is likely to allow pesticide analysis based on immunochemistry to
be even more specific and sensitive.

Although one must ultimately rely upon the immune system of

an animal to determine the specificity of a given antibody population, methods were discussed earlier which can be used to
predict the antibody specificity when a given hapten is used.
Antisera, which will detect parent pesticide plus toxic metabolites, could be used in combination with one or more highly
specific antisera to quantitate several molecules of interest.
An assay of moderate specificity will be of greater use in some
analytical applications than a highly specific assay. The more
general assay may be very useful in screening for the presence of
a class of compounds or the presence of a specific functionality
in a metabolite. Such assays can be used very effectively by
coupling them with chromatographic techniques such as thin-layer
chromatography (TLC) or open column chromatography. The use of
immunochemical tests as sensitive detectors for high performance
liquid chromatography certainly offers promise in pesticide
analysis. A nice example of such a procedure is the analysis of
N,N-dimethylindolealkylamines in biological fluids (80).

   Speed of Analysis. The speed with which many immunochemical
analyses can be completed illustrates a major advantage of
immunochemical procedures. Immunochemical assays are most time
and cost effective when the sample load is large. Parker (4)
estimated that a single technician could perform 100-5000 radioimmunoassays per day with little or no assay automation in comparison to 20-40 GLC assays (3). Numerous inexpensive systems
are available to decrease analysis time. These systems may
include solid phase separation techniques, automatic dispensers,
test tube racks which will fit directly into a centrifuge and/or
scintillation counter, and data handling systems. Alternatively,
there are fully automated systems based on RIA or ELISA which
require very little operator attention and which handle 25-240
samples/hr. Gochman and Bowie (81) have outlined the basis of
operation and summarized the features of automated RIA systems
and extensive literature is available from the manufacturers.
   As with many analytical procedures, the most time-consuming
part of the assay is sample preparation. The high specificity
and sensitivity of immunoassays may tremendously reduce the
workup needed before actual analysis of the sample. For example,
analysis of allethrin in milk by the accepted analytical procedure based on electron capture GLC required 4-8 hour per sample
in our hands (82). Similar sensitivity and higher specificity
could be realized using an immunochemical assay requiring 15-30
minutes per sample. With some loss of sensitivity, immunoassays
may be very rapid. Turbidity measurements can be made so quickly
and quantitatively that they may be very useful for field
analyses of pesticides. Such rapid procedures might prove very
useful in determining pesticide coverage on specific areas of a
plant immediately after application, detecting drift, or monitoring the safety of a field for worker reentry.
   In this chapter we have discussed the advantages of immuno-

chemical methods as a supplement to more classical analytical
techniques. Possibly among the most important contributions of
immunochemistry to future pesticide analysis will be its use as a
tool to open new areas of pesticide analytical chemistry. If
very rapid, inexpensive assays can be developed, pesticide
analysis may be increasingly employed to enhance effective pesticide use rather than as simply an enforcement or residue tool.

    Cost Effectiveness. As with the other advantages of immuno-
chemical analysis, cost may be quite variable. Reagent costs for
several automated systems have been estimated at under $1.25 per
sample. The cost is obviously much lower for less sophisticated
assay systems, especially if some reagents are prepared in house.
A major consideration is the expense of new instrumentation. For
dedicated or automated instrumentation for either RIA or ELISA
procedures, the cost may be $50-100,000. However, most analyti-
cal laboratories already have the basic instrumentation needed
for immunoassays. Moderate sensitivity can be obtained through
the use of numerous procedures such as radial immunodiffusion and
hemagglutination. These procedures require no expensive equipment or reagents and they may be very useful in areas where
equipment acquisition or maintenance is a problem.
    The expense of an analytical procedure depends upon much
more than the cost of the final analysis. Much of the expense of
an assay is related to sample preparation, and for many applications immunoassays have tremendously reduced the time needed for
sample preparation. Another consideration is the amount of time
needed for the development of an assay. The additional expertise
which must be developed in an analytical laboratory before
immunoassays can be used with confidence may seem formidable, and
waiting for an animal to develop antibodies may lead to unacceptable delays in assay development. On the other hand, once a
usable antibody titer is obtained, the development of a workable
assay is usually straightforward. It is also likely, if immunoassays become accepted for some aspects of pesticide analysis,
immunoassay kits or at least critical reagents will become commercially available. Such kits already exist for many pharma-
ceutical products and hormones, and numerous companies will
supply antibodies to a user supplied hapten on a contract basis
(83).

    Applicability. Parker (4) points out that one can assume
that workable radioimmunoassays can be developed "with all except
the smallest or most unstable molecules." Once a useful antibody
titer is obtained, often only very small changes in a generalized
procedure are needed to obtain a workable assay. Although
immunoassays would appear to be generally applicable to pesticide
analytical problems they may be most useful in solving specific
problems which appear intractable when classical procedures are
used. Immunoassays are often most sensitive and specific when

several polar functionalities exist. Such compounds may be
rather nonvolatile or heat labile and difficult to analyze by
classical methods. Although sensitive, specific immunoassays
have been developed for nonpolar compounds, such compounds may
be most readily analyzed by GLC procedures. For laboratories not
interested in the development of their own antisera, each Fall
"Lab World" (83) lists suppliers of immunochemical reagents.

   Problems with Immunoassays. As with any analytical technique, there are numerous problems associated with the use of
immunochemical technology. Most of these problems are common to
any analytical procedure, but some are relatively unique to
immunoassay and have been covered by Parker (4). The parameters
which should be monitored to maintain quality control of the
assay have been discussed by Rodbard et al. (50). A major con-
cern discussed earlier is cross reactivity or interference,
especially if it is unexpected. One can guard against this
problem by employing well characterized antiserum, by using
sample blanks, and by running standard curves in the presence of
extracts. One must rely upon the equipment and reagents used in
analytical procedures. Antibodies are certainly not as stable as
many chemical reagents; however, the guaranteed shelf life of
many commercial lyophilized preparations is over 5 years at 4°C.
The integrity of the reagents must be periodically reestablished,
especially if the assays are only performed sporadically.
   Immunoassays lend themselves to the processing of a large
number of samples. The same number of control and standard
assays are required whether one or a large number of samples are
assayed. For an analytical laboratory faced with analyzing a
large number of samples for the presence of a few pesticides,
immunochemical procedures are likely to offer many advantages
over some more clasical analytical methods. If the same laboratory were faced with quantitating the residues of a large number
of differing chemicals in a few samples, immunochemical procedures are likely to be less cost and time efficient than an
equally sensitive GLC based assay.

   Possible Contributions of Immunochemical Methods to
Pesticide Analysis. As Ercegovich (3) pointed out, it is unlikely that immunochemical methods will replace current, established analytical methods of pesticide analysis. However, the
analytical chemist who carefully compares the attributes and
deficiencies of immunochemical methods of analysis with other
procedures is likely to find applications for which immunochemical methods offer distinct advantages.
   In many cases, those compounds which are most difficult to
assay by classical procedures because of numerous polar functionalities and poor volatility are the very compounds which lend
themselves most readily to immunochemical analysis (4). One can
also predict that the number of pesticides marketed with a high

degree of optical purity will increase, and immunochemical methods lend themselves to the analysis of chirality at the residue level (6,7). Thus, there will probably be some pesticides for which immunochemical methods will provide the future enforcement procedures of choice for residue analysis.

It is envisioned that immunochemical procedures can be more commonly used as a supplement to classical methods of pesticide analysis. Since samples can often be analyzed without expensive and time-consuming cleanup procedures usually required of most methods, the immunological assays can rapidly screen many samples at significantly lower cost. When the immunoassays indicate that samples contain appreciable pesticide residues, the samples can be further analyzed by GLC or other methods. Alternatively, an immunochemical assay may provide a confirmatory test. Specific antisera also can be used to concentrate pesticides and to clean up extracts by means of affinity chromatography procedures, thereby, permitting greater sensitivities of GLC, HPLC or other methods. It is expected that immunochemical and especially the ELISA procedures may contribute to field reentry and human exposure problems where simple, rapid, inexpensive procedures are desired. Finally, the possible usage of these methods in developing countries could be of practical importance due to the simplicity of the procedures, the ease with which they are interfaced with thin-layer analysis, and the use of relatively simple laboratory apparatus.

Acknowledgements

This review is dedicated to the memory of C. D. Ercegovich, who pioneered the application of immunochemical methods to pesticide residue analysis. A. Karu (Dept. of Biochemistry, University of California, Riverside) provided advice on monoclonal antibody technology, and Siong Wie (UCR) critically reviewed the manuscript. The original research presented here was supported by EPA Grant R806447-01, NIEHS Grant 5 R01 ES01260-03 and the California and Pennsylvania Agricultural Experiment Stations. B.D. Hammock was supported by NIEHS Research Career Development Award 1 K04 ES00046-01. This is a publication from the Pennsylvania Agricultural Experiment Station.

Literature Cited

1. Yalow, R. S. and Berson, S. A. J. Clin. Invest., 1960, 39, 1157.
2. Yalow, R. S. Science, 1978, 200, 1236.
3. Ercegovich, C. D. in "Pesticide Identification at the Residue Level," Advances in Chemistry Series #104, R. F. Gould, Ed. American Chemical Society, Washington, D.C., 1976, p. 162.
4. Parker, C. W. "Radioimmunoassay of Biologically Active Compounds;" Foundations of Immunology Series, Prentice-Hall, Inc., Englewood Cliffs, N.J., 1976, 239 pp.
5. Flynn, E. J. and Spector, S. J. Pharmacol. Exp. Ther. 1972, 181, 547.
6. Wing, K. D., Hammock, B. D. and Wustner, D. A. J. Agric. Food Chem., 1978, 26, 1328.
7. Wing, K. D. and Hammock, B. D. Experentia, 1980, 35, 1619.
8. Lemieux, R. U., Bundle, D. R., Baker, D. A. J. Am. Chem. Soc., 1975, 97, 4076.
9. Lemieux, R. U., Baker, D. A. and Bundle, D. R. Can. J. Biochem., 1977, 55, 507.
10. Turková, J. Affinity Chromatography, Elsevier Scientific Publishing Co., New York, 1978, p. 151.
11. Langone, J. J. and Van Vunakis, H. Res. Commun. Chem. Pathol. Pharmacol., 1975, 10, 163.
12. Anderson, G. W., Zimmerman, J. E. and Callahan, F. M. J. Am. Chem. Soc., 1964, 86, 1839.
13. Sylwester, A., Wing, K. D., and Hammock, B. D. "Immuno-chemical Analysis of Insecticides: Haptens and Antigens for the IGR Diflubenzuron," Paper #30, Pacific Branch Meeting, Entomological Society of America, June 26, 1979. Publication in preparation.
14. Lensky, Y., Baehr, J.-C. and Porcheron, P. C. R. Acad. Sc. Paris, Series D., 1978, 287, 821.
15. Lauer, R. C., Solomon, P. H., Nakanishi, K. and Erlanger, B. F. Experentia, 1974, 30, 560. Ibid. 558.
16. Maróy, P., Vargha, J., Horváth, K. FEBS Letters, 1977, 81, 319.
17. Rudinger, J. and Ruegg, U. Biochem. J., 1973, 133, 538.
18. Bolton, A. E. and Hunter, W. M. Biochem. J., 1973, 133, 529.
19. Sheehan, J. C. and Hlavka, J. J. J. Org. Chem., 1956, 21, 439.
20. Goodfriend, T. H., Levine, L. and Fasman, G. D. Science, 1964, 144, 1344.
21. Spindler, K.-D., Beckers, C., Gröschel-Stewatt, U., Emmerich, H. Z. Physiol. Chem., 1978, 359, 1269.
22. deReggi, M. L., Hirn, M. H., Delaage, M. A. Biochem. Biophys. Res. Commun., 1975, 66, 1307.

23. Vallejo, R. P. and Ercegovich, C. D. "Analysis of Potato
    for Glycoalkaloid Content by Radioimmunoassay (RIA),"
    Methods and Standards for Environmental Measurements,
    National Bureau of Standards Publication 519, Wash-
    ington, D.C., April 1979.

24. Haas, G. J., Guardia, E. J. Proc. Soc. Exp. Biol. Med.,
    1968, 129, 546.

25. Centeno, E. R., Johnson, W. J. and Sehon, A. H. Int. Arch.
    Allergy, 1970, 37, 1.

26. Lukens, H. R., Williams, C. B., Levison, S. A., Dandliker,
    W. B. and Murayama, D. Environ. Sci. Technol., 1977,
    11, 292.

27. Kado, N. Y., Wei, E. T. J. Natl. Cancer Inst., 1978, 61,
    221.

28. Chae, K., Cho, L. K., and McKinney, J. D. J. Agric. Food
    Chem., 1977, 25, 1207.

29. Pauling, L., Pressman, D., Campbell, D. H., Ikeda, C. and
    Ikawa, M. J. Am. Chem. Soc., 1942, 64, 2994.

30. Landsteiner, K. The Specificity of Serological Reactions,
    Harvard University Press, Cambridge, Mass., 1945 (Dover
    Publications, Inc., New York, 1962, 330pp.).

31. Al-Rubae, A. Y., Ph.D. Thesis, "The Enzyme-Linked Immuno-
    sorbent Assay, A New Method for the Analysis of Pesti-
    cide Residues," The Pennsylvania State University,
    1978.

32. Williams, C. A. and Chase, M. W., Ed. Methods in Immunology
    and Immunochemistry. Academic Press, New York, N.Y.,
    1967, 479pp.

33. O'Connor, J. D. and Borst, D. W. Science, 1972, 178, 418.

34. Borst, D. W. and O'Connor, J. D. Steroids, 1974, 24, 637.

35. Porcheron, P., Foucrier, J., Gros, Cl., Pradelles, P.,
    Cassier, P. and Dray, F. FEBS Letters, 1976, 61, 159.

36. Rao, P. N., Moore, P. H. Jr., Peterson, D. M. and Tcho-
    lakian, R. K. J. Steroid Biochem., 1978, 9, 539.

37. Rao, P. N. and Moore, P. H. Jr. Steroids, 1977, 29, 461.

38. Rao, P. N. and Moore, P. H. Jr. Steroids, 1976, 28, 101.

39. Beiser, S. M., Butler, V. P. Jr., Erlanger, B. F. "Hapten-
    Protein Conjugates: Methodology and Application," in
    Textbook of Immunopathology, 2nd Ed., V. 1, p. 15. Ed.
    P. A. Miescher and H. J. Miller-Eberhard, Grune and
    Stratton, Inc., New York, N.Y., 1976.

40. Habeeb, A. F. S. A. Anal. Biochem., 1966, 14, 328.

41. Böhlen, P., Stein, S., Dairman, W. and Udenfriend, S. Arch.
    Biochem. Biophys., 1973, 155, 213.

42. Tigelaar, R. E., Rapport, R. L., Inman, J. K. and Kupfer-
    berg, H. J. Clin. Chim. Acta, 1973, 43, 231.

43. Melchers, F., Potter, M. and Warner, N., Eds. "Lymphocyte
    Hybridomas - Second Workshop on 'Functional Properties
    of Tumors of T and B Lymphocytes'," Current Topics in
    Microbiology and Immunology, 81. Springer-Verlag, New
    York, 1978, 246 pp.

44. Nieschlag, E., Wickings, E. J. Z. Klin. Chem. Klin. Bio-
    chem., 1975, 13, 261
45. Fox, J. L. Chem. Eng. News, 1979, 57, 15.
46. Rogers, R. C. Radioimmunoassay Theory for Health Care
    Professionals, Hewlett Packard Corp. Loveland, Colo-
    rado, 1974, 55pp.
47. Cais, M. Dani, S., Eden, Y., Gandolfi, O., Horn, M., Isaacs,
    E. E., Josephy, Y., Saar, Y., Slovin, E. and Snarsky,
    L. Nature, 1977, 270, 534.
48. Cais, M., Slovin, E. and Snarsky, L. J. Organometallic
    Chem., 1978, 160, 223.
49. Ekins, R. P., Newman, G. B. and O'Riordan, L. H. "Satura-
    tion Assays" in Statistics in Endocrinology, J. W.
    McArthur and T. Colton, Eds. MIT Press, Cambridge,
    Mass., 1970, p. 345.
50. Rodbard, D., Rayford, P. L. and Ross, G. T. "Statistical
    Quality Control of Radioimmunoassays," in Statistics in
    Endocrinology, J. W. McArthur and T. Colton, Eds. MIT
    Press, Cambridge, Mass., 1970, p. 411.
51. Yalow, R. S. and Berson, S. A. Principles of Competitive
    Protein-Binding Assays. W. D. Odell and W. H. Daugha-
    day, Eds. J. B. Lippincott Co., Philadelphia, 1971, p.
    374.
52. Rodbard, D. Principles of Competitive Protein-Binding
    Assays, W. D. Odell and W. H. Daughaday, Eds. J. B.
    Lippincott Co., Philadelphia, PA, 1971, p. 204.
53. Rodbard, D. Anal. Biochem., 1978, 90, 1.
54. Rodbard, D. and Tacey, R. L. Anal. Biochem., 1978, 90, 13.
55. Seaton, B., Lusty, J. and Watson, J. J. Steroid Biochem.
    1976, 7, 511.
56. Schöneshöfer, M. Clin. Chim. Acta, 1977, 77, 101.
57. Horrocks, D. L. Nucl. Instr. Meth., 1976, 133, 293.
58. Ashcroft, J. Anal. Biochem., 1970, 37, 268.
59. Fraker, P. J. and Speck, J. C. Jr. Biochem. Biophys. Res.
    Commun., 1978, 80, 849.
60. Avrameas, S. Immunochemistry, 1969, 6, 43.
61. Engvall, E. and Perlmann, P. J. Immunology, 1972, 109, 129.
62. Engvall, E., Jonsson, K. and Perlmann, P. Biochem. Biophys.
    Acta, 1971, 251, 427.
63. Engvall, E. and Perlmann, P. Immunochemistry, 1971, 8, 871.
64. Belanger, L., Sylvestre, C. and DuFour, D. Clin. Chim.
    Acta, 1973, 48, 15.
65. Belanger, L., Hamel, D., DuFour, D. and Pouliot, M. Clin.
    Chem., 1976, 22, 198.
66. Kirkpatrik, A., Wepsic, H. T. and Nakamura, R. M. Clin.
    Chem., 1977, 23, 50.
67. Ruitenberg, E. J., Steerenberg, P. A., Brosi, B. J. M. and
    Buys, J. J. Immunological Methods, 1976, 10, 67.

68. Voller, A., Bidwell, D. E. and Bartlett, A. Bull. Wld.
       Hlth. Org., 1976, 53, 55.
69. Voller, A. La Ricerca Clin. Lab., 1978, 8, 289.
70. Avrameas, S. Int. Rev. Cytol., 1970, 27, 349.
71. Ruitenberg, E. J., vanAmstel, J. A., Brosi, B. J. M. and
       Steerenbert, P. A. J. Immunological Methods, 1977, 16,
       351.
72. Ouchterlony, O. Prog. Allergy, 1958, 5, 1.
73. Crowle, A. J. Immunodiffusion, Academic Press, New York,
       1961, 333pp.
74. Jalanti, R. and Henney, C. S. J. Immunological Methods,
       1972, 1, 123.
75. Rowe, D. S. Bull Wld. Hlth. Org., 1969, 40, 613.
76. Köhler, G. and Milstein, C. Nature, 1975, 256, 495.
77. Sutherland, G. L. Residue Reviews, 1965, 10, 85.
78. Landsteiner, K. and van der Scheer, J. J. Exp. Med., 1928,
       48, 315.
79. Karush, F. J. Am. Chem. Soc., 1956, 78, 5519.
80. Riceberg, L. J. and Van Vunakis, H. V. J. Pharmacol. and
       Exp. Therapeutics, 1978, 206, 158.
81. Gochman, N. and Bowie, L. J. Anal. Chem., 1977, 49, 1183A.
82. Moore, J. B. "Residue and Tolerance Considerations," in
       Pyrethrum the Natural Insecticide, J. E. Casida, Ed.,
       Academic Press, N.Y., 1976, p. 293.
83. Sweeney, M. Lab World, 1978, 29, 48.

RECEIVED March 3, 1980.

19

Negative Ion Mass Spectrometry

E. C. HORNING, D. I. CARROLL, I. DZIDIC, and R. N. STILLWELL

Institute for Lipid Research, Baylor College of Medicine, Houston, TX 77030

Negative ion mass spectrometry will come into use in the future in many applications, and it is likely to prove particularly valuable in pesticide work and in toxicological studies. Most manufacturers of mass spectrometers are now prepared to supply instruments that can be used for either positive or negative ion mass analysis, or for concurrent operation by rapid switching, and numerous older instruments are being modified for negative ion studies. Current investigations are usually directed to gaining additional information about basic processes involved in negative ion formation and to exploring applications.

Source conditions for both positive and negative ion mass spectrometry are usually discussed in terms of source pressure and style of ionization. The source conditions that have been used for negative ion formation are in Table I.

The pressure in the source, while an important technological variable, is not a determinant of the mass spectrum. It is possible, for example, to duplicate electron impact ionization (EI) mass spectra of steroids under chemical ionization (CI) conditions, with nitrogen as the charge and energy transfer reagent, by using a short residence time for the ions in the source. In general, the internal energy of product ions is inversely related to the degree of equilibration attained in the source, and this can be varied by varying the residence time of ions in the source. The usual mode of operation for atmospheric pressure ionization (API) (1,2) leads to highly stable ions which have been thermally and chemically equilibrated with the carrier gas and other ions in the source, and very few fragment ions are usually observed. The highest degree of fragment ion formation, short of electron impact ionization reactions, is observed under the low pressure chemical ionization (LPCI) conditions employed by Brandenberger (3).

Negative ions are formed under a variety of conditions. The best-known reactions are those of electron attachment and dissociative electron attachment, since these are the reactions

which occur in electron capture detectors. Low energy electrons are the reagent, and these are present when gases such
as nitrogen, argon and argon/methane are used under API or CI
conditions. Proton removal or proton transfer to a basic ion
will also lead to anion formation. This reaction can be used
to classify organic compounds in terms of gas phase acidity.
For example, picric acid is a strong gas phase acid; aliphatic
alcohols are weak gas phase acids. Oxygen substitution reactions leading to phenolate ions occur for some substances.
Many chlorine-substituted compounds and some aromatic hydrocarbons react with O_2^- ions; this type of reaction is discussed
later. Adduct formation, involving a halide ion, leads to
negative ions under some circumstances. There are also a few
addition reactions of specialized reagents that will lead to
negative ions.

    Much of our experience in negative ion studies is based
upon atmospheric pressure ionization mass spectrometry (1).
The conditions used by Brandenberger (3) for identification
studies involve low pressure chemical ionization with nitrous
oxide or methane. Ordinary CI (0.3-1 Torr) conditions have
been used in a number of laboratories. Current evidence suggests that there may not be a single condition that can be
recommended as being best in all applications of negative ion
mass spectrometry, and that at least two or perhaps more
conditions will ultimately find wide use. Our approach to the
question of "best" conditions is to use low energy conditions
in quantitive analytical work, and to use higher energy conditions in identification and structural studies. There are a
variety of reasons for this, but in general the effect of
higher energy conditions is that fragmentation is increased,
and this increases the amount of data available for interpretation. In the future, it is possible that two ionization
steps will be used in the same analytical system for both
qualitative and quantitative analysis. At present, with
positive ions, it is possible to convert an initially formed
ion to fragment ions by collision with neutral gas molecules.
When two mass analyzers are used, the general process has
been called MS/MS by McLafferty. Much less is known about the
stability of negative ions under varying conditions.

    Our recent API negative ion mass spectrometry studies (4)
have been carried out with a source (Figure 1) having the
physical dimensions of a commercially available electron
capture detector. With an electrode in place, and with a
nickel-63 foil in place, the source chamber functions as an
electron capture detector. A very small nickel-63 foil is
used in the aperture region. The gas stream from a gas
chromatograph is split before entering the source, so that
detection is achieved in three ways. The split stream is
directed in part to a standard electron capture detector and
in part to the API source, which allows both a source electron

capture response and mass analysis of ions from the source be recorded simultaneously.

Electron attachment studies have, with some exceptions, given the expected results. Table II contains a list of compounds or compound types known to give molecular negative radical ions by electron attachment, in order of decreasing electron affinity. Figures 2 and 3 show the experimental result of ionizing benzil with thermalized electrons; the only product is an $M^{\overline{\cdot}}$ ion, and all three expected responses are observed. Azulene, one of the compounds in Table II, shows the expected response under API conditions, but benzophenone does not. When these experiments were repeated under CI conditions, all compounds in Table II gave molecular negative ions.

Electron attachment reactions may end with $M^{\overline{\cdot}}$ formation or with the formation of a fragment ion(s) through cleavage to a stable anion and a neutral radical (which is not observed). These reactions, when halogenated (Cl,Br,I) compounds are involved, give the corresponding halide ion. Many examples of

$$M + e \longrightarrow M^{\overline{\cdot}}$$

$$[M^{\overline{\cdot}}] \longrightarrow (M-R)^- + R\cdot$$

$$[M^{\overline{\cdot}}] \longrightarrow (M-X)\cdot + X^-$$

dissociative electron attachment reactions are known. Figures 4 and 5 show the mass analysis and detector responses for methyl parathion. An $M^{\overline{\cdot}}$ ion is not observed. The fragment ions at m/z 138, 141, 154 and 248 correspond to stable anions which are formed by the elimination of a neutral radical with or without rearrangement. Quantification can be carried out by selected ion detection of $(M-15)^-$ ions, or by use of one of the other ions. The structure is evident from the fragment ions.

Proton transfer reactions are not at present useful in pesticide studies. These reactions are gas phase analogs of reactions occurring in solution which result in the ionization of acids. Basic ions are used as reagents. The ions which are employed under API conditions (5) are Cl^- and $O_2^{\overline{\cdot}}$, for the ionization of relatively strong acids, while under CI conditions it is possible to use the strongly basic ions F^-, $O^{\overline{\cdot}}$ or HO^- for the ionization of very weak acids.

Oxygen substitution reactions occur for many halogenated compounds and for aromatic hydrocarbons. The best known reactions are those of aromatic polychloro and nitrochloro compounds (6).

TABLE I.   SOURCE CONDITIONS FOR NEGATIVE ION FORMATION

<u>Source Condition</u>

API	760	Torr	equilibrated
CI	0.3-1	Torr	not equilibrated
CI	<0.01	Torr	not equilibrated
EI	$<10^{-5}$	Torr	not equilibrated

TABLE II.   ELECTRON AFFINITY

Benzoquinone

1,4-Naphthoquinone

Maleic anhydride

m-Chloronitrobenzene

Phthalic anhydride

Nitrobenzene

Azulene

Polycyclic aromatic hydrocarbons

Benzophenone

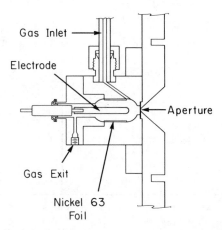

Journal of Chromatography

Figure 1. This API source is designed to serve both as an electron capture detector and as a source for the mass spectrometer (4).

The stream from a gas chromatograph is split, with one segment directed to a commercial electron capture detector and the other to this source. Three responses can be observed: (1) an electron capture response from the usual gas chromatograph–electron capture detector combination; (2) an electron capture response from the API source, which should duplicate the normal response; and (3) a mass spectrometric response based on ions generated under API conditions.

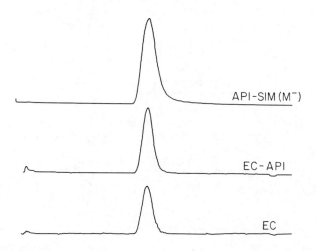

Journal of Chromatography

Figure 2. The responses due to negative ion formation from benzil are: (EC) the GC-electron capture response; (EC–API) the corresponding response from the API source; and (API–SIM) the selected ion response for the M⁻ ion of benzil (4). (benzil, C_6H_5-CO-CO-C_6H_5, Ar/CH_4, 240°C, PZ179 column)

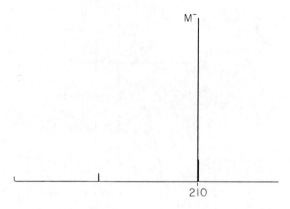

Journal of Chromatography

Figure 3.   Scanned API negative ion mass spectrum for benzil.   This compound ionizes to form M⁻ ions (4).

The carrier gas was argon/methane, and the source temperature was 240°C.   A PZ179 column was used in the gas chromatograph.   This mass spectrum was obtained while the compound was in the source, as indicated in Figure 2.

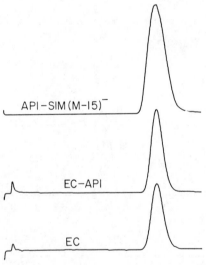

Journal of Chromatography

Figure 4.   The responses due to negative ion formation from methyl parathion are: (EC) the GC-electron capture response; (EC–API) the corresponding response from the API source; and (API–SIM) the selected ion response for the (M-15)⁻ ion derived from methyl parathion by loss of a methyl radical (4). (Ar/CH₄, 240°C, PZ179 column)

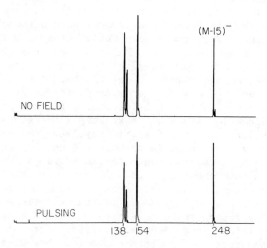

Figure 5. Scanned API negative ion mass spectrum for methyl parathion (4).

This compound ionizes to form four ions by cleavage and rearrangement. The loss of a methyl radical leads to the ion at m/z 248; the ion at m/z 154 amu is the p-nitrothio-phenolate ion; the ion at m/z 138 is the p-nitrophenolate ion; and the ion at m/z 141 is due to loss of the p-nitrophenyl radical. The carrier gas was argon/methane, and the source temperature was 240°C. A PZ179 column was used in the gas chromatograph. This mass spectrum was obtained while the compound was in the source, as indicated in Figure 4. Under these conditions, variations in the field in the source did not alter the mass spectrum.

$$O_2 \qquad\qquad\qquad\qquad\qquad MX$$

$$\downarrow e \qquad\qquad\qquad\qquad\qquad \downarrow e$$

$$O_2^{\cdot -} + MX \longrightarrow [MXO_2]^{\cdot -} \longleftarrow O_2 + MX^{\cdot -}$$

$$\downarrow$$

$$(M{+}O{-}X)^- + XO^{\cdot}$$

As far as is known, an intermediate ion $[MXO_2]^{\cdot -}$ is formed from
aromatic halides, but this ion has never been observed because
it cleaves immediately to form a phenolate ion and a neutral
radical (which is not observed, but which leads to a halide
ion as an observed product). Under API conditions, with a
relatively high concentration of oxygen, $MX^{\cdot -}$ ions are not
observed, and the final reaction products are a phenolate ion
and a halide ion. When the concentration of oxygen is reduced
so that only a trace is present, $MX^{\cdot -}$ ions are observed, and
the reaction proceeds from the right to give the same intermediate and final product ions. Under CI conditions, which are
not equilibrated, $O_2^{\cdot -}$ and $MX^{\cdot -}$ ions are observed at the same
time, and the substitution reaction presumably can proceed
from either direction since the next step is irreversible. An
element of structural specificity is involved; o- and p-nitro-
chlorobenzene are converted to phenolate ions, but the chief
product from m-nitrochlorobenzene is the $M^{\cdot -}$ ion. Polychloro
compounds (hexachlorobenzene, 2,3,4,5,6-pentachlorobiphenyl)
give chlorophenolate ions. This reaction has been studied by
Hass and his colleagues (7). Substitution reactions with
aromatic hydrocarbons are not observed under API conditions,
but under conditions where ions with higher internal energy
are present the corresponding phenolate ions are detected as
products. It is probable that an analogous reaction also
occurs for aliphatic compounds, since Grimsrud (8) found that
an enhanced electron capture response resulted for methyl
chloride when oxygen was present. The effect of oxygen on the
negative ion product from Tetradifon is shown in Figure 6.
Thermalized electrons will react with this compound to form
$(M{-}HCl)^{\cdot -}$ ions; a ring closure is assumed to occur for $M^{\cdot -}$
ions (not observed) with elimination of hydrogen chloride. If
oxygen is present, a substitution reaction occurs, leading to
$(M{+}O{-}Cl)^-$ ions. The phenolate ion which is formed is highly
stable, and the observed response is greater for the substitution reaction than for the cyclization reaction. In most
instances where halogenated pesticides are under study, it
would be desirable to examine the possibility of using this
reaction for analytical purposes. Oxygen should not, of course,
be added to a carrier gas stream before a GC column, but it

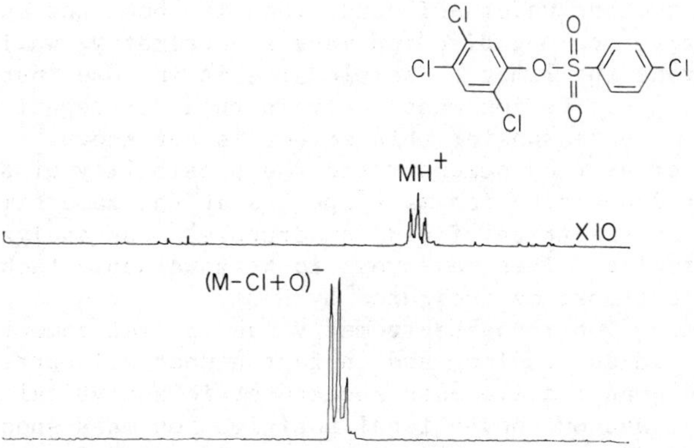

Journal of Chromatography

Figure 6. Scanned API positive and negative ion mass spectra for Tetradifon (4). The nitrogen carrier gas contained a very low concentration of oxygen (10^{-3} torr O_2 N_2); the source temperature was 200°C. The positive ion mass spectrum shows only MH^+ ions resulting from protonation. The negative ion mass spectrum shows $(M - Cl + O)^-$ ions resulting from a substitution reaction of oxygen for chlorine. If oxygen is excluded from the source, the negative ion product is $(M - HCl)^-$, which is probably formed from $M^{\cdot-}$ by ring closure with elimination of hydrogen chloride.

can be added in low concentration to gas streams entering MS
sources. When nonfilament sources are used, oxygen can be
added in higher concentrations.

Adduct ions are formed under some circumstances. Polyha-
lides will add Cl⁻ to give negative ions, for example; this
mode of negative ion formation has been used for analytical
purposes by Dougherty et al (9,10,11).

The limiting sensitivity of detection for negative ions
has been determined with several different types of instruments.
There is general agreement that subpicogram sensitivity of
detection can be achieved under both API and CI conditions, and
the limits of sample size for different instruments are in the
range 10-100 fg. A comparison of API and CI positive ion data,
however, leads to a different conclusion. The limiting sensitivity of detection under API conditions is about the same for
both positive and negative ion mass spectrometry, while under
CI conditions the limit of sample size is in some instances much
higher for positive ion mass analysis than for negative ion mass
analysis. The reason for this effect is not known.

Hunt et al (12) demonstrated the possibility of obtaining
positive and negative ion mass spectra at the same time by
employing an electrical field (quadrupole) mass analyzer with
rapid switching. This may prove to be a valuable technique in
many applications, as indicated by Hunt.

Negative ion mass spectrometry has not yet come into general
use in pesticide studies, and in fact almost all current qualitative and quantitative mass spectrometric analytical methods
are still based on conventional positive ion mass spectrometry
under CI or EI conditions. Identification procedures based upon
negative ion mass spectra, or upon both positive and negative
ion spectra, will probably be used in the future for many
classes of compounds, along with methods based on ion transformations. The condition used by Brandenberger (3) for identification studies is low pressure CI; the reactions which occur
are due to electrons or to O⁻ ions. Characteristic ions are
found for certain drug classes which are important in forensic
chemistry. Several quantitative applications of negative ion
mass spectrometry have been published. For example, Markey
and Lewy (13) described an analytical procedure for melatonin
in biologic samples which involved derivative formation followed
by CI negative ion mass spectrometry. Studies of related
methods for catecholamines are under way in several laboratories.
These applications require very high sensitivity of detection.

A recent toxicologic application of API negative ion mass
spectrometry involved the analysis of human urine samples for
2,3-dibromopropanol. This problem arose because of the use of
tris(2,3-dibromopropyl) phosphate as a flame retardant of
children's sleepwear. Both tris and its metabolite, 2,3-
dibromopropanol, are mutagenic and carcinogenic agents, and the

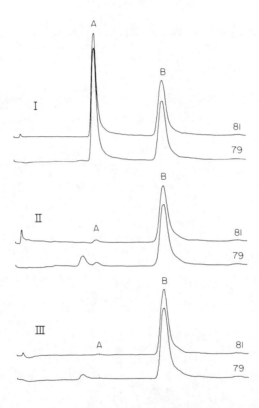

Science

Figure 7.   Urinary analyses for 2,3-dibromopropanol for: (I) a child wearing new tris-treated sleepwear; (II) the same child wearing old tris-treated sleepwear before use of the new sleepwear, and (III) a child who had never worn tris-treated sleepwear (14).

The observed responses for bromide ions at m/z 79 and 81 are for (A) 2,3-dibromopropanol and (B) 1,4-dibromo-2-butanol used as an internal standard. These analyses were carried out by atmospheric pressure ionization mass spectrometry with negative ion detection. Samples were introduced by gas chromatography (5% SE-30 on 100–120 mesh Gas Chrom Q; 1.8 m × 2 mm glass column; 110°C isothermal; nitrogen 40 mL/min).

TABLE III.   ANALYSES OF URINE SAMPLES FOR 2,3-DIBROMOPROPANOL

Day	New treated pajamas	2,3-Dibromopropanol ng/ml
1	No	0.4
2	No	0.4
3	Yes	11
4	Yes	29
5	Yes	(sample lost)
6	Yes	21
7	Yes	18
8	No	9
9	No	14
10	No	6
11	No	6
12	No	8

use of <u>tris</u> was terminated, but not before a very large number
of children were exposed to treated cloth as a consequence of
a federal requirement for flame-proofing. There was a tendency
to minimize the seriousness of the circumstances by pointing
out that there was no evidence of the absorption of <u>tris</u> through
the skin of exposed children, but this fact was due to the use
of inadequate analytical methods. Figure 7 shows GC/MS analyses
for 2,3-dibromopropanol by API negative ion mass spectrometry
in a bioanalytical system (<u>14</u>), for three samples from (I) a
child wearing new <u>tris</u>-treated sleepwear, (II) the same child
wearing old <u>tris</u>-treated sleepwear, before wearing the new
sleepwear, and (III) a child who had never worn treated sleepwear. The chart records are for bromide ions at m/z 79 and 81.
The results are in Table 3. Similar data were obtained by
Dougherty et al, using adduct ion formation (<u>14</u>). This problem,
in a program organized by Ames and his colleagues (<u>14</u>), required
the use of negative ion techniques for its solution.

The chief delaying factor in the use of negative ion mass
spectrometry in pesticide studies is the relatively high cost
of mass spectrometric methods. Analytical procedures based
upon gas chromatography with electron capture detection have
been in use for many years, and technological improvements
in gas chromatography continue to be made. The relatively high
possibility of making errors in identification and quantification with an electron capture detector, however, suggests that
all definitive procedures should be based upon mass spectrometry,
and that validation of methods involving chromatography and
chromatographic detectors of the usual type should be based upon
mass spectrometric evidence rather than upon chromatographic
evidence alone.

LITERATURE CITED

1.   Carroll, D. I.; Dzidic, I.; Stillwell, R. N.; Horning,
     E. C. in "Trace Organic Analysis: A New Frontier
     in Analytical Chemistry"; Hertz, H. S.; Chesler,
     S. N., Eds., National Bureau of Standards
     Special Publication 519, 1979; pp. 655-671.

2.   Horning, E. C.; Horning, M. G.; Carroll, D. I.; Dzidic, I.;
     Stillwell, R. N. Anal. Chem., 1973, 45, 936.

3.   Brandenberger, H. in "Instrumental Applications in
     Forensic Drug Chemistry"; Klein, M.; Kruegel, A. V.;
     Sobol, S. P., Eds., U. S. Government Printing Office:
     Washington, D.C., 1979; pp. 4147.

4.   Horning, E. C.; Carroll, D. I.; Dzidic, I.; Lin, S-N.;
     Stillwell, R. N.; Thenot, J-P. J. Chromatogr., 1977, 142,
     481.

5.   Dzidic, I.; Carroll, D. I.: Stillwell, R. N.;
     Horning, E. C. J. Amer. Chem. Soc., 1974, 96, 5258.

6.   Dzidic, I.; Carroll, D. I.; Stillwell, R. N.; Horning,
     E. C. Anal. Chem., 1975, 47, 1308.

7.   Busch, K. L.; Norstrom, A.; Bursey, M. M.; Hass, J. R.;
     Nilsson, C. A. Biomed. Mass Spectrom., 1979, 6, 157.

8.   Grimsrud, E. O.; Miller, D. A. Anal. Chem., 1978, 50, 1140.

9.   Dougherty, R. C.; Dalton, J.; Biros, F. J. Org. Mass
     Spectrom., 1972, 6, 1171.

10. Tannenbaum, H. P.; Roberts, J. D.; Dougherty, R. C.
     Anal. Chem., 1975, 47, 49.

11. Dougherty, R. C.; Roberts, J. D.; Biros, F. J. Anal. Chem.,
     1975, 47, 54.

12. Hunt, D. F.; Stafford, G. C., Jr.; Crow, F. W.; Russell,
     J. W. Anal. Chem., 1976, 48, 2098.

13. Lewy, A. J.; Markey, S. P. Science, 1978, 201, 742.

14. Blum, A.; Gold, M. D.; Ames, B. N.; Kenyon, C.; Jones,
     F. R.; Hett, E. A.; Dougherty, R. C.; Horning, E. C.;
     Dzidic, I.; Carroll, D. I.; Stillwell, R. N.; Thenot,
     J-P. Science, 1979, 201, 1020.

RECEIVED March 3, 1980.

Analysis of Organotin Pesticide Residues by Gas Chromatography/Mass Spectrometry

T. E. STEWART and R. D. CANNIZZARO

Thompson–Hayward Chemical Company, Kansas City, KS 66106

Organotin compounds (tetravalent Sn) were found in the 1950's to possess certain fungicidal and biocidal activity.-(1) This activity along with the fact that organotins were much less toxic than other organometallics, such as organomercurials or organoarsenicals, established their agricultural importance.

Triaryltin compounds such as triphenyltin hydroxide (TPTH, Fentin hydroxide, Du-Ter, I), or triphenyltin acetate (TPTOAC, Brestan, II) have been shown to be effective as protectant fungicides. These compounds have demonstrated other activity such as antifoulants, bactericides and algicides. Due to their low phytotoxicity they have been developed for plant protection. In several cases they have been reported to exhibit the unusual property of deterring insects from feeding.(2)

Trialkyltin compounds have been shown to be effective agriculturally. Tributyltin oxide (III) has insecticidal and fungicidal activity, but generally has not been useful for plant protection due to phytotoxicity and volatility problems. Tricyclohexyltin hydroxide (TCTH, IV) has been an extremely effective acaricide.(3) Developed under the trade name Plictran, this compound was one of the first commercial acaricides. Another trialkyltin compound under development is hexakis (β,β-dimethylphenylethyl)distannoxane (Vendex, V), which is also an effective acaricide.

The earliest methods for the determination of organotin compounds involve acidic digestion of the compound to inorganic tin, which is then determined spectrophotometrically with several suitable colorimetric reagents.(4,5,6,7) These methods suffer from a lack of specificity since there may be organotins present within the plant. These methods also cannot distinguish among the different organotin pesticides and are very difficult to perform.

Recent methodology overcomes a number of these problems. Several methods are available for the analysis of tricyclo-hexyltin hydroxide.(8,9) These methods are not applicable to the triphenyltin compounds because of their thermal instability. Work by Eposto, et. al.(10) has shown that TPTH and TCTH can be derivatized with phenylmagnesium bromide (PhMgBr) to form the tetraphenyl and tricyclohexylphenyl derivatives, respectively, which can then be quantitated by GLC. Recent work by Soderquist(11) involves the reduction of the triphenyltin hydroxide to triphenyltin hydride, which is then determined by GLC. This method also allows for the formation of hydrides of any metabolites such as the diphenyltin dihydride, which may be quantitated. Wright(12) has shown that the methyl derivative of TPTH can be chromatographed on a capillary column and detected with a FP detector. At Thompson-Hayward, a method has been developed for TPTH that possibly may be used for a variety of other organotin pesticides. This method involves the conversion of TPTH to the methyl derivative. The flow scheme is shown in Figure 1. TPTH is first converted to the triphenyltin chloride (TPTCl, VI) and then the chloride is reacted with methylmagnesium chloride (CH_3MgCl) to form the methyl derivative, triphenylmethyltin (TPTM, VII).

The methyl derivative can be gas chromatographed and quantitated by a mass spectrometer set to monitor characteristic ions (m/e) of the compound (selective ion monitoring).

Experimental:

All solvents and reagents used in this method were pesticide quality. Methylmagnesium chloride 2.9M in THF was obtained from Aldrich Chemical Company. Triphenyltin hydroxide and triphenylmethyltin were obtained from Ventron. Tricyclohexyltin hydroxide and Vendex were analytical standards. Triphenylethyltin was synthesized and characterized at Thompson-Hayward Chemical Company.

Gas chromatographic and mass spectral data were compiled by a Finnigan 6110 data system coupled to a Finnigan 4000 mass spectrometer and 9600 gas chromatograph. Some results were obtained using a Finnigan PROMIM (Programmable Multiple Ion Monitoring) system, coupled to a 4-pen strip chart recorder and Laboratory Data Control integrator.

Methods and Procedures:

Prior to extraction, samples to be analyzed are ground or chopped until they are of uniform consistency. The prepared sample is weighed into a blender; 150 ml of suitable solvent is added and the sample is blended at a high speed

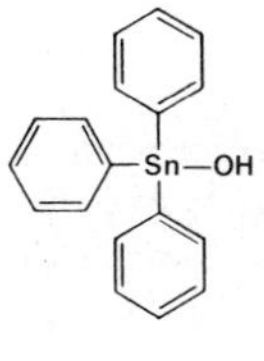

I TPTH
DU·TER

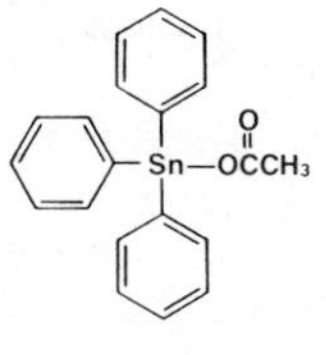

II TPTOAc
BRESTAN

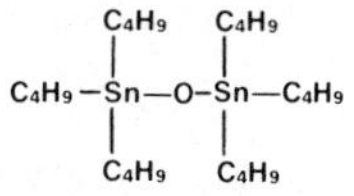

III

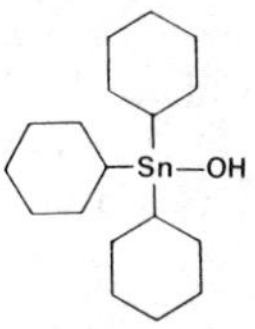

IV TCTH
PLICTRAN

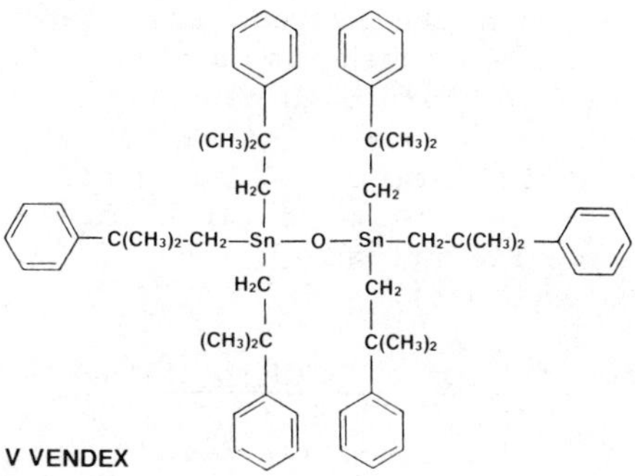

V VENDEX

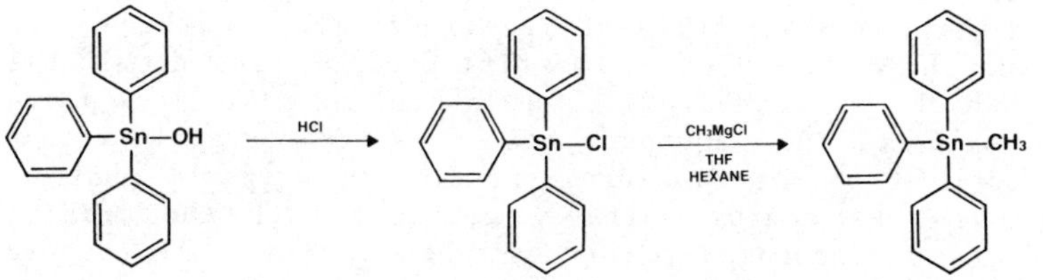

I TPTH VI TPTCl VII TPTM

Figure 1. Reaction scheme for the conversion of TPTH to TPTM

for 5 minutes. The extract is vacuum filtered through a
Buchner funnel into a flat-bottom flask. The filter cake is
again blended and filtered. The two extracts are combined,
dried over sodium sulfate and then evaporated to dryness on a
rotary flash evaporator.

Florisil Column Clean-up:

The extract is dissolved in 10 ml of dichloromethane and
transferred to a 19 mm i.d. X 300 mm glass chromatographic
column packed as follows: 15 ml of 10% deactivated Florisil
(10 ml H_2O + 90 grams Florisil) followed by a 2 cm layer of
sodium sulfate which has been prewashed in order with 50 ml
methanol, 50 ml acetone, and 50 ml dichloromethane. The
sample flask is rinsed twice with additional 10 ml and 5 ml
portions of dichloromethane, which are transferred to the
column. When the solvent has reached the top of the sodium
sulfate, the flask is rinsed with two 15 ml portions of
acetone which are then transferred to the column. The solvent
is allowed to drain to the top of the sodium sulfate layer.
These washes are discarded. A 200 ml volume of methanol is
added to the column to elute the TPTH, which is collected in
a flask. The methanol is evaporated to dryness on a rotary
flash evaporator.

Conversion of TPTH to TPTCl:

The cleaned-up residue is dissolved in 25 ml of 6N HCl
by sonication and the resulting solution transferred to a
separatory funnel. The flask is then rinsed with 50 ml of
dichloromethane which is likewise transferred to the separatory funnel. The TPTCl is partitioned into the dichloromethane, which is removed, filtered and dried through a pad of
sodium sulfate. The aqueous phase is extracted again with an
additional 50 ml of dichloromethane. The organic phase is
dried over the sodium sulfate, combined with the original
extract, and evaporated to dryness on a rotary flash evaporator.

Conversion of TPTCl to TPTM:

The TPTCl is dissolved in 6 ml of hexane and 4 ml
tetrahydrofuran and reacted with 2.0 ml of 2.9M CH_3MgCl in
THF for one half hour at room temperature. Excess CH_3MgCl is
deactivated with 50 ml saturated ammonium chloride solution
and the entire reaction mixture is transferred to a separatory

funnel. The flask is rinsed with 50 ml of hexane which is then transferred to the separatory funnel. The TPTM is extracted into hexane, which is filtered through a sodium sulfate pad into a flat-bottom flask. The aqueous portion is extracted with an additional 50 ml of hexane, the hexane is filtered and combined with the original extract. The hexane extract is evaporated to dryness on a rotary flash evaporator.

Alumina Column Clean-up:

A 19 mm i.d. X 300 ml glass chromatographic column is prepared as follows: 1 cm of sodium sulfate is added to a closed column; 50 ml of hexane is then added, followed by 21 ml of 6% deactivated neutral alumina (94 grams alumina + 6 grams H_2O). After the alumina settles 1 cm of sodium sulfate is added and the hexane is drained to the top of the sodium sulfate. The sample containing the TPTM is dissolved in 10 ml of hexane and transferred to the column. The flask is rinsed with an additional 5 ml of hexane which is transferred to the column. The hexane is allowed to drain to the top of the sodium sulfate and is discarded. 200 ml of hexane is added to the column and the TPTM eluate is collected in a flat-bottom flask.

GC/MS Analysis:

A sample is diluted in an appropriate volume of internal standard stock solution and an aliquot of the sample is injected into a gas chromatograph/mass spectrometer monitoring m/e 349 and 351. The peak areas of the TPTM and internal standard are integrated and expressed as the ratio of TPTM:-Internal Standard and the ratio of those areas respectively are used to construct a standard curve. The amount of TPTM present is then calculated from the standard curve.

Results and Discussion:

As mentioned previously, the formation of the methyl derivative is similar to several methods that already have been developed for organotin pesticides. The utilization of this derivative in conventional gas chromatographic residue procedures presents many challenges. Electron capture detection is sensitive enough but requires extensive clean-up while with FID and FPD the sensitivity is lower than desired. The development of a routine GC/MS procedure (single or multiple ion monitoring) was chosen then as the only possible alternative. A SIM (selective ion monitoring) method is specific for retention time and as well as for characteristic ions (m/e).

The mass spectrum of triphenylmethyltin is presented in Figure 2 and shows the predominent tin isotopes, ^{116}Sn, ^{118}Sn, and ^{120}Sn. The various tin isotopes are listed in Table I.

TABLE I

TIN ISOTOPE MASSES AND ABUNDANCES

	Mass (13)	Abundance (14)%
^{112}Sn	111.90494	0.95
^{114}Sn	113.90296	0.65
^{115}Sn	114.90353	0.34
^{116}Sn	115.90211	14.24
^{117}Sn	116.90306	7.57
^{118}Sn	117.90179	24.01
^{119}Sn	118.90332	8.58
^{120}Sn	119.90213	32.27
^{122}Sn	121.90341	4.71
^{124}Sn	123.90524	5.98

The fragmentation pattern of TPTM reveals no parent molecular ion, but a tin isotope cluster at m/e 347, 349, and 351, which corresponds to the triphenyltin moiety. The spectrum also shows a tin isotope cluster at m/e 193, 195, and 197 corresponding to the monophenyltin moiety. This fragmentation is similar to that observed by Chambers(15).

The mass spectrometer can be set to monitor several characteristic ions (m/e) of the triphenylmethyltin derivative. The most logical m/e values to be scanned would be 347, 349 and 351 because they would indicate the presence of a triphenyltin moiety, while allowing fewer interfering compounds. In this fashion the method distinguishes between a triphenyltin compound and any other organotin pesticide. To enhance the reproducibility of the method, triphenylethyltin (TPTE) is added as an internal standard before injection. The TPTE has a slightly longer retention time on 3% OV-1 (as shown in Figure 3) and gives a similar fragmentation pattern as shown in Figure 4.

A SIM series of TPTM standards with the internal standard TPTE, (TPTE concentration:200 pg/ul) is shown in Figure 5. A standard curve is constructed by plotting the ratio of the peak areas of TPTM/TPTE vs. the amount of TPTM. A typical standard curve is shownin Figure 6. The curve is linear from 25 pg to 1 ng of TPTM. Most of our anlayses are conducted in the range of 50-400 pg TPTM.

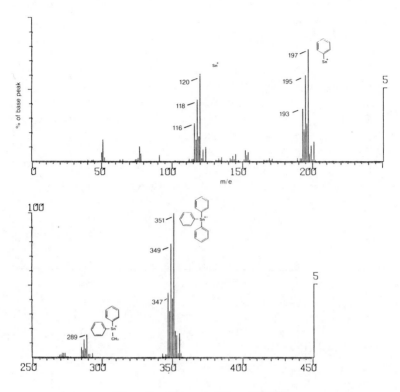

Figure 2.   Mass spectrum of TPTM

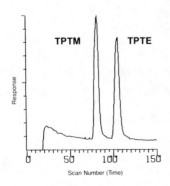

Figure 3.   Reconstructed gas chromatogram of TPTM and TPTE on 3% OV-1 at 200°C

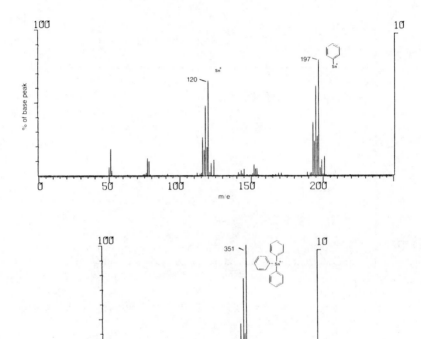

Figure 4.   Mass spectrum of TPTE

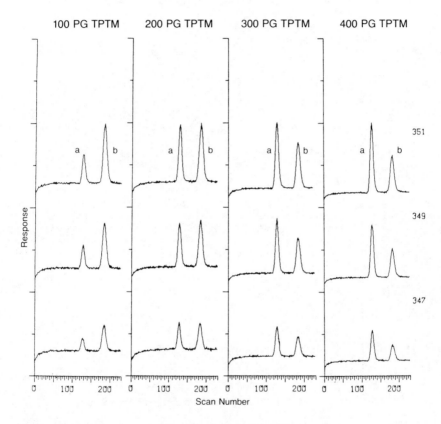

Figure 5.   SIM chromatograms of TPTM (a) and TPTE (b) (200 pg/μL), monitoring m/e 347, 349, and 351

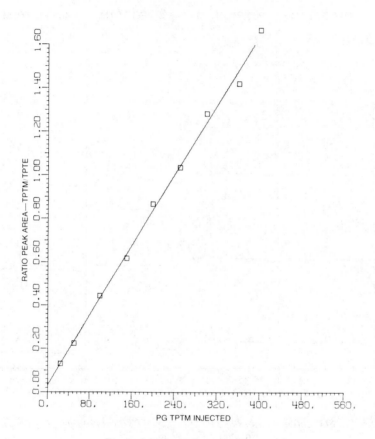

Figure 6.   Typical standard curve

Figure 7 presents typical PROMIM chromatograms from the analysis of TPTH in soybeans. The chromatograms are very clean showing none of the interferences which normally would be encountered with an EC or FI detector. Analyses of TPTH in soybean foliage, radish tops, barley and other crops have very similar chromatograms. None of the 63 pesticides registered for use on soybeans are observed to interfere with the analysis. Typical recovery data for soybean seed and soybean foliage are listed in Table II.

In our research to improve the method we have noted that the conversion of TPTH to TPTCl is not necessary for the derivatization with CH_3MgCl. The conversion of TPTH directly to TPTM has been shown to be quantitative, based on recovery efficiency data and studies involving standards. This reaction scheme is shown in Figure 8.

Analyses done using this procedure have very similar results. Usually water samples that do not require extensive clean-up are analyzed with this method.

Other Applications:

The application of this type of method to other organotin compounds is a very real possibility, as indicated by the following investigations:

Tricyclohexyltin Hydroxide (TCTH IV):

Tricyclohexyltin hydroxide was derivatized by the reaction scheme in Figure 9 to form the methyl derivative, TCTM. The methyl derivative may be detected by monitoring its characteristic ions (m/e). The mass spectrum of TCTM is shown in Figure 10. The fragmentation shows a tin isotope cluster at m/e 301 which corresponds to the dicyclohexylmethyltin moiety. A comparison of the mass spectra of TPTM and TCTM reveals divergent fragmentation pathways. TPTM displays a very intense ion at m/e 351 (M-15) while TCTM exhibits a base peak at m/e 219 which corresponds to the monocyclohexylmethyltin moiety. TCTM also displays a tin isotope cluster at m/e 301 which corresponds to the dicyclohexylmethyltin moiety. This tin isotope cluster appears to be appropriate for quantitative purposes. TPTM and TCTM can easily be separated on a 3% OV-7 column as shown in Figure 11. Figure 12 demonstrates that SIM chromatograms of TPTM and TCTM present no problems in distinguishing between the two derivatives even when complete separation is not achieved.

TABLE II

Recovery of Triphenyltin Hydroxide From
Soybeans and Soybean Foliage

Soybeans

Level of Fortification (ppm)	Number of Trials	% Recovery Average
0.05	34	76
0.10	18	83
0.20	2	77

Mean = 78.3
Standard Dev. = 11.43

Soybean Foliage

Level of Fortification (ppm)	Number of Trials	% Recovery Average
0.05	5	96
0.10	5	98
0.50	15	82
1.00	2	68
25.00	7	86
50.00	1	81
100.00	2	76

Mean = 84.5
Standard Dev. = 18.6

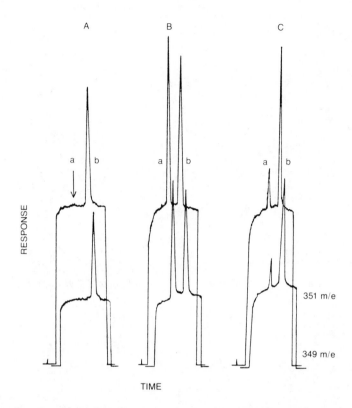

Figure 7. Typical PROMIM chromatograms of TPTM (a) and TPTE (b) in soybean extracts: (A) soybean control sample (10 mg sample aliquot injected, < 0.05 ppm); (B) soybean control sample fortified with 0.05 ppm TPTH (10 mg sample aliquot injected, 0.045 ppm found, 90% recovery); (C) soybeans treated at 0.5 lb a.i./acre (10 mg sample aliquot injected, < 0.05 ppm)

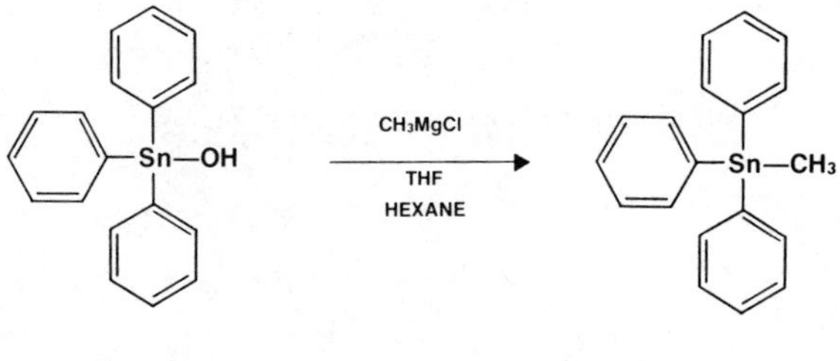

Figure 8. Alternate reaction scheme for the conversion of TPTH to TPTM

IV TCTH

TCTM

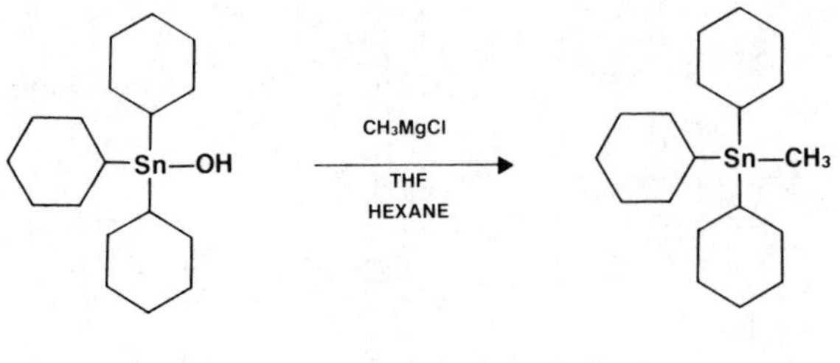

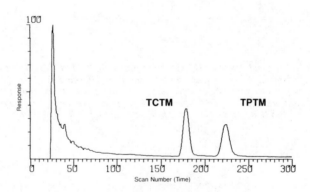

Figure 9. Reaction scheme for the conversion of TCTH to TCTM

Figure 10. Mass spectrum of TCTM

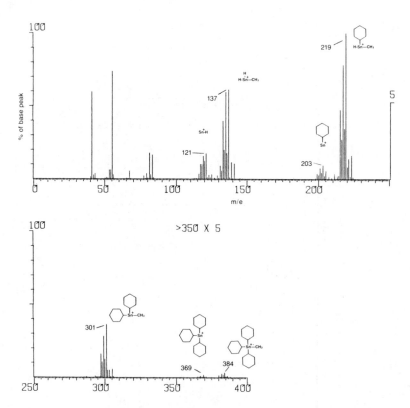

Figure 11.   Reconstructed gas chromatogram of TPTM and TCTM on 3% OV-1 at 210°C

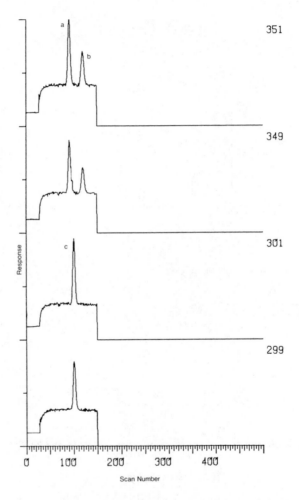

*Figure 12.   SIM chromatograms of TPTM (a), TPTE (b), and TCTM (c) on 3%
OV-1 at 210°C*

If greater analytical specificity is desired, monitoring the tricyclohexyltin moiety would be indicated. Our studies have shown that the ethyl derivative of TCTH exhibits an enhanced tin isotope cluster at m/e 369 which satisfies these requirements. A mass spectrum of TCTE is shown in Figure 13.

Vendex:

Vendex is a related organotin compound which exists in a bis-oxide form. Therefore to employ a derivatization scheme similar to that utilized on organotins already described we had to consider cleaving the oxide bond. Our initial derivatization scheme, similar to that employed with TPTH is shown in Figure 14. Later work revealed the same results regardless of whether or not HCl was used to cleave the oxide bond. The gas chromatogram of the methyl derivative of Vendex is shown in Figure 15, along with a mass spectrum in Figure 16. The tin isotope cluster at m/e 401 corresponds to the di-(β,β -dimethylphenylethyl)methyltin. These ions (m/e) could be monitored to quantitate this derivative. The mass spectrum in Figure 16 reveals a tin isotope cluster at m/e 345 which could possibly interfere with the analysis of TPTH.

Figure 17 shows SIM chromatograms of TPTM and TPTE with a large amount of the Methylvendex derivative. A small potential interference was observed. However, the isotopic abundance of the Methylvendex derivative is not similar to that observed with TPTM. Therefore we conclude that the possibilities for confusion are minimal.

Conclusion:

The GC/MS methodology presented for the determination of TPTH residues in crops provides for greater specificity than any other currently available, and exhibits excellent sensitivity. The recovery efficiency data generated demonstrate that this method of residue analysis is a reliable, routine tool for the determination of TPTH residues in a wide range of agricultural crops. In addition, the results of our research suggest that the procedures developed and reported here could be adapted for the analysis of other organotin residues.

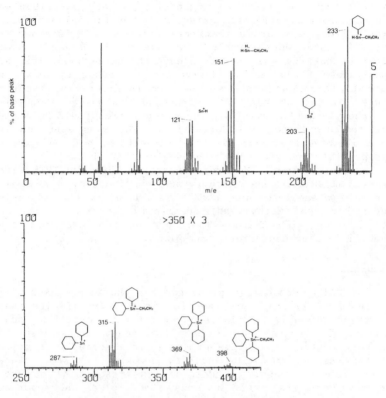

Figure 13.   Mass spectrum of TCTE

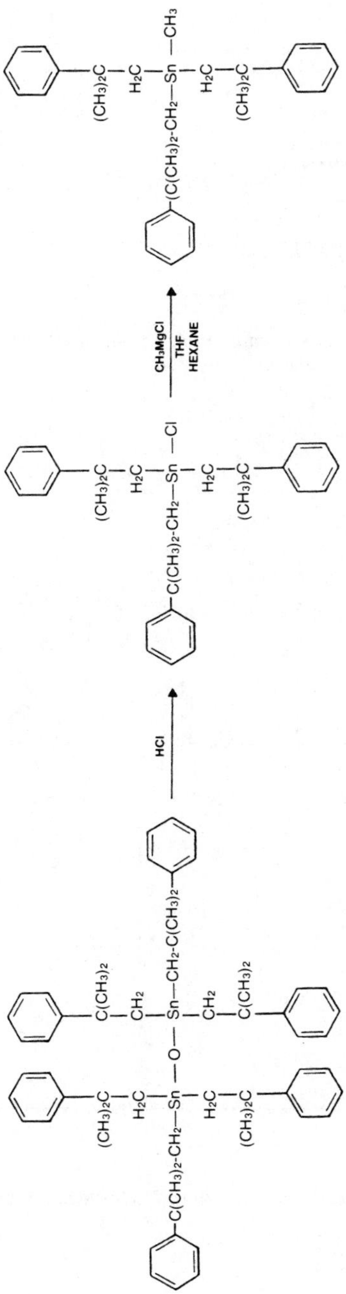

Figure 14. Reaction scheme for the conversion of vendex to the methylvendex derivative

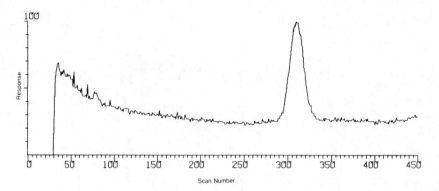

Figure 15.   Reconstructed gas chromatogram of the methylvendex derivative on 3% OV-7 at 250°C

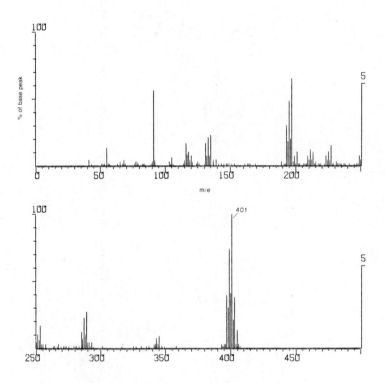

Figure 16.   Mass spectrum of methylvendex derivative

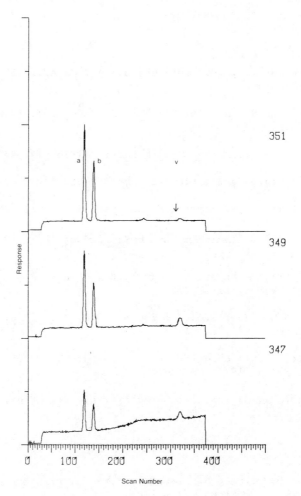

Figure 17. SIM chromatograms of TPTM·(a), TPTE (b), and methylvendex derivative (v) on 3% OV-7 with temperature programming to 265°C

ACKNOWLEDGEMENT

    We would like to acknowledge the assistance of Mr. Loren
Dunham and Ms. Virginia Corbin at Stoner Laboratories in the
development of the methyl derivative of TPTH.   We would also
like to acknowledge the assistance of Ms. Ann B. Grow and Mr.
Ken Shearer at Thompson-Hayward Chemical Company.

Literature Cited:

1. van der Kerk, G.J.M., Luijten, J.G.A., J. Appl. Chem.,
    1954, 4, 314.

2. Ascher, K.R.S., Nissim, S., Int. Pest Control, 1965,
    7, 21.

3. Hunter, R.C., Environ. Health Persp., 1976, 14, 47.

4. Newman, E.J., Jones, P.D., Analyst, 1966, 91, 406.

5. Analytical Method Committee, Analyst, 1967, 92, 320.

6. Corbin, H.B., J. Assoc. Off. Anal. Chem., 1970, 53,
    140.

7. Trombotti, G., Maini, P., Cavmo, B., Kovacs, A.,
    Pesticide, Sci., 1970, 1, 144.

8. Gauer, W.O., Seiber, J.N., Crosby, D.G., J. Agric.
    Food Chem., 1974, 22, 252.

9. Camoni, I. Chiacchierini, E., Iachetta, R., Margi, A.,
    Ann. Chimica, 1975, 68, 267.

10. Esposto, R., Camoni, I., Chiacchierini, E., Margi, A.,
    Ann. Chimica, 1978, 68, 235.

11. Soderquist, C.J., Crosby, D.G., Anal. Chem., 1978,
    50, 1435.

12. Wright, B.W., Lee, M.L., Booth, G.M., J. High Resol.
    Chrom. & Chrom Comm. 1979, 189.

13. Konig, L.A., Mattauch, J.H.E., Wapstra, A.E., Nuclear
    Phys., 1962, 31, 18.

14. Beynon, J.H., "Mass Spectrometry and its Applications
    to Organic Chemistry" Elsevier, Amsterdam, 1960.

15. Chambers, D.B., Glockling, F., Weston, M., J. Chem
    Soc. (A), 1967, 1759.

RECEIVED March 15, 1980.

INDEX